Quality and Preservation of Fruits

Editor

N. A. Michael Eskin, Ph.D.
Professor
Department of Foods and Nutrition
University of Manitoba
Winnipeg, Manitoba, Canada

CRC Press
Taylor & Francis Group
Boca Raton London New York

CRC Press is an imprint of the
Taylor & Francis Group, an **informa** business

First published 1991 by CRC Press
Taylor & Francis Group
6000 Broken Sound Parkway NW, Suite 300
Boca Raton, FL 33487-2742

Reissued 2018 by CRC Press

Library of Congress Cataloging-in-Publication Data

Quality and preservation of fruits / editor, N. A. Michael Eskin.
p. cm.
Includes bibliographical references and index.
ISBN 0-8493-5561-3
1. Fruit—Preservation. 2. Fruit—Quality. I. Eskin, N. A. M.
(Neason Akivah Michael)
TX612.F7Q35 1990
664'.8—dc20 90-2298

A Library of Congress record exists under LC control number: 90002298

Publisher's Note
The publisher has gone to great lengths to ensure the quality of this reprint but points out that some imperfections in the original copies may be apparent.

Disclaimer
The publisher has made every effort to trace copyright holders and welcomes correspondence from those they have been unable to contact.

ISBN 13: 978-1-315-89705-9 (hbk)
ISBN 13: 978-1-351-07615-9 (ebk)

Visit the Taylor & Francis Web site at http://www.taylorandfrancis.com and the
CRC Press Web site at http://www.crcpress.com

PREFACE

The main focus of this book is to review those factors that influence the quality of fruits for both the fresh market as well as for the food processing industry. The production of a successful commercial crop depends on the development of suitable varieties in addition to agronomic and climatic conditions. While high yields are important to the farmer, the crop must be of sufficiently high quality to meet the demands of the consumer market and the requirements of the food processing industry. Such quality factors include appearance, texture, and flavor together with nutritional quality and wholesomeness. Increased concern for a healthier diet has resulted in a larger consumption of fresh produce in North America over the past decade. The greater urbanization of our society has placed enormous demands on the food industry for a greater variety of processed fruit products in the convenience food market. This has led to the development of new processing technologies that more effectively preserve the quality attributes associated with the original fresh produce.

This volume is concerned with major fruit crops grown in North America for the consumer market and/or the food processing industry.

THE EDITOR

N. A. Michael Eskin, Ph.D., is professor of food chemistry at the University of Manitoba, Canada. He was born in England in 1941. He graduated with a B.Sc. degree with honors in biochemistry from the University of Birmingham in 1963 and received his Ph.D. in Physiological Chemistry from the same university in 1966. Professor Eskin taught food chemistry at the Polytechnic of the Southbank in London for 2 years before coming to the University of Manitoba in 1968. He has taught courses in flavor chemistry, lipid chemistry, nutrition related to agricultural crops, as well as research methodology. He has authored over 60 research papers and reports and has been chairman and speaker at many national and international scientific meetings.

Professor Eskin is the author of *Plant Pigments Flavors and Textures: The Chemistry and Biochemistry of Selected Compounds* published in 1979 and is co-author of *Biochemistry of Foods* published in 1971. Both books were translated into Japanese while the latter was also published in German and recently also in Malay. A second edition of *Biochemistry of Foods* is currently in press. Professor Eskin is considered an authority on canola oil and co-authored *Canola Oil: Properties and Performance* published in 1987. He has reviewed a number of books as well as many research papers for both journals and granting agencies.

Professor Eskin was the recipient of the Winnipeg Rh Institute Award in 1980 for outstanding contributions to research and scholarship in the natural sciences. He was a Lady Davis Fellow at the Department of Biotechnology and Food Engineering at the Technion, Israel from 1982 to 1983. Professor Eskin is a Fellow of the Institute of Food Science and Technology of the U.K. He was a visiting professor at the Faculty of Agriculture of the Hebrew University from 1973 to 1974. Professor Eskin is a lecturer on canola oil for courses offered by the Canadian International Grains Institute. He has guest lectured at a number of universities in Canada (including Toronto, Guelph, and McGill) and abroad as well as the Weizmann Institute, Hebrew University, Volcani Institute, and Bar Ilan University in Israel. He is a member of the Canadian Institute of Food Science and Technology, Institute of Food Technologists, Institute of Food Science and Technology (U.K.), and serves on several committees of the American Oil Chemists' Society.

CONTRIBUTORS

Béla S. Buslig, Ph.D.
Research Scientist
U.S. Citrus and Subtropical Products Laboratory
Winter Haven, Florida

Gary A. Couvillon, M.S., Ph.D.
Department Head
and
Division Chairman
Department of Horticulture
University of Georgia
Athens, Georgia

Stephen R. Drake, Ph.D.
Research Food Technologist
Irrigated Agriculture Research and Extension Center
Prossey, Washington

N. A. Michael Eskin, Ph.D., F.I.F.S.T.
Professor
Department of Foods and Nutrition
University of Manitoba
Winnipeg, Manitoba, Canada

David V. Fisher, Ph.D., F.A.S.H.S.
Former Head Pomology Section
and
Director
Agriculture Canada Research Station
Summerland, B.C., Canada

John A. Kitson, F.I.F.S.T.
Former Head Food Processing Station
Agriculture Canada Research Station
Summerland, B.C., Canada

Gerard Kewer, Ph.D.
Associate Professor
Department of Horticulture
University of Georgia
Tifton, Georgia

Porter B. Lombard, Ph.D.
Professor
Department of Horticulture
Oregon State University
Corvallis, Oregon

Justin R. Morris, Ph.D.
Professor
Department of Food Science
University of Arkansas
Fayetteville, Arkansas

Manuel G. Moshonas, Ph.D.
Research Chemist
U.S. Citrus and Subtropical Products Laboratory
Florida Department of Citrus
Winter Haven, Florida

Klayton E. Nelson, Ph.D. (Deceased)
Professor
Department of Viticulture and Encology
University of California, Davis
Davis, California

Daryl G. Richardson, Ph.D.
Associate Professor
Department of Horticulture
Oregon State University
Corvallis, Oregon

Philip E. Shaw, Ph.D.
Research Leader
U.S. Citrus and Subtropical Products Laboratory
Winter Haven, Florida

William A. Sistrunk, Ph.D.
Professor Emeritus
Department of Food Science
University of Arkansas
Fayetteville, Arkansas

Ronald E. Wrolstadt, Ph.D.
Professor and Head
Department of Food Science and Technology
Oregon State University
Corvallis, Oregon

This book is dedicated to seven men in my life:

My late father Joshua Eskin
My brother-in-law Percy Wilenski
My brother-in-law Adrian Becker
My son Katriel Eskin
My son Joshua Eskin
My son Ezra Eskin
My son Daniel Eskin

and

To the Memory of Professor Klayton E. Nelson

TABLE OF CONTENTS

Chapter 1*

ORANGES

Béla S. Buslig

TABLE OF CONTENTS

* Mention of a trademark or proprietary product is for identification only and does not imply an endorsement or warranty of the product by the Florida Department of Citrus over other products.

I. INTRODUCTION

A. HISTORY

The sweet orange, *Citrus sinensis* [L.] Osbeck, is probably native to southeastern Asia, most likely southern China, and Indo-China. References to its cultivation date back to early Chinese literature. The earliest existing book on the orange, dated in 1178 A.D., makes several clear references to sweet oranges in ancient Chinese literature of the period around 300 B.C. Although there seems to be some evidence of sweet oranges in Roman gardens as early as the 1st century A.D., there is no written evidence for the cultivation of sweet oranges in Europe until the 15th century. The first importation of sweet orange trees to Europe is attributed to Genoese traders by some references, and to Portuguese explorers by others, but considerable uncertainty exists about the time of introduction.[4,17,38] In any case, following the appearance of sweet oranges during the 15th and 16th centuries in Europe, it was cultivated extensively around the Mediterranean and became quite popular.

With the exploration of the Americas, sweet oranges, along with other citrus fruits (or in actuality, seeds of the same), were brought first to the West Indies and shortly afterwards to the North and South American continents late in the 15th and early in the 16th centuries. By the latter part of the 16th century numerous citrus varieties were well established in the southern part of North America and in Central America and were extremely popular among colonists and natives alike. Settlers in Florida during the early 18th century found many wild citrus groves throughout the state.[4,17] Introduction of orange culture in the Western portion of the present U.S. was by missionaries, early in the 18th century in Arizona, and by the middle of the century in California.[4] The popularity of the sweet orange resulted in widespread distribution of plantings throughout tropical and subtropical regions of the world and created considerable demand for the fresh fruit, and with the invention of frozen concentrated orange juice (FCOJ),[25] additional demand for production.

During the early years of commercial production the primary outlet for sweet oranges was the fresh market. The established growing areas in Europe, the Middle East, and North Africa adequately served the main market areas in Europe. Plantings in the Americas primarily served the Western Hemisphere. The predominant U.S. producer in the early part of the century was California, which as late as 1945 produced about 60% of the total U.S. crop. California's dominance was enhanced in these early years by several devastating freezes in the Eastern U.S., which severely reduced production in this area and drove the Eastern production area further south, essentially almost exclusively into Florida. Replantings in Florida after the freezes in the late 19th century established a vigorous citrus industry, but it was only after the development of FCOJ that Florida surpassed California in orange production. By the early 1950s Florida became the major citrus producer in the world, despite some setbacks due to freezes. Development of a strong demand for orange juice and the need for additional, less costly supplies of oranges led to the plantings of large acreages in some of the developing countries, mainly Brazil. This development, coupled with another series of freezes in the early 1980s in Florida, enabled Brazil to take over the number one position in orange production (Table 1).

B. COMMERCIALLY IMPORTANT VARIETIES

Sweet oranges are usually classified into four types — the common, acidless, pigmented, and navel oranges.[4,17] The most widely grown and commercially the most important, are the common sweet oranges. Approximately three-fourths of the total sweet oranges produced fall in this category. These include the major varieties used for the fresh fruit market and also the most important varieties used for processing. Total production of oranges by the major producing countries exceeded 30,000,000 metric tons during the 1986 to 87 season.

TABLE 1
Orange Production by Selected Countries (1978/79 to 1986/87)[a]

Country	1978/79	1980/81	1982/83	1984/85	1986/87
Argentina	706	668	639	653	630
Australia	399	352	392	489	470
Brazil	8154	9302	9210	11715	10975
Cuba	200	360	400	408	410
Cyprus	113	123	148	155	164
Egypt	843	921	1201	1182	1235
Greece	480	527	665	775	881
Israel and Gaza	1113	879	1039	1081	1042
Italy	1619	1735	1498	1960	2260
Mexico	1280	1600	1380	1000	1480
Morocco	630	685	670	686	650
South Africa[b]	579	569	510	486	625
Spain	1633	1693	1652	1365	2023
Turkey	656	695	656	760	700
U.S.	8502	9694	8827	6241	7158

Note. Values shown represent 1000s of metric tons.

[a] Includes tangerines and Temples in some countries.
[b] Includes some fruit produced in Swaziland, Botswana, and Mozambique, marketed through the South African Citrus Board.

Adapted from Reference 12.

Table 1 shows production by selected countries for the period, 1978 to 87. It is estimated that in excess of 50% of the total world production, and more than two-thirds of the output of the major producing countries, is processed, primarily into various juice products. The major varieties used in processing are described below.

1. Valencia

This variety[16,17] is the most widely grown among all of the sweet oranges. The origin of the Valencia orange is quite obscure, possibly originating from Portugal via the Azores. In the U.S. both California and Florida strains seem to have been supplied from the same source during the mid-19th century. The trees are quite adaptable to a variety of climates ranging from desert-like as in California and Arizona to hot and humid semitropical areas, such as Florida. The time from bloom to maturity can vary from 6 months in the warm, tropical areas near the equator to 18 months in the colder temperate areas of commercial production. Due to its high heat requirement, the Valencia is the latest maturing sweet orange variety.[27] The fruit is medium-large, oblong to spherical, containing few if any seeds. The internal color at maturity is quite good, although the peel under mild winters can be pale or can regreen. The juice flavor is excellent, but somewhat acid. The fruit holds well on the tree without deterioration and ships well. The Valencia orange is an excellent fresh fruit and is the most desired variety for the manufacture of FCOJ. Valencia oranges grown in the U.S. and the Americas are not identical with oranges grown around the region of Valencia, Spain, or Spanish varieties bearing the name, Valencia.

2. Pineapple

The Pineapple variety[16,17] originated as a chance seedling around Citra, Florida around 1860. The fruits are medium size, largely spherical, moderately seedy, juicy, sweet, possess good color, and excellent flavor. The name, Pineapple, was given to this variety due to its delicate fragrance. It is a mid-season maturing fruit, possessing good fresh fruit and processing qualities. In Florida its popularity increased recently, along with the early maturing Hamlin variety, due to the earlier maturity which can be an advantage in freeze years. This variety does not hold up as well on the tree as does the Valencia.

3. Hamlin

The Hamlin[16,17] is the earliest maturing variety grown in Florida. It originated as a chance seedling in the late 1800s near Glenwood, Florida. The fruit is medium-small, relatively round, sometimes oblate, contains few seeds, is juicy and sweet, with a mild flavor, low in acidity. The juice, particularly at the earliest harvests, has the least color of all Florida varieties. Although another variety, Parson Brown, has similar early maturing characteristics and good flavor, Hamlin largely supplanted it due to its lack of seeds. This variety has become quite important in Florida as it matures and can be harvested prior to any serious chance of a freeze.

4. Pera

By far the major variety grown in Brazil,[16,17] the Pera orange trees bear fruit which are medium to small and ovoid to ellipsoid in shape with few seeds. Juice is of good color, flavor, and sweetness. Pera is late maturing, holds well on the tree with little deterioration in quality, and stores and ships well. Nearly three-fourths of Brazil's orange and juice exports are of this variety.

C. OTHER COMMERCIALLY GROWN VARIETIES FOR PROCESSING

The varieties above are the most commonly grown in the major processing areas of the world. Other commercial varieties, such as the Shamouti or Jaffa, widely grown in Israel, the Cadenera and the Berna of Spain and North Africa, and the Calabrese of Italy are of considerable importance in Europe, both for fresh fruit and for processing. Their production volume is relatively small in comparison with the others produced by the major producers, Brazil and the U.S. Several additional varieties are grown in other citrus-growing regions and are undoubtedly important to the local economy. Excellent descriptions of the varieties grown globally can be found in the *Citrus Industry*.[16]

II. CULTURAL AND ENVIRONMENTAL CHARACTERISTICS

Except for the mandarin and the sour oranges, the sweet orange tree is the most hardy of all of the commercial citrus varieties. Cool temperatures are required for optimal peel color and the development of tolerance of low temperatures. Sweet orange trees and mature fruit can tolerate short periods of 28°F reasonably well, but extended periods below this temperature cause fruit and tree damage. Green fruit freezes at higher temperatures. When dormant, due to exposure to cool temperatures, sweet orange trees, but not the fruit, can withstand 18 to 20°F for short periods without injury.[17] In areas prone to periods of temperatures below freezing, various frost protection methods are employed to reduce or prevent damage to the trees and to a certain extent to the fruit. Such approaches rely on tree wraps and banking, wind machines, irrigation, heaters, or the combination of such methods. Certain rootstocks also afford some protection by increasing hardiness.

Under optimal growth conditions, the sweet orange tree can reach a height of 25 to 40 ft if not pruned. These favorable conditions, mild winters, adequate moisture and mineral nutrition, and sufficient total heat exposure (the requirement depending on variety), produce trees which yield the most desirable fruit-quality characteristics. However, under the usual weather circumstances in the citrus-producing regions of the world, such conditions are rarely found year after year.[27] Fertilization, irrigation,[24] growth regulators,[10] nutrition, and pesticide sprays,[1,22] as well as pruning are necessary to produce healthy trees of manageable shape and size for efficient production.[17]

ARID DESERT
cool nights
low rainfall

high sugar, high juice → low sugar, high acid
thin peel, poor color ← thick peel, brilliant color
fungal blemishes — minimal blemishes

HUMID TROPICS
warm nights
high rainfall

Oranges, as well as all other citrus fruits, are of tropical origin. Under typical hot and humid tropical conditions, green fruit color, high juice yield, and high sugar content are the norm. Most commercial citrus growing is in areas having more temperate conditions, higher elevations, or aridity. Such conditions can have dramatic effects on the external appearance and internal quality factors of the fruit.[23,27] The effects of climatic extremes can be summarized in the following diagram.[14,36]

As the fruit reaches a final growth plateau, warm days and cool nights cause a color change from green to yellow towards orange, indicating the onset of maturity. In areas with warmer nightly temperatures, peel color change is a poor indicator of the onset of maturity. Fruits, such as apples, bananas, avocados, etc., produce a burst of ethylene as the fruit approaches maturity. This in turn triggers a large increase of respiratory activity, known as the climacteric, with a concurrent color change indicating the onset of maturity. Oranges and other citrus fruits do not show such a climacteric phase triggering the final phase of maturity. There is some increase in ethylene production as the fruit color starts to change, but no significant changes in respiratory metabolism occur to indicate the onset of maturity. In oranges, maturity is primarily indicated by slowing of the accumulation of soluble solids and the slow decrease of titratable acidity.[3] Due to the lack of easily defined physiological maturity, the ratio of total soluble solids to titratable acidity (Brix/acid ratio) is used to define legal maturity in California[11] and in Florida.[31] Table 2 shows representative values of internal quality factors in Florida oranges.

The harvesting season, in light of the above considerations, depends on the variety and the climate in which the particular variety is grown. Local standards for maturity and marketing of the fruit largely capitalize on the quality factors achievable in the area, such as good external color and blemish-free skin in arid summer and cool winter regions as opposed to internal factors, such as high juice yield, high sugar content, and lower acidity in humid tropical and semi-tropical areas.[14,36] For purposes of fresh fruit sales, external appearance is more important, therefore the internal maturity factors are deemphasized, although in most areas a minimum Brix/acid ratio is established to maintain acceptable quality. For processing, internal quality is the overriding factor, therefore juice yield and higher Brix/acid ratio is emphasized. Not all varieties do equally well in all growing areas. Certain varieties, grown in several areas, show dramatically different external and internal characteristics, so much so that universal standards of maturity may not be appropriate.[14]

III. FACTORS INFLUENCING THE QUALITY OF THE CROP FOR PROCESSING

The orange varieties discussed earlier constitute the vast majority of the fruit processed. Characteristics desirable for the fresh market, such as good color and blemish-free skin, as less important for fruit destined for processing; as a result cosmetic rejects from packing houses are often sent to processing plants as long as they are wholesome and meet minimal standards established for maturity. These standards as discussed earlier involve internal quality factors such as required soluble solids, acidity, Brix/acid ratio, and juice color, low level of bitter components, and good flavor.[31,32] While the external appearance of fruit is less critical, control of fungal, yeast, and bacterial infections or insect damage which all can cause internal quality changes, reduced fruit, and juice yield are quite necessary.

A. ROOTSTOCK

Oranges are normally grown as specific scion and rootstock combinations for a variety of reasons, the most important being to reduce the period of juvenility associated with seedlings grown on their own roots. In addition, the selection of the appropriate rootstock for the specific soil conditions of the growing area is quite important. Effects of various rootstocks have been observed on juice yield, soluble solids, acidity, juice color, peel oil contents, and bitterness.[18,40] Table 3 shows observations of rootstock effects on juice yield, soluble solids, and titratable acidity,[40] and Table 4 shows effects on juice color.[40] As these factors are the basis of minimum standards for maturity of grade, judicious selection of the appropriate rootstock with consideration given to suitability for a given soil type, and disease resistance can be of great importance.

B. FERTILIZATION AND IRRIGATION

Fertilization and irrigation play important roles in the production of fruit for the fresh market and processing. Nitrogen and potassium are the two principal nutrients in citrus production. The other macronutrients and most of the micronutrients do not play significant roles except in case of severe deficiency.[24] In general nitrogen applications increase the juice content, soluble solids, acidity, and juice color. Excessive applications will lower fruit production. Potassium at high levels has a negative effect on juice quality, although its effect on increased fruit production per acre leads to increased soluble solids per acre. High rates of phosphorus causes lower acid content, thus increasing the Brix/acid ratio. On application to soils low in phosphorus increased fruit production is observed. Table 5 shows the observed effects of nitrogen and potassium fertilization.[24]

If moisture is limited, irrigation promotes tree growth and increases fruit production. A major role of irrigation is the enhancement of the effects of fertilization. Adequate irrigation will increase juice yield, soluble solids/acidity ratio, and fruit production. Excessive irrigation causes adverse effects in fruit quality.[24]

C. HARVESTING

Harvesting of oranges is primarily accomplished by manual labor. Although the technology is available for mechanical harvesting of citrus fruit, the tendency of the fruit to tightly adhere to the branches impedes efficient harvesting by this means. The force required for removal of oranges is usually provided by air blast or mechanical shakers which cause various degrees of peel injury, and possible tree damage due to the vigorous shaking. Fruit harvested this way is not usually suitable for fresh use, and often, depending on the delay between harvest and delivery to the processing plants, is prone to microbial decay. Research programs have focused on the development of abscission promoting chemicals which cause significant loosening of the fruit to enable tree shakers to remove fruit more efficiently, without serious tree damage. Most of the active compounds were found to induce the formation of ethylene by superficial damage to the fruit peel, which in turn causes the formation of an abscission layer at the attachment of the fruit to the branch. The ideal compound needs to be selective so as to act only on mature fruit and not on leaves and immature fruit. This last requirement is due to the differing growth habits of the varieties grown. Early and mid-season oranges normally have fruit only at the same stage of growth on the trees. Late season oranges, in contrast, have both mature and very young fruit present at the same time, and therefore need the selective action on the mature fruit. Currently 4 compounds are available under experimental label, each showing somewhat different characteristic action. Weather conditions have a profound influence on the action of these agents. With the aid of these abscission agents mechanical harvesting could become more practical, although the fruit will probably be used for processing due to the peel damage.[39]

TABLE 2
Representative Quality Factors of Florida Oranges[a]

Total soluble solids (TSS), °Brix	9.1—13.6
Total sugars, g/100 ml	7.17—10.72
Sucrose, g/100 ml	3.15—5.75
Titratable acidity, % by wt.	.69—1.17
Ratio (TSS/TA)	9.6—17.7
pH	3.3—4.0
Oil, % by vol.	.005—.031
Chemical oxygen demand, ppm	80—560
Aldehydes, ppm	9.2—39.6
Oxygenated terpenes, ppm	3.5—14.3
Glycosides, mg/100 ml	70—155
Limonin, ppm	0—12.5
Sinking pulp, % by vol.	8—22
Cloud, % light transmission	14—39
Alcohol insoluble solids, mg/100 ml	270—584
Water insoluble solids, mg/100 g	99.6—377.8
Total pectin, mg/100 g	47.3—119.5
Water soluble pectin, mg/100 g	12.0—65.2
Hydroxide soluble pectin, mg/100 g	23.3—65.2
Serum viscosity, cps	1.39—3.50
Protein, mg/100 ml	93—192
Ash, mg/100 ml	367—542
Potassium, mg/ 100ml	202—320
Magnesium, mg/100 ml	8.8—13.6
Calcium, mg/100 ml	1.3—6.2
Sodium, mg/100 ml	.22—1.04
Color score, CN	32.3—38.9
Citrus red, CR	16.3—41.5
Citrus yellow, CY	61.2—86.2
β-Carotene[b], I.U./100 ml	190—400
Thiamine[b], μg/100 ml	60—145
Riboflavin[b], μg/100 ml	11—90
Pyridoxine[b,c] μg/100 ml	25—80
Folic acid[b], μg/100 ml	1.2—3.3
Ascorbic acid[b], mg/100 ml	35—56

[a] Values from Hamlin, Pineapple, and Valencia oranges.
[b] Values from Reference 33.
[c] Includes pyridoxal and pyridoxamine.

Adapted from Reference 2.

IV. PROCESSING

A. ORANGE JUICE

During the 1986 to 87 season more than half of all of the oranges produced by the major citrus growing countries were processed (Table 6). The two primary producers, Brazil and the United States, were responsible for over 85% of the total processing output. Most of this was in the form of juice, primarily FCOJ. Other products manufactured, seemingly minor in comparison to juice production, but nevertheless significant to local economies, are cold-pressed peel oil, dried peel and pulp for cattle and poultry feed, chilled or canned sections, candied peel, and various specialty items.

TABLE 3
Effect of Rootstocks on Selected Juice Characteristics

Rootstock	Juice yield (%)	TSS[a] (%)	TA[b] (%)	TSS/TA
Hamlin orange				
Troyer citrange	44	10.4	0.69	15.2
Citrumelo F 80-8	47	10.4	0.72	14.5
Sour orange	44	10.3	0.80	12.9
Citrumelo F 80-18	47	10.2	0.72	14.2
Trifoliate orange	46	10.2	0.63	16.4
Swingle citrumelo	45	10.1	0.72	14.0
Hybrid FF 1-131-20	46	10.1	0.60	16.9
Rough lemon	25	8.3	0.60	13.8
C. volkameriana	29	8.2	0.55	15.0
Valencia orange				
Citrumelo F 80-8	64	12.3	0.95	13.0
Swingle citrumelo	64	12.0	0.99	12.2
Smooth flat Seville	63	11.5	0.90	12.7
Rangpur × Troyer	59	11.5	0.84	13.6
Carrizo citrange	61	11.3	0.88	12.9
Rough lemon	58	10.3	0.81	12.7

[a] Total soluble solids.
[b] Titratable acidity.

Adapted from Reference 40.

Upon delivery of fruit from the field a sample of each delivered load is tested for maturity.[14,31,32] The delivered fruit may be temporarily placed in storage bins. Fruit is then washed, sorted, with that intended for the fresh market receiving fungicide and wax treatment, and the defective fruit removed. The fruit to be juiced is sent by conveyors to mechanical extractors. In the U.S. either the FMC In-Line (FMC Corp.) or the Brown (AMC Corp.) extractors are used in most processing plants. Both types of extractors require fruit which are carefully sized.[5]

The FMC In-Line extractor consists of 5 pairs of one fixed and one moving cup, constructed with intermeshing fingers and a central perforated tube with a cutting edge. After the fruit is deposited in the lower cup, the upper cup descends and the sharp edge of the tube cuts a hole in the fruit. As the fingers of the upper and lower cups mesh together, the fruit is squeezed against the perforated tube, expressing the juice from the fruit. A restrictor in the lower end of the tube prevents loss of juice and forces it through the perforations. This also provides for the initial removal of seeds and larger pulp particles. When the upper cup is firmly seated, the perforated tube with the restrictor rises to compress the contents of the tube, which results in the recovery of the remaining juice and ejection of the plug of seeds, pulp, and peel. As the fruit is compressed, a spray of water is applied to the outside of the fruit to recover the expressed peel oil as an aqueous oil slurry. Adjustment of cup type, size, and arrangement of perforations, restrictors, and the clearance can determine the extent of the squeeze and therefore the type of the juice produced. Capacity of the extractors depends on the number of strokes per minute, usually 75 to 100.[5]

TABLE 4
Effect of Rootstocks on Juice Color (2 Year Average

Rootstock	Color numbers (CN)
	Hamlin orange
Trifoliate orange	34.3
Troyer citrange	34.0
C. macrophylla	34.0
Swingle citrumelo	33.8
Sweet orange	33.7
C. volkameriana	33.7
Rough lemon	33.2
Sour orange	32.9
	Valencia orange
Trifoliate orange	37.2
Cleopatra mandarin	37.0
Swingle citrumelo	37.0
Carrizo citrange	36.9
Sour orange	36.7
Palestine sweet lime	36.2
C. volkameriana	35.6

Adapted from Reference 40.

The Brown extractors operate by halving the fruit and applying the cut half against rotating reamers. The extent of extraction can be varied by changing the depth of penetration of the rotating reamers. With the Brown extractors if peel oil recovery is desired it is performed as a preliminary step by pricking the oil glands in the fruit peel prior to juice extraction on a series of needle equipped rollers, with the oil collected as an aqueous emulsion, or by cold pressing the peel after juice extraction with a screw-type press.[31]

Both types of extractors have capabilities of controlling the amount of peel oil in the extracted juice. The juice from both types of extractors is sent to finishers to remove pieces of peel, seeds, and excess pulp particles. Finishers are essentially screening devices, and are either screw or paddle-type. They may be adjusted by varying the screen perforation size to set the nominal particle size of the finished juice and by changing the pressure on the restrictor device of the peel and pulp discharge. The pressure applied determines the type of juice produced.[2] Juice yield obtained is directly related to extraction and finishing pressures and has an important bearing on the sensory quality of the juice produced. In Florida yield control is voluntary but yield monitoring and reporting is required by law.[31,32] In Florida the yield is determined using official state test extractor and analysis systems operated in each processing plant by State of Florida, Department of Agriculture and Consumer Services inspectors. Representatives of growers, processors, and regulatory agencies meet at regular intervals during the processing season to determine a fair and equitable voluntary yield control factor and to make adjustments depending on changes in conditions during the season.[5,14]

The various types of orange juice produced commercially can be classified as FCOJ, fresh unpasteurized, single-strength chilled, and canned. Each category can be further subdivided into several variations of the type of the product and furthermore, blends with other juices are also produced.

TABLE 5
Effects of Rates of Nitrogen and Potassium on Juice Quality[a]

	N (lb/A/yr)				K (lb/A/yr)			
	70	**140**	**210**	**280**	**70**	**140**	**210**	**280**
Juice yield (%)	58.6	59.2	59.2	59.6	59.5	59.3	59.2	58.7
TSS[a] (%)	11.3	11.7	11.9	12.4	12.0	11.9	11.8	11.6
TA[c] (%)	0.71	0.75	0.78	0.80	0.76	0.79	0.83	0.83
TSS/TA	16.0	15.7	15.3	15.4	12.0	11.9	11.8	11.6
Color (CN)	38.5	38.8	38.9	38.8	38.8	38.7	38.7	38.7

[a] Florida Valencia oranges.
[b] Total soluble solids.
[c] Titratable acidity.

Adapted from Reference 24.

TABLE 6
Oranges Processed by Selected Countries
(1983—84 to 1987—88; 1000s of metric tons)

	Production year				
Country	**1983—84**	**1984—85**	**1985—86**	**1986—87**	**1987—88**
Argentina	100	120	100	195	165
Australia	272	333	340	274	309
Brazil	7162	8650	6936	7548	7507
Cuba	10	90	100	110	115
Cyprus	13	22	17	44	20
Egypt	7	7	7	8	8
Greece	90	165	60	148	79
Israel and Gaza	521	559	329	484	297
Italy	514	450	600	575	320
Mexico	320	70	220	343	400
Morocco	131	121	145	46	227
South Africa[a]	120	140	155	172	160
Spain	175	230	163	140	118
Turkey	80	96	70	75	70
U.S.	4957	4503	4971	5201	5836

[a] Includes some fruit produced in Swaziland, Botswana, and Mozambique, marketed through the South African Citrus Board.

Adapted from Reference 34.

1. Frozen Concentrated Orange Juice (FCOJ)

Frozen concentrated orange juice (FCOJ) was developed by researchers of the Florida Citrus Commission at the U.S. Department of Agriculture Laboratory in Winter Haven, Florida during the early and mid-1940s. Although there were various techniques available to concentrate and/or dehydrate citrus juices prior to this time, the products were seriously defective in appearance and in flavor. The loss of volatile flavor components and heat damage during manufacture made these products barely palatable. Advances in engineering design led to the use of a high vacuum, low temperature evaporator for juice concentration, which significantly reduced the heat damage plaguing earlier attempts. The overconcentrated product was added to a small quantity of fresh ''cut-back'' juice to achieve the desired concentration and most significantly, to restore the fresh flavor and aroma lost during the concentration process. The resulting product was packaged and stored frozen, retaining the desirable flavor, appearance and nutrients for long periods of time. The developers, L. G. MacDowell, E. L. Moore and C. D. Atkins, were granted U.S. Patent Number 2,543,109 in 1948,[25] and assigned it to the U.S. Government. The Winter Haven Laboratory and the cooperating research groups received the Distinguished Service Award from the U.S. Department of Agriculture in 1956 for the development of FCOJ. The success of the FCOJ is evidenced by the fact that from its first commercial season in 1946 to 1987/88, production grew from 852,000 l to 647,160,000 l in Florida alone, despite several series of freezes in the interim.

The original low temperature evaporators for FCOJ production have been replaced by modern multi-effect, multi-stage, high temperature, short time evaporators, which eliminated some of the problems due to incomplete inactivation of some of the pectic enzymes. In Florida, Brazil, and several other producing countries the most commonly used evaporator is the thermally accelerated short term evaporator (TASTE), consisting of multiple, cascaded sets of tubular heat exchangers. Juice flow and the temperatures of the successive stages are controlled to optimize evaporation while reducing heat exposure of the processed juice.[7] In the initial stages of evaporation, the volatile flavor and aroma components are collected. This product is an aqueous and oil mixture, which is subsequently separated into flavor fractions called aqueous essence or aroma and essence oil. After refining and standardization, these products are used to restore flavor to the concentrate before packaging. In recent developments, both the control of the evaporator heating and the juice flow is performed by microprocessor controlled equipment.[7] Plate type heat exchangers, operating on a similar high temperature, short term heating basis are employed mainly by European producers.[15]

FCOJ produced today is a highly standardized and regulated product, where numerous parameters have to meet certain federal and state requirements.[9] During manufacture of the final product, juices of various origin are concentrated and stored in refrigerated "tank farms". The frozen concentrates are blended to achieve the desired product, flavor is restored by added peel oil and essence. Fresh "cut-back" juice is rarely, if ever, used today to restore flavor to FCOJ. The blend is packaged and distributed in frozen form.

One of the newer developments in the production of orange juice concentrate employs freeze concentration. With this method the juice is subjected to controlled freezing and the ice crystals are allowed to grow to sufficient size to permit removal by differential screening from the slowly circulating slurry. The quality of the product is superior to most conventionally produced FCOJ. The equipment and operating cost, the stringent requirement for optimum oil content, pulp removal, and pasteurization for stability prior to concentration makes this method considerably more expensive than the FCOJ produced by the TASTE evaporator. In addition, the size of the equipment is far larger than the equivalent capacity conventional evaporator.[8] The superior quality product, however, is attractive for enhancement of the quality of conventional FCOJ.

2. Chilled Orange Juice

The availability of refrigerated storage in the home created considerable demand for chilled juice due to its convenience and high quality. When produced as a pasteurized product from the juice lines directly, chilled juice comes closest to fresh juice in flavor and aroma. Initially chilled juice products were primarily distributed to restaurants and through dairy distribution channels, as dairies already had the sanitary processing equipment and the low temperature delivery systems. The ever increasing demand caused processors to increase their involvement in this area and it also led to the development of new, improved variations on this product. Chilled juices are distributed in paper cartons, plastic and glass bottles, and plastic jugs of assorted sizes. Production of chilled juices can be directly from the fresh juice lines in processing plants or reconstituted from concentrated juices. Chilled juice is either produced from fresh juice or reconstituted from FCOJ. When produced from fresh juice, it may be pasteurized and the containers (mostly glass and plastic) are either filled hot, or with heat sensitive packaging, the pasteurized juice is rapidly cooled and the sterilized containers are filled cold under aseptic conditions. Blending of various juices may be employed with early season fruit to achieve the required minimum color score. Chilled juice reconstituted from FCOJ is pasteurized, peel oil is added for flavor, and packaged similarly. Recently there has been a marked increase in production of a form of chilled juice, which receives minimal treatment after extraction before packaging and distribution. It is sold as fresh unpasteurized orange juice[6] and has the best flavor of all commercially marketed orange juices. Such juice has a very short storage life and is intended for rapid consumption, since clarification and microbial spoilage occurs within a few short days. Nevertheless, the high flavor quality of this product causes sufficient demand which numerous producers are willing to satisfy. Several varieties of orange and other fruit juice blends as well as fortified juices are also packaged and distributed in chilled form. Specifications for the grades of these products exist in U.S.[35] and Florida regulations.[32] These products require refrigerated storage, recommended to be below 10°C, but preferably near freezing for maximum stability of flavor and appearance. Although some of the products, particularly those packed in glass, seem to be stable at ambient temperatures, deterioration of flavor, color, and the loss of ascorbic acid is accelerated under those conditions.[5] Juices packed in plastic containers or plastic lined cartons stored under refrigerated conditions show loss of flavor by migration of the compounds responsible for the typical orange flavor into the plastic container wall or into the liner, thus reducing a desirable attribute.[28] Chilled orange juice is also delivered to distribution points, usually dairies, in refrigerated tankers where the juice is packaged in cartons or bottles.[5]

3. Canned Orange Juice

Canned single-strength orange juice is one of the oldest products of the processed citrus juice industry. Although it has been mainly supplanted by chilled and concentrated orange juices due to their higher quality and the ever increasing availability of home refrigeration, considerable quantities are still made. Canned juices are produced by reconstituting FCOJ or by blending the juices extracted from several varieties of oranges, followed by deaeration, deoiling, pasteurization, and canning.[5] Certain canned orange juices are produced with added sugar, although any added ingredients must be declared on the label.[31,32,35] Deaeration serves primarily to reduce flavor deterioration, to prevent lowering of ascorbic acid levels due to dissolved oxygen, and to reduce frothing during the filling step. Deoiling is performed to control the peel oil level of freshly extracted juices prior to packaging. It is done under vacuum with the application of mild heat to produce vapors which carry off about 90% of the volatile peel oil. The U.S. Standards for orange juice specify an upper limit for peel oil content, measured by titration as D-limonene.[13,26]

As citrus juices in their raw state contain various active enzymes as well as possible microbial contaminants, pasteurization is necessary to stabilize these juices. Microorganisms are inactivated at orange juice pH at approximately 71°C, but higher temperatures are needed to prevent loss of cloud due to enzymic action.[5] The juice is rapidly heated to about 92°C in tubular or plate heat exchange type pasteurizers for up to about 30 s and from there it is pumped hot into filler bowls and rapidly filled into cans. Live steam may be injected into the headspace followed by closing of the can. The cans are inverted to sterilize the inside of the lid, and rotated on conveyors with a spray of cold water to minimize flavor changes caused by long-term heating. The cans emerge with sufficient residual heat to facilitate drying and thereby prevent rusting. Plain tinned or enameled cans are usually used, with tin playing a significant role in the prevention of discoloration (browning) and the loss of ascorbic acid.[5]

Canned orange juices are normally stored at ambient temperature. However, studies indicate that it is desirable to store the juices at the lowest practical temperature, as temperatures greater than about 20°C cause accelerated deterioration of flavor, appearance, and ascorbic acid content.[19] Although tin coated cans are suitable and even desirable for long term storage of canned orange juices due to the protective action on ascorbic acid and appearance, excessive leaching of tin at elevated storage temperatures can be cause for concern.[5]

Quality control of the production steps and the individual products described above is performed by state and federal inspectors located in the various processing plants.[13,26,37] In Florida the legislature enacted the Florida Citrus Code of 1949, the most comprehensive piece of food quality control legislation ever adopted. Through this code, the Florida Citrus Commission (created by legislative action in 1935), now the Florida Department of Citrus, was empowered to develop strict regulations in the manufacture of FCOJ and other processed citrus products. Quality regulations are monitored by a continuous inspection program of the Florida and U.S. Departments of Agriculture. This program covers all phases of the manufacturing process, from the fruit through equipment to warehousing.[13] The various orange juice products are defined in the Standards of Identity under the Federal Food, Drug and Cosmetic Act.[35]

B. BY-PRODUCTS

1. Orange Flavor Volatiles

During the initial stages of evaporation the volatile components found in orange juice are removed with the water vapor, as mentioned earlier. These volatile fractions contain most of the compounds responsible for orange flavor. These flavor fractions, collected throughout the season from the different varieties of fruit processed, may be standardized by the producer or sent to flavor companies specializing in the refining and standardization of the aqueous essences and essence oils. Aqueous essence, which consists mainly of ethanol, with aldehydes, esters and terpenoid hydrocarbons contributing to the characteristic orange flavor, is distilled and standardized on certain aldehydes and esters deemed important to good flavor and aroma.[28,30] The standardized essences are used to restore flavor to the concentrates prior to packaging. The essence oils, containing mainly terpenoid compounds, are also refined and used for various flavoring purposes.

2. Orange Peel Oils

Orange peel oils are an important by-product of the production of orange juice. Peel oil is usually collected as an aqueous slurry either simultaneously with the extraction of juice (FMC), prior to extraction (AMC) or by cold pressing the peel residues after the extraction, as described earlier. The oil-water slurry is put through a finishing operation, through a conventional finisher, followed by centrifugation to remove particulate matter and separate the aqueous phase. This product is known as cold pressed oil and is used as flavoring for juices and various products desiring orange flavor. The cold pressed oils are often distilled to produce oil products reduced in terpene hydrocarbons. The distillation step, known in the industry as "folding", enriches the oil in oxygenated components and higher boiling sesquiterpenes, removing most of the d-limonene. The folded oils are used as flavoring materials for various products. The hydrocarbon fraction, mostly d-limonene, is used as a solvent or base material for various synthetic products.[21,29]

3. Water Extracted Soluble Solids

The finisher pulp residues from the extracted juices contain considerable amounts of soluble solids trapped in the particulate material. Some processors employ a series of finishers with water flowing through the finisher pulp in a countercurrent configuration, sometimes with pectic enzymes added. The trapped soluble material thus extracted is then concentrated in normal juice evaporators and used in beverages. This material, known as water extracted soluble solids (WESOS) or more commonly as "pulp-wash", cannot be added to any product labeled as orange juice which is packaged in Florida.[31,32]

4. Waste Products

The production of orange juice results in considerable amounts of remaining waste materials in the form of peel and pulp residues. Initially, these wastes were dumped on pastures and fed wet to livestock. By 1974 to 75, in excess of 3.5 million tons of such residues were produced for disposal. Most of this material, which includes grapefruit residues, was converted into dried citrus pulp and pellets, used primarily as livestock feed supplements.[20] In the initial processing of the wet material, calcium hydroxide is added to the peel material which is subjected to continuous presses to produce a peel juice. This press liquor contains about 9 to 15% soluble solids, which are 60 to 75% sugars. The press liquor is concentrated to a syrup, molasses, used as feed additive to cattle or for the production of alcohol.[20] Sweet oranges are also utilized in numerous specialty products, including flavonoids, pectin, citrus sections and salads, candied peel, syrup, and wines.[20] Commercial importance of these products, however, is minor compared with orange juice and are primarily produced by small producers. An exhaustive compilation of manufacturing of various by-products and specialty products is discussed in Reference 20.

REFERENCES

1. **Albrigo, L. G.,** Stress factors affecting external and internal quality of citrus fruit, in *Factors Affecting Fruit Quality. Proc. 1988 Citrus Short Course,* Lake Alfred, FL. Ferguson, J. J. and Wardowski, W. F., Eds., 1988, 69.
2. **Attaway, J. A., Barron, R. W., Blair, J. G., Buslig, B. S., Carter, R. D., Dougherty, M. H., Fellers, P. J., Fisher, J. F., Hill, E. C., Huggart, R. L., Maraulja, M. D., Petrus, D. R., Ting, S. V., and Rouse, A. H.,** Some new analytical indicators of processed orange juice quality, 1971-72, *Proc. Fla. State Hort. Soc.,* 85, 192, 1972.
3. **Bain, J. M.,** Morphological, anatomical and physiological changes in the developing fruit of the Valencia orange, *Citrus sinensis* (L.) Osbeck. *Aust. J. Bot.,* 6, 1, 1958.
4. **Batchelor, L. D. and Sinclair, W. B.,** World Production and Important Commercial Varieties of Oranges, in *The Orange,* Sinclair, W. B., Ed., University of California, Division of Agricultural Sciences, 1961, chap. 1.
5. **Berry, R. E. and Veldhuis, M. K.,** Processing of oranges, grapefruit and tangerines, in *Citrus Science and Technology,* Vol. 2, Nagy, S., Shaw, P. E., and Veldhuis, M. K., Eds., AVI Publishing, Westport, CT. 1977, chap. 4.
6. **Carter, R. D.,** Florida orange juice. Commonly called fresh-squeezed Florida orange juice. Production/Packaging/Distribution. Florida Department of Citrus, Lake Alfred, FL. 1987.
7. **Carter, R. D.,** Heating steam control in a TASTE evaporator using a microprocessor controller, *Proc. Int. Soc. Citriculture,* 1988, 1717.
8. **Carter, R. D.,** Some recent advances in the citrus processing industry in Florida, *Proc. Int. Soc. Citriculture,* 1988, 1697.
9. **Chadwell, K. C.,** Citrus processing regulations: Florida standards. *The Citrus Industry* 69(12), 4, 1988.
10. **Davies, F. S. and Jackson, L. K.,** Plant growth regulators and fruit quality in citrus, in *Factors Affecting Fruit Quality. Proc. 1988 Citrus Short Course,* Lake Alfred, FL. Ferguson, J. J. and Wardowski, W. F. Eds., 1988, 49.
11. **Erickson, L. C.,** The General Physiology of Citrus, in *The Citrus Industry,* Vol. 2, Reuther, W., Webber, H. J., and Batchelor, L. D., Eds., University of California, Division of Agricultural Sciences, 1968, chap. 2.
12. Florida Agricultural Statistics Service. Citrus Summary, 1982, 1984 and 1987-88. Florida Department of Agriculture and Consumer Services. 1983, 1985 and 1989.

13. Florida Department of Citrus. Reconstituted Florida orange juice. Production/Packaging/Distribution. 2nd rev., Carter, R. D., Ed., Florida Department of Citrus, Lake Alfred, FL, 1985.
14. **Grierson, W. and Ting, S. V.,** Quality standards for citrus fruits, juices and beverages. *Proc. Int. Soc. Citriculture* 21, 1978.
15. **Heid, J. L. and Casten, J. W.,** Vacuum concentration of fruit and vegetable juices, in *Fruit and Vegetable Juice Processing Technology*, Tressler, D. K., and Joslyn, M. A., Eds., AVI Publishing, Westport, CT. 1961, chap. 9.
16. **Hodgson, R. W.,** Horticultural varieties of citrus, in *The Citrus Industry*, Vol. 1, Reuther, W., Webber, H. J., and Batchelor, L. D., Eds., University of California, Division of Agricultural Sciences, 1967, chap. 4.
17. **Hume, H. H.,** *Citrus Fruits*, Macmillan, NY, 1957.
18. **Kefford, J. F. and Chandler, B. V.,** The influence of rootstocks on the composition of oranges, with special reference to bitter principles, *Aust. J. Agric. Res.*, 12, 56, 1961.
19. **Kefford, J. F., McKenzie, H. A., and Thompson, P. C. O.,** Effects of oxygen on quality and ascorbic acid retention in canned and frozen orange juices, *J. Sci. Food Agric.*, 10.51, 1959.
20. **Kesterson, J. W. and Braddock, R. J.,** By-products and specialty products of Florida citrus, *Bull. 784*, Agric. Experiment Stations, Inst. Food and Agric. Sciences, University of Florida, Gainesville, FL, 1976.
21. **Kesterson, J. W., Hendrickson, R., and Braddock, R. J.,** Florida citrus oils, *Bull. 749*, Agric. Experiment Stations, Inst. Food and Agric. Sciences, University of Florida, Gainesville, FL, 1971.
22. **Knapp, J. L., Tucker, D. P. H., Noling, J. W., and Vandiver, V. V., Jr.** Florida Citrus Spray Guide - 1988. *Circ. 393-N*, Florida Cooperative Extension Service, University of Florida, Gainesville, 1988.
23. **Koch, K. E.,** Production and environmental factors affecting the Brix/Acid ratio, in *Factors Affecting Fruit Quality, Proc. 1988 Citrus Short Course*, Lake Alfred, FL. Ferguson, J. J., and Wardowski, W. F., Eds., 1988, 78.
24. **Koo, R. C. J.,** Irrigation and fertilization effects on fruit quality, in Factors Affecting Fruit Quality, *Proc. 1988 Citrus Short Course*, Lake Alfred, FL. Ferguson, J. J., and Wardowski, W. F., Eds., 1988, 35.
25. **MacDowell, L. G., Moore, E. L., and Atkins, C. D.,** Method of preparing full-flavored fruit juice concentrates, U.S. Patent 2,453,109, Nov. 9, 1948.
26. **Redd, J. B., Hendrix, C. M., Jr., and Hendrix, D. L.,** Quality control manual for citrus processing plants, Intercit, Inc. Safety Harbor, FL 1986.
27. **Reuther, W.,** Climate and fruit quality, in *Factors Affecting Fruit Quality, Proc. 1988 Citrus Short Course*, Lake Alfred, FL. Ferguson, J. J., and Wardowski, W. F., Eds., 1988, 9.
28. **Shaw, P. E.,** Citrus essential oils, *Perfumer and Flavorist*, 3(6).35., 1979.
29. **Shaw, P. E.,** Review of Quantitative analyses of citrus essential oils, *J. Agric. Food Chem.*, 27.246, 1979.
30. **Shaw, P. E., and Wilson, C. W., III.,** Importance of selected volatile components to natural orange, grapefruit, tangerine and mandarin flavors, in *Citrus Nutrition and Quality, ACS Symp. Ser.*, No. 143., Nagy, S., and Attaway, J. A., Eds., American Chemical Society, Washington, D. C., 1980, 167.
31. State of Florida, *Florida Citrus Fruit Laws*, Chapter 601, Florida Statutes, State of Florida, Department of Citrus. 1980 ed. and subsequent revisions.
32. State of Florida, *Official Rules Affecting the Florida Citrus Industry*, Pursuant to Chapter 601, Florida Statutes, State of Florida, Department of Citrus, Effective January 1, 1975 and subsequent revisions.
33. U.S. Department of Agriculture. Chemistry and technology of citrus, citrus products and byproducts, *Agric. Handbo. No. 98*, U.S. Department of Agriculture, Agricultural Research Service, Washington, D. C., 1956.
34. U.S. Department of Agriculture, *Horticultural Products Review, Circ. Ser.*, FHORT 7-85, 1-86, 1-87, 12-88, Foreign Agricultural Service, 1985, 1986, 1987, 1988.
35. U.S. Department of Agriculture, U. S. Standards for Grades of Orange Juice, Effective January 10, 1983, Agricultural Marketing Service, Fruit and Vegetable Division, Processed Products Standardization and Inspection Branch., incl. in Code of Federal Regulations, Title 7, part 52, rev. Jan. 1, 1989.
36. **Wardowski, W. F.,** Factors affecting fruit quality, in *Factors Affecting Fruit Quality, Proc. 1988 Citrus Short Course*, Lake Alfred, FL. Ferguson, J. J., and Wardowski, W. F., Eds., 1988, 5.
37. **Wardowski, W. F., Soule, J., Grierson, W., and Westbrook, G.,** Florida citrus quality tests, *Bull. 188*, Florida Coop. Extension Service, Inst. Food and Agric. Sciences, University of Florida, Gainesville, FL. 1979.
38. **Webber, H. J.,** History and development of the citrus industry, in *The Citrus Industry*, Vol. 1., Reuther, W., Webber, H. J., and Batchelor, L. D., Eds., University of California, Division of Agricultural Sciences, 1967, chap. 1.
39. **Wilson, W. C., Coppock, G. E., and Attaway, J. A.,** Growth regulators facilitate harvesting of oranges, *Proc. Int. Soc. Citriculture* 1981, 278.
40. **Wutscher, H. K.,** Rootstock effects on fruit quality, in *Factors Affecting Fruit Quality, Proc. 1988 Citrus Short Course*, Lake Alfred, FL. Ferguson, J. J., and Wardowski, W. F., Eds., 1988, 24.

Chapter 2*

THE LEMON

Manuel G. Moshonas and Philip E. Shaw

TABLE OF CONTENTS

* Mention of a trademark or proprietary product is for identification only and does not imply a warranty of the product by the U.S. Department of Agriculture over other products which may also be suitable.

I. INTRODUCTION

The lemon came into existence so long ago that its origin cannot be accurately determined. It is fairly well established, however, that the lemon was first grown in southeastern Asia, probably in southern China. By the 10th century, lemons were cultivated in Egypt and by the 12th century were grown extensively in Persia and Palestine. During that period the Arabs (and later the Crusaders) took the lemon to Spain, northern Africa, and Italy from where lemon plantings spread throughout southern Europe. The Europeans, led by the Italians, cultivated lemon and expanded the industry to the point where they became the suppliers of the world markets. Columbus brought lemon seeds to Haiti which led to introduction of the lemon to America. The lemon's first home in North America was Florida. By the mid-18th century the commercial importance of the lemon was enhanced when James Lind, a British naval doctor discovered that the disease ''scurvy'', which often struck sailors who remained at sea for long periods, was due to a lack of an unknown substance that could be provided by a daily ration of lemons or limes. This unknown substance later proved to be vitamin C. During this period, the lemon emerged as an important commercial crop in Florida. However, freezes, diseases, and effects of high humidity dealt a heavy blow to the Florida lemon industry and greatly reduced the commercial importance of the lemon in Florida. As the Florida lemon industry declined, the California (and to a much lesser extent, the Arizona) industry which began in 1841, developed into one of the world's major producers and distributors of lemons. These states supply fresh fruit to nearly every household in the United States and export just as many fresh lemons to the world markets. In recent years, lemon production in Florida has increased, but the primary outlet for the fruit is processed products. Italy is the only other lemon producer that rivals U.S. production. Other nations that produce lemons as a commercial crop include Argentina, Australia, Brazil, Chile, Cyprus, Greece, Israel, Lebanon, Morocco, Spain, South Africa, Tunisia, and Turkey.

A. HISTORICAL DEVELOPMENT OF MAJOR COMMERCIAL VARIETIES

1. Eureka

The Eureka variety, grown extensively in California and Arizona, originated in California around 1858 when it was selected by W.C. Workman from seedlings grown from lemon seeds that originated in Italy. Two decades later this variety was propagated from budwood by T.A. Carey who later introduced the Carey's Eureka. The Eureka tree's ability to bear early, together with high yield, has made it one of the most widely grown lemon trees. Careful selection of seedlings from Californian Eureka clones has yielded closely related, but more hardy, varieties (Allen, Cascade, Cook, Meek, and Thorton) which have gained wide acceptance. Further development of these varieties through selection and nucellar budlines is currently being developed by California nurserymen.

2. Lisbon

The Lisbon variety originated in Portugal and is believed to have been a clonal seedling of the Gallego variety. In the mid 1800s the Lisbon lemon was brought to California where it became increasingly popular in the interior areas of that State and in Arizona. Along with its desirable characteristics of vigorous growth, uniform fruit size and prolific production, the tree is very hardy and less susceptible to cold than other varieties. The cloning and selection process has led to related varieties such as the Bradbury, Cavors, Deaver, Hall, Jameson, Ledig, Limoneira-8A, Monroe, Prior, Prospect, Strong, and USDA. The Bearss variety, currently popular with Florida growers, is believed to be a clone selection of the Lisbon. One known nucellar variety of Lisbon developed by H. B. Frost and named Frost Lisbon, was released in 1950. Selection and use of nucellar budlines of other popular Lisbon clones are currently underway.

3. Femminello Ovale

The Femminello Ovale variety was a clone of the Femminello Stusato, one of the oldest Italian varieties. This variety accounts for over 75% of Italy's production and is especially adaptable to the "forcing" treatment widely used by Sicilian lemon growers. The "forcing" or "verdelli" practice begins with withholding water from lemon groves in June and July until the trees begin to wilt. The trees are then generously fertilized and irrigated, "forcing" an out-of-season bloom which results in a large lemon crop from April through August when demand for lemons is high because of its greater resistance to disease. A clonal selection of the Femminello Ovale called Santa Teresa has emerged as the most popular new lemon variety being planted in Italy.

4. Genoa

The Genoa lemon was named after the city in Italy where it was first grown. In 1875 this variety was brought to California by Don J. Rubio. After the Genoa lemon was introduced in Argentina and Chile it became the dominant variety for commercial production in those countries.

5. Berna

The Berna lemon is believed to have originated in Spain where it accounts for over 90% of the commercial production there. The similarity to the fruit and tree of the Lisbon, suggests that the Berna may have been a Lisbon clone. The Berna is also grown commercially in Morocco and Algeria.

II. AGRONOMIC CHARACTERISTICS

The lemon tree, the most cold-sensitive of all citrus trees, can be grown commercially only in subtropical regions. Under the most favorable conditions, which include mild winters, cool summers, light rainfall, and low humidity, the tree grows vigorously and can reach a height of 20 ft if not pruned. Although lemon trees are similar to orange trees in cultivation, longevity, appearance, and propagation, they derive more benefit from heavy pruning. In high humidity, warm regions, such as Florida, lemon trees are more susceptible to fungal diseases—particularly citrus scale. The trees flower and produce lemons throughout the year but reach a maximum output either in late spring or summer in the coastal regions of Spain, Northern Africa, Sicily, Greece, and Southern California, or, in the fall and winter in the interior growing regions. Lemons grown commercially are harvested while still green and picked according to size rather than maturity. When the fruit no longer passes through a 2 1/4 in. diameter ring carried by pickers, it is harvested. The fruit are then washed, waxed, graded, and sorted for maturity based on color. Those closest to maturity are separated and placed in cold rooms until the proper color is attained. The major portion of the freshly picked fruit are immature and are stored, treated, and controlled in curing rooms.

Twelve countries produce approximately 97% of the lemons sold in commercial markets of the world. These countries and the quantity of lemons they produced from 1976 to 1982 are listed in Table 1.

All commercially grown lemon varieties have evolved from mixed parentage and thus do not come true to seed. Trees used in commercial plantings are propagated by budding or grafting lemon scion cuttings on seedlings (rootstock) judged to be superior and hardier than lemon seedlings. The rootstocks most commonly used for lemons are chosen due to resistance of Tristasia, Phytophthora, Exocortis, or vigorous transmission of water nutrients. They include sour orange *(C. aurantium)* which also has a high cold tolerance and low incidence of foot rot or mal di gomma; grapefruit *(C. paradisi):* sweet orange *(C. sinensis);* mandarin orange *(C. reticulata);* tangelo *(C.* reticulata × *C. paradisi);* and rough lemon

TABLE 1
Production of Lemons in Specified Countries
(1977/78 to 1981/82)

Country	1977/78	1978/79	1979/80	1980/81	1981/82
U.S.	53,600	41,700	44,500	65,000	51,000
Italy	46,479	44,622	44,506	39,922	45,899
Spain	18,162	14,391	18,394	19,497	26,112
Argentina	16,247	17,466	22,978	23,849	22,630
Brazil	21,064	21,296	NA	NA	NA
Turkey	14,507	9,864	12,766	12,186	13,346
Greece	11,257	7,776	8,356	10,445	11,605
Chile	4,004	4,294	4,178	4,352	4,468
Israel	1,741	2,669	3,075	2,379	3,482
South Africa	1,973	1,973	2,089	2,959	3,191
Cyprus	1,741	1,857	2,147	2,031	2,785
Australia	2,785	2,263	2,785	2,205	2,437
All others	812	5,454	5,629	4,874	4,120
Total	194,372	175,625	171,403	189,699	191,075

Note: Values shown represent 1000s of cartons (38 lbs).

From *Citrus Fruit Industry Statistical Bulletin*. 1983. Sunkist Growers, Inc. With permission.

(C. jambhiri) which is highly important in certain kinds of soil and has been widely used in Argentina, Brazil, Florida, South Africa, and Australia. Recently horticulturists have developed nucellar clone budlines that are more vigorous in growth and more resistant to viruses. Although the only known nucellar clone selections are Frost Eureka and Frost Lisbon, nucellar lines of most popular clonal sections of major varieties are being developed.

Lemons are picked according to size rather than color. The size is such that most fruit have a thick, firm, green rind, are immature, and have a high acid content, all of which tend to maximize their storage life. Those lemons considered for storage are graded and those found to be highly colored, of improper size, scarred or misshapen are removed and used in processed products. The remaining fruit are washed, sorted into four color grades, and may be waxed for storage. The length of storage, determined by color grade, ranges from a few weeks for fruit that are yellow to 6 months for those that are dark green.

To these handling practices, the California industry has added a 2,4-D (2,4-dichlorophenoxy acetic acid) treatment which is applied during the final stages (washing or waxing) of preparation for storage. 2,4-D makes the fruit more resistant to fungal diseases, while attached, by delaying the abscission process of the fruit calyx (button). Since most lemons have Alternaria citri spores under the button the extended life of the button prevents fungus from entering the fruit. The treatment also delays aging and reduces external decay.[1-4] The treatment does not, however, inhibit fungal growth and once fungus enters the fruit it continues to grow as it would in untreated fruit.[5,6] Table 2 shows how 2,4-D affects "black buttons", Alternaria decay and other miscellaneous decay in stored lemons that have reached different degrees of color-based maturity. Lemons infected internally with Alternaria suffer adverse flavor effects but, because no external damage occurs, these fruit are very difficult to detect and discard. Significantly reducing Alternaria and other fungi through the application of 2,4-D, therefore, improves the general quality of lemons for both fresh and processed markets. Quality of lemons treated with 2,4-D is further improved because the aging process is delayed, which in turn reduces external decay.

Lemons which have been accurately separated by color are carefully packed in boxes and placed in storage rooms. These rooms are closely controlled so that the temperature

TABLE 2
Effect of 2,4-D on Storage of Yellow, Greenish-Yellow, and Light-Green Lemon Fruits

Treatment	Months of storage	% Black buttons			% Alternaria decay			% Miscellaneous decay		
		Yellow	Greenish-yellow	Light-green	Yellow	Greenish-yellow	Light-green	Yellow	Greenish-yellow	Light-green
Wax emulsion, no growth regulator	2	—	5.4	—	—	0.5	—	—	0.5	—
	4	—	58.5	44.0	—	2.7	0.0	—	1.9	3.7
	6	—	73.2	64.0	—	9.5	5.2	—	14.5	10.5
Wax emulsion, 500 ppm of 2,4-D	2	0.4	0.0	—	0.0	0.0	—	0.0	0.0	—
	4	2.9	1.1	0.0	0.5	0.0	0.0	0.4	1.1	1.1
	6	2.5	0.0	0.0	1.5	0.1	0.0	2.7	0.0	1.7

From Coggin, C. W., Jr. and Hield, H. Z., In *The Citrus Industry* Vol. II, Reuther, W., Batchelor, L. D., and Webber, J. H., Eds., University of California Press, 1968, 383. With permission.

will range between 55°F ± 2°F and the relative humidity will range between 88% and 92%. Careful temperature control is necessary to prevent damage to the peel and segment membrane which occurs at lower temperatures and to deter the growth of decay-producing organisms which occur at higher temperatures. The boxes are stacked according to color grade and in such a manner as to expose them to a generous quantity of fresh air. During the storage period, also referred to as the "curing process", the rind becomes thinner and tougher due to loss of moisture in the albedo and membranes. Curing also produces an increase in the size of juice sacs resulting in higher juice yield and juice quality than that obtained from similar fruit allowed to yellow on the tree.[7] By controlling the curing conditions the final stage of fruit yellowing can be delayed or accelerated depending on market demand and crop production. In times of very high demand, lemons can be cured to a yellow color in less than ten days by exposing them to ethylene gas and slightly higher temperatures.

Although lemon storage problems have been significantly reduced there are continual efforts to further reduce factors which adversely affect fruit quality. Researchers in Israel, Australia, Japan, Florida, and Texas have investigated the benefits of wrapping individual citrus fruit, including lemons, by placing them in plastic bags and applying heat to shrink the plastic tightly around the fruit.[8,9] The plastic which is permeable to oxygen and carbon dioxide allows the fruit to breathe. However, because the film is impermeable to moisture, the plastic impedes moisture loss thus helping the fruit retain juice and stay firm. Results show that aging and shrinking are greatly reduced with some reduction in decay by lessening the spread decay and less contact-dusting of mold spores. The wrap packing may also replace the waxing of lemons which reduces fruit breathing and can lead to development of a disagreeable off-flavor during long storage.[8] The rate of respiration (which has an impact on "keeping" quality of lemons in storage) is influenced by cultural practices, age of fruit at the time of storage, pre-storage preparation, and curing conditions.[10] Careful control of factors that influence respiration rate of stored lemons can minimize physical and physiological changes that adversely affect fruit quality. Changes that can occur during storage include: (1) albedo browning, (2) pitting, (3) peteca, (4) red block, (5) membrane darkening, and (6) water breakdown.[10] Studies made by the California Fruit Growers Exchange Laboratory in 1943 (unpublished) showed increases in the juice/fruit ratio, total soluble solids, and total acids in all lemons during storage.[10] Lemons at three levels of maturity (based on color) were examined with the following results:

1. Yellowish-green lemons stored for six weeks showed an increase in (a) juice/fruit between 9.6% and 12.1%, (b) total soluble solids between 2.8% and 3.7% and (c) total acids between 7.1% and 8.5%.
2. Light green lemons stored for eight weeks showed an increase in: (a) juice/fruit between 23.9% and 28.2%, (b) total soluble solids between 3.3% and 11.0% and (c) total acids between 13.1% and 21.2%.
3. Dark green lemons stored for 11 weeks showed an increase in: (a) juice/fruit between 21.1% and 34%, (b) total soluble solids between 8.8% and 13.4%, (c) total acids between 15.6% and 24.2%. Another important compositional change occurs during storage. The ascorbic acid (vitamin C) content increases. Although a number of factors influence the quantitative change, the ascorbic acid level in lemons continually increases during the entire storage period.[10] Vitamin C increases in stored lemons were found to range from 6.5% to 27.0% in yellowish-green fruit, 10.0% to 40% for light green fruit, and 11.9% to 34.9% for dark green fruit.

Lemon juice composition and methods for compositional analysis have been well documented.[11,12] Although the composition and flavor of lemon juice is quite different from most citrus fruit, it compares favorably to other citrus fruit in flavor acceptance, nutritional value, stability, and is superior in storage qualities. A general listing of lemon juice components are shown in Table 3.

TABLE 3
Composition of Lemon Juice

Constituent	Content per 100g
Protein (total N × 6.25)	0.26—0.77 g
Amino nitrogen	0.019—0.046 g
Fat (ether extract)	None—0.6 g
Soluble solids, total (°Brix)	7.1—11.9 g
Acid, total, as anhyd. citric	4.20—8.33 g
Malic acid	0.15—0.41 g
Sugar, total, as invert	0.77—4.08 g
Reducing sugar	0.78—2.63 g
Sucrose	0.03—0.63 g
Minerals, total ash	0.15—0.35 g
Calcium	5.6—27.9 mg
Phosphorus	5.3—16.6 mg
Iron	0.14—0.69 mg
Magnesium	5.8—11.3 mg
Potassium	99—128 mg
Sodium	1.0—5.0 mg
Sulfur	2.0—8.0 mg
Chlorine	2.3—4.0 mg
Vitamin A (as carotene)	None or trace
Thiamine (B-1)	0.004—0.125 mg
Riboflavin (B-2)	0.005—0.073 mg
Niacin	0.056—0.196 mg
Inositol	56—76 mg
Folic acid	0.00082—0.00094 mg
Flavanones	46—54 mg
Ascorbic acid (vitamin C)	31—61 mg
pH	2.11—2.48

From Swisher, H. E. and Swisher, L. H., in *Fruit and Vegetable Juice Processing Technology*, 3rd ed., Nelson, P. E. and Tressler, D. K., Eds., AVI Publishing, Westport, CT, 1980, 163.

III. FACTORS INFLUENCING THE QUALITY OF THE CROP FOR PROCESSING

Lemons grown specifically for processing should possess several characteristics different from those of lemons grown primiarly for the fresh fruit market. The production of blemish-free fruit, so important for the fresh fruit market, is less important for fruit destined to be processed. Citrus scab (also known as lemon scab)[13] caused by a fungus and russett caused by mites are two serious surface blemishes to lemons produced in Florida that are of little concern when the fruit is to be processed, although russett can decrease yields.[14]

Factors should be present in fruit grown primarily for processing that are important in the major products that will result, primarily concentrate for lemonade and cold-pressed lemon peel oil. Thus, the juice should have a good flavor plus a high content of citric acid,[15] and the fruit should furnish a high yield of good quality peel oil.[16] In addition, desirable tree characteristics such as longevity, high fruit yield, and freedom from disease are also important.[14] Since the fruit is usually processed shortly after harvest, keeping quality is not as important as with lemons sold as fresh fruit. Fruit size is not of major concern in fruit used for processing. Juice extractors can be adapted to various sizes of fruit for processing, whereas uniformity of size is essential to good marketing of fresh lemons.

The California and Florida lemon crops illustrate the differences in fruit raised primarily for the fresh fruit market (California) and for processing (Florida). The major lemon cultivar grown in California is the Eureka. This cultivar is small, of superior quality, and is produced in most areas of California.[10] Although lemons bloom year-round, much of the Eureka fruit matures in the spring which is the optimum time to meet the peak demand for lemons in the summer months for use in lemonade and iced tea. A second major cultivar, Lisbon, is preferred in southern California and Arizona, because the fruit is less susceptible to sunburn and to freeze injury.[10]

Bearss (a Lisbon-type) is the major cultivar grown in Florida, where the fruit is primarily destined for processed products.[17] In studies on 42 lemon selections as potential fruit to be grown and harvested in Florida, several selections of the Bearss cultivar rated highest in peel oil yield and aldehyde content (a measure of oil quality).[16] Most selections produced acceptable yields of juice and citric acid.[15] The one selection that did not give a good qualtiy juice was a Meyer lemon sample. The off-flavor in the juice was caused by the presence of peel oil and this undesirable flavor was removed by vacuum concentration. Thus, all samples including Meyer could be processed into good quality concentrated lemon juice products. Meyer lemon oil contains a high level of thymol that causes an off-flavor in juice products made from that oil, and an aroma that renders it unsuitable for marketing as a typical lemon oil.[18] Meyer lemons are processed separately and the oil is usually not collected.

Since oil yield and quality together with citric acid content are of primary consideration in selecting lemons to be processed, factors that affect these parameters are considered important to growers and processors. Kesterson et al.[16,19] showed that oil quality (aldehyde content) and yield were related. Low oil content in the peel resulted in a high water/oil ratio during processing that caused a low aldehyde content in the resulting oil after separation from the water phase.[19] These workers also showed that oil yield for lemons processed in Florida varies considerably from season-to-season.[20]

The amount of total soluble soilds, a measure of the citric acid content, varies with the rootstock used in propagation.[10] In a study with Eureka and Lisbon cultivars on 5 different rootstocks, trifoliate orange produced fruit with the highest and rough lemon the lowest soluble solids content. Many other factors such as disease resistance, fruit yield, susceptibility to cold, and other stress factors enter into the choice of rootstock.

Freeze-damage to lemon fruit caused a noticeable decrease in juice content (decrease in specific gravity of whole fruit), but the specific gravity of the juice remained unchanged.[10] These results indicated that the loss of water from the fruit was compensated for by loss of total soluble solids so that the total soluble solids content of juice from freeze-damaged and undamaged fruit were similar. However, up to 75% of the total acid could be lost due to freeze damage, thereby decreasing the value of the crop.

IV. PROCESSING

More than half of the lemons produced in the U.S. during the 1980 to 1981 season were processed (Table 4). Other major lemon producing countries and the portion of the fruit used in processing are listed in Table 4.

The two major products made from lemon fruit are cold-pressed peel oil and concentrated juice. Cold-pressed lemon oil brings a premium price (over $6/lb.) and is widely used in flavoring synthetic cola flavored and lemon-lime flavor drinks, desserts, and is added to many cleansers compounded for home use. The concentrated juice used in the production of frozen concentrated lemonade is standardized on the basis of acidity, because citric acid is the most important major constituent. Since sugar is added to achieve the desired sugar/acid ratio in the finished lemonade product, the sugar content of the concentrated lemon juice is relatively unimportant. Lemon juice contains about twice as much citric acid as total sugars, whereas lemonade contains 15 to 16 times more sugar than citric acid.[21,22]

TABLE 4
Quantities and Countries Processing Lemons

Country	Production		Lemons processed			
	1979/80	1980/81	1979/80	% of Total	1980/81	% of Total
U.S.	716	1096	347	48.5	703	64.1
Italy	767	688	238	31.0	180	26.2
Argentina	396	411	180	45.4	175	42.6
Spain	317	342	18	5.7	20	5.8
Turkey	220	210	23	10.5	20	9.5
Australia	48	38	35	72.9	19	50.0
Greece	144	180	15	10.4	18	10.0
South Africa	36	51	8	22.2	15	29.4
Israel	53	42	16	30.2	9	21.4
Cyprus	37	36	4	10.8	3	8.3

Note: Values shown represent 1000s of metric tons.

From *Citrograph,* 67, 268, 1982. With permission.

Techniques for processing lemons are basically the same whether the fruit is primarily grown for processing or is rejected by a fresh fruit packinghouse based on size, maturity or the presence of peel blemishes. Fresh fruit for processing is washed, graded for removal of unwholesome fruit, stored in bins to be processed as soon as possible, and regraded before processing. In the U.S. juice is usually expressed with an extractor made by either Brown International Corporation or the FMC Corporation. In the FMC In-Line Extractor, juice and peel oil are separated by the extractor in a single step.[23] Water is used to wash the peel oil from the mashed peel by forming a water-oil emulsion during this step; the ratio of water to peel oil partly determines the yield and quality of the peel oil obtained from the emulsion.[19] Large pulp particles are kept out of the juice during extraction by a screen, but a paddle finisher or screw press is used to further reduce the pulp content prior to concentration. Flavor deterioration and ascorbic acid loss is minimized by deaeration of the juice, storage at cold temperatures (less than 10°C), and concentration or pasteurization as rapidly as possible after extraction.

A relatively new oil recovery system that is used extensively in California for separation of lemon oil from whole fruit prior to juice extraction was developed by Brown International Corp. The oil extractor uses a bed of rolls covered with needle-like projections that are mounted in a shallow tank. Whole fruit advances in a single layer across the rolls which rotate in the same direction as the fruit movement. The rolls pierce the oil cells located in the outer portion of the peel to release the oil. An overhead paddle conveyor maintains uniform flow of fruit. Water enters at the discharge end and flows countercurrent to the fruit flow, forming an emulsion with the released oil. The oil-water emulsion discharges over a weir which maintains liquid level. The emulsion is screened in a paddle finisher and then centrifuged in two stages. The low solids content of the emulsion permits high efficiency in the first stage of centrifugation. Thus, water can be recycled from the centrifuge back to the oil extractor without further purification. Recycling of water results in high aldehyde content in the final oil and improved disposal of waste water by decreasing the volume generated during oil recovery.[24] This process also produces one of the highest oil yields currently obtainable.

High oil yield is especially important in lemon oil production because of its relatively high value compared to other citrus oils, and several methods to increase oil yield are available. Addition of enzymes to increase yields in the centrifugation steps have been described.[27] Experimental use of bioregulators sprayed on lemon trees have increased the oil yield in the fruit.[26] Aqueous discharges from centrifuges used to recover peel oil normally contain considerable peel oil (0.25%) that is usually lost. Lund and Bryan[27] used steam distillation to recover this oil and showed that it had the same citral content as the cold-pressed oil removed by centrifugation. Cold-pressed oil was not significantly preferred over distilled oil in an aroma panel test. However, some panel members noted the cold-pressed oil to have a "fuller" or "more-rounded" aroma, probably because some of the higher-boiling components were in greater concentration in the cold-pressed oil.

The Italian lemon industry markets the bulk of the crop as fresh fruit, but about 25% of the crop is processed.[28] The fruit is processed mainly for the peel oil and the juice is of secondary importance. Both large, modern, well-equipped plants and small, family-run operations are involved in processing lemons. The large plants use Italian made processing equipment which either deoils the fruit first and then recovers the juice or extracts the fruit first and then recovers the peel oil; FMC In-Line equipment that recovers both oil and juice in one step is rarely used. Use of the fumatrice machine has been the preferred method for extraction of lemon peel oil because of the high quality of oil obtained.[28]

Lemon juice is pasteurized both for sterilization and for inactivation of enzymes to stabilize the cloud. For maximum cloud stability and flavor in unfrozen concentrated products, the juice is pasteurized first and then concentrated in a vacuum evaporator.[23] Frozen concentrated juice is generally prepared from unpasteurized or partially pasteurized juice (to stabilize the cloud). The concentrated juice is standardized on the basis of its acidity and stored at −18°C or lower.[23]

Lemon juice processed in California is generally concentrated in an APV evaporator designed for medium production runs of heat-sensitive materials.[23] In Florida, the same temperature-accelerated short-time (TASTE) evaporator used to concentrate orange and grapefruit juices[29] is used to concentrate lemon juice, but the maximum temperature is lowered from the 93° to 100°C normally used to 80°C for lemon juice to avoid loss of cloud in the finished product.[30] Lemon juice is concentrated to a standard level of 400 g of anhydrous citric acid/l.

Lemon peel oil may be added back to the concentrated juice to provide optimum lemon flavor. For juice products that are to be stored at room temperature, the peel oil content should be low to avoid development of off-flavors. Frozen concentrated juices are generally stored without added peel oil. For optimum flavor, the upper limit of peel oil concentration in single-strength juice is about 0.025%. Aqueous lemon essence has been produced during concentration of juice. The aqueous essence has an aroma characteristic of fresh lemon juice and contains volatile flavor components from juice, but as yet has not been used commercially to flavor lemon juice products.[31] Sulfur dioxide and sodium benzoate are sometimes added to lemon juice products as preservatives.

Concentrated lemon juice can either be used to prepare frozen concentrated lemonade by the addition of sugar and other flavor ingredients,[32] it can be canned directly under sterile conditions, or it can be reconstituted to single-strength lemon juice. Because of the high acidity, use of low temperatures for storage of lemon juice products is more important than for other citrus juices, particularly concentrated juice. High storage temperature can destabilize the cloud and cause loss of ascorbic acid.

Apart from its use in lemonade, lemon juice is much more widely used as a flavoring in other foods than are other citrus juices. Its major household use is in iced tea, but use on fish and vegetables, in salad dressings and salads, in pies, cakes, cookies, and alcoholic beverages all attest to the universal popularity of lemon flavor. A book devoted to lemon fruit was recently published.[33]

REFERENCES

1. **Stewart, W. S.,** Effects of 2,4-dichlorophenoxyacetic acid and 2,4,5-trichlorophenoxyacetic acid on citrus fruit storage, *Proc. Am. Soc. Hortic. Sci.*, 54, 109, 1949.
2. **Erickson, L. G.,** Plant growth regulators for lemons, *Calif. Citrog.*, 37, 201, 1952.
3. **Stewart, W. S., Palmer, J. E., and Hield, H. Z.,** Packinghouse experiments on the use of 2,4-dichlorophenoxyacetic acid and 2,4,5-trichlorophenoxyacetic acid to increase storage life of lemons, *Proc. Am. Soc. Hortic. Sci.*, 59, 324, 1952.
4. **Iwasaki, T., Nishiura, M., and Shichijo, T.,** Effects of 2,4-D and 2,4,5-T on fruit drop and abscission of stem button in some citrus varieties, *Bull. Hort. Div. Nat. Tokai-Kinki Agr. Exp. Sta.*, 3, 17, 1956.
5. **Erickson, L. C., DeWolfe, T. A., and Brannaman, B. L.,** Growth of some citrus-fruit pathogens as affected by 2,4-D and 2,4,5-T, *Bot. Gaz.*, 120, 31, 1958.
6. **DeWolfe, T. A., Erickson, L. C., and Brannaman, B. L.,** Retardation of alternaria rot in stored lemons with 2,4-D, *Proc. Am. Soc. Hortic. Sci.*, 74, 367, 1959.
7. **Swisher, H. E., and Swisher, L. H.,** Lemon and lime juices, Ch. 4. in *Fruit and Vegetable Juice Process. Tech.*, 3rd Ed., Nelson, P. E., and Tressler, D. K., Eds., Avi Publishing, Westport, Conn., 144, 1980.
8. **Anon.,** Texas A&I tests wraps for grapefruit, *Citrograph*, 68, 131, 1983.
9. **Anon.,** Decay control: the endless war, *Citrograph*, 68, 236, 1983.
10. **Bartholomew, E. T. and Sinclair, W. B.,** *The Lemon Fruit*, University of California Press, Los Angeles, CA, 1951.
11. **Joseph, G. H., Stevens, J. W., and MacRill, J. R.,** Nutrients in California lemons and oranges, *J. Am. Diet Assoc.*, 38, 552, 1961.
12. **Birdsall, J. J., Derse, P. H., and Teply, L. J.,** Nurtients in California lemons and oranges, II. Vitamin, mineral and proximate composition, *J. Am. Diet Assoc.*, 38, 555, 1961.
13. **Pratt, R. M.,** *Florida Guide to Citrus Insects, Diseases and Nutritional Disorders in Color*, Agricultural Experiment Station, Gainesville, FL, 1958, 106.
14. **Knorr, L. C.,** The growing of lemons in Florida: historical, varietal and cultural considerations, *Proc. Fla. State Hortic. Soc.*, 71, 123, 1958.
15. **Wenzel, F. W., Olsen, R. W., Barron, R. W., Huggert, R. L., Patrick, R., and Hill, E. C.,** Use of Florida lemons in frozen concentrate for lemonade, *Proc. Fla. State Hortic. Soc.*, 71, 129, 1958.
16. **Kesterson, J. W. and Hendrickson, R.,** Evaluation of cold-pressed Florida lemon oil and lemon bioflavonoids, *Proc. Fla. State Hortic. Soc.*, 71, 132, 1958.
17. **Burns, R. M., Canhom, A. E., and Becker, D. H.,** Observations of lemon men's Florida tour, *Citrograph*, 61, 300, 1976.
18. **Moshonas, M. G., Shaw, P. E., and Veldhuis, M. K.,** Analysis of volatile constituents from Meyer lemon oil, *J. Agric. Food Chem.*, 20, 751, 1972.
19. **Kesterson, J. W., Hendrickson, R., and Braddock, R. J.,** Florida citrus oils, *University Fla. Inst. Food Agric. Sci. Bull.*, 749, 1971.
20. **Kesterson, J. W. and Braddock, R. J.,** Total peel oil content of the major Florida citrus cultivars, *J. Food Sci.*, 40, 931, 1975.
21. **McCready, R. M.,** Carbohydrates, in *Citrus Science and Technology*, Vol. 1, Nagy, S., Shaw, P. E., and Veldhuis, M.K., Eds., AVI Publishing, Westport, CT, 1977, 74.
22. **Vandercook, C. E.,** Organic acids, in *Citrus Science and Technology*, Vol. 1, Nagy, S., Shaw, P. E., and Veldhuis, M. K., Eds., AVI Publishing, Westport, CT, 1977, 208.
23. **Swisher, H. E. and Swisher, L. H.,** Lemon and lime juice processing, in *Citrus Science and Technology*, Vol. 2, Nagy, S., Shaw, P. E., and Veldhuis, M. K., Eds., AVI Publishing, Westport, CT, 1977, 253.
24. **Shaw, P. E.,** Citrus essential oils, *Perfumer and Flavorist*, 3, 35, 1979.
25. **Braddock, R. J. and Kesterson, J. W.,** Use of enzymes in citrus processing, *Food Technol.*, 33(11), 78, 1979.
26. **Yokoyama, H., Gold, S., and DeBenedict, L.,** Effect of bioregulators on the essential oil of lemon, Citrus Research Conference, Pasadena, CA, December 8, 1983.
27. **Lund, E. D. and Bryan, W. L.,** Composition of lemon oil distilled from commercial mill waste, *J. Food Sci.*, 41, 1194, 1976.
28. **Safina, G.,** Citrus processing around the world, Part 5—Italy, in *Citrus Science and Technology*, Vol. 2, Nagy, S., Shaw, P. E., and Veldhuis, M. K., Eds., Avi Publishing, Wesport, CT, 1977, 623.
29. **Berry, R. E. and Veldhuis, M. K.,** Processing of oranges, grapefruit and tangerines, in *Citrus Science and Technology*, Vol. 2, Nagy, S., Shaw, P. E., and Veldhuis, M. K., Eds., Avi Publishing, Westport, CT, 1977, 177.
30. **Taylor, L. E.,** personal communication, 1983.
31. **Moshonas, M. G. and Shaw, P. E.,** Analysis of flavor constituents from lemon and lime essence, *J. Agric. Food Chem.*, 20, 1029, 1972.

32. U.S. Standards for Grades of Frozen Concentrated Lemonade, 1983 Code of Federal Regulations 7-52:1421. U.S. Government Printing Office, Washington, D.C.
33. **Sinclair, W. B.,** *The Biochemistry and Physiology of the Lemon and other Citrus Fruits,* University of California Press, Oakland, CA, 1984.

Chapter 3*

THE GRAPEFRUIT

Béla S. Buslig

TABLE OF CONTENTS

* Mention of a trademark or proprietary product is for identification only and does not imply an endorsement or warranty of the product by the Florida Department of Citrus over other products.

I. INTRODUCTION

A. HISTORY

The origin of grapefruit *(Citrus paradisi* Macfadyen) is relatively unclear. At present some citrus historians consider it indigenous to the Malayan and Indian archipelagos,[41] while others definitely exclude China or the adjacent southeastern portion of Asia,[30] and still others consider it originating in the West Indies.[18,33] It is reasonably well established, however, that by the 12th century the shaddock or pummelo (*C. grandis* [L.] Osbeck), which is possibly the progenitor of the present day grapefruit, was grown in Palestine, and shortly thereafter in Spain, being brought there by the Arabs. Records indicate that by the 17th century the shaddock was grown in the West Indies. The present day grapefruit is suggested to have been derived from a variety of pummelo or shaddock by mutation[18,41] or it may be a hybrid of the shaddock and the sweet orange.[33] The first definite description of the grapefruit, under the name "forbidden fruit", was made in 1750 from Barbados.[41] The name "grapefruit" probably originated in Jamaica, derived from its habit of growing in bunches. In the early 19th century, grapefruit was introduced into Florida by Count Odette Philippe, originally in the Indian River region, and later around the west coast of Florida.[18] Essentially all commercial grapefruit cultivars grown in the U.S. can be traced to this original introduction.

B. MAJOR COMMERCIAL VARIETIES

The grapefruit varieties grown commercially fall into two distinct categories: seedy and seedless. These categories can be further subdivided into white and pigmented (pink and red) types. The origins of most commonly cultivated commercial varieties can be traced to Florida, where the grapefruit industry began.[30] In Florida, approximately 60% of the total grapefruit production during the 1987 to 88 season was of the white seedless varieties, the remainder were mainly the seedless colored varieties,[16] with only about 5% of the total consisting of the seedy varieties (mainly Duncan). About half of the world grapefruit production (Table 1) is consumed as fresh fruit, while the rest of the fruit is utilized and consumed as processed products, mainly juice (Table 2).

1. Duncan

Duncan, the oldest defined clone grown in Florida and around the world, is a white seedy variety.[30] The fruit is large, oblate to globose, with short radiating basal furrows. Color is pale to light yellow, with medium thick, smooth rind. Flesh is buff to yellowish in color, tender, and very juicy. Flavor is excellent. Medium early, in Florida from late October to February, in maturity. The tree is large and vigorous, very productive and reportedly the most cold resistant of the grapefruit varieties.[17,18] Although seedless varieties have largely replaced it on the fresh market, its excellent flavor and large size makes it the most desirable grapefruit variety for processing.[17]

2. Marsh

The Marsh is the most important seedless variety grown around the world. The fruit is medium in size, oblate to spherical, containing few or no seeds. Rind is medium thin, with smooth, tough surface, pale to light yellow in color. The flesh is buff colored, tender, and very juicy. Flavor is good, although not as pronounced as the Duncan. The fruit holds well on the tree, stores, and ships well. It is the latest maturing of all the commercial varieties. The tree is large, spreading and vigorous, producing well. It is more cold sensitive than the Duncan variety.[17] The Marsh variety originated as a chance seedling around 1860 on a farm near Lakeland, Florida. This variety is of great horticultural interest, not only because of its seedlessness, but as the originator of most commercially important pigmented varieties.[17]

TABLE 1
Grapefruit Production by Selected Countries (1000s of metric tons)

Country	Production year 1983—84	1984—85	1985—86	1986—87	1987—88
Argentina	147	157	178	160	150
Australia	30	37	39	30	33
Brazil	35	24	21	27	29
Cuba	150	155	237	250	284
Cyprus	98	96	95	111	96
Egypt	3	2	2	2	2
Greece	4	5	5	6	5
Israel and Gaza	415	390	382	406	327
Italy	8	8	8	8	3
Mexico	78	53	82	91	105
Morocco	5	5	6	5	4
South Africa[a]	89	102	99	121	115
Spain	13	8	13	19	18
Turkey	25	21	22	30	27
U.S.	1974	2046	2122	2330	2523

[a] Includes some fruit produced in Swaziland, Botswana, and Mozambique marketed through the South African Citrus Board.

Adapted from Reference 35.

3. Other White Varieties

The Triumph is an early to mid-season variety. It produces medium-small fruit, and is oblate, globose, or ellipsoid, or flattened at both ends. The rind is medium-thick, smooth, pale to light yellow. The flesh is tender and very juicy, and very seedy. Flavor is lacking bitterness and exceptionally good. The tree is productive, but less vigorous than other grapefruits.[17] Although the flavor and the lack of bitterness are desirable characteristics, the seediness and the lower tree vigor limited its commercial importance. The Walters is a mid-season variety which produces medium-large, seedy fruits. The rind is relatively thin, the juice has a pronounced bitter but excellent flavor.[17] It is only important as the progenitor of some of the pink varieties. Neither the Triumph, nor the Walters is planted extensively.

4. Pink Varieties

Commercially grown pink varieties of grapefruit, the Thompson (Pink Marsh) and the Foster (Foster Pink), originated as chance mutations of the Marsh seedless and the Walters (a minor, white-fleshed, seedy variety) respectively.[17] The Thompson fruit is similar in appearance to the Marsh, with the exception of a light pink color in the juice sac and segment membrane walls. The fruit holds well on the tree, but the color tends to fade on prolonged on-tree storage. The tree is large, productive and vigorous. It is somewhat earlier in maturity than the Marsh. The Foster is medium-large, very seedy, with the rind exhibiting a pink blush under favorable circumstances. The color extends into the albedo. It is no longer important commercially, having been largely supplanted by another variety originating from it, the Foster seedless, which has a similar appearance but better flesh color.[17] Neither variety, though, is used as extensively for processing as the deeper colored varieties, which are planted more extensively.

5. Red Varieties

The major red fleshed varieties, the Redblush (Ruby, Red Marsh, Red seedless) and

TABLE 2
Grapefruit Processed by Selected Countries (1000s of metric tons)

	Production year				
Country	**1983—84**	**1984—85**	**1985—86**	**1986—87**	**1987—88**
Argentina	70	80	77	90	80
Australia	18	22	22	18	20
Brazil	18	18	18	22	24
Cuba	20	20	20	45	30
Cyprus	23	14	17	26	22
Greece	1	3	3	3	3
Israel and Gaza	246	253	234	249	193
Mexico	35	19	33	39	51
Morocco	1	1	2	2	2
South Africa[a]	17	33	34	39	35
Spain	1	1	1	2	2
Turkey	1	1	2	3	3
U.S.	1037	1227	1147	1252	1317

[a] Includes some fruit produced in Swaziland, Botswana, and Mozambique marketed through the South African Citrus Board.

Adapted from Reference 35.

Star Ruby, originated as limb sports of the Thompson. The red fleshed varieties are the primary grapefruit planted in Texas and Arizona, and to a lesser extent in California and Florida. With the exception of the deeper flesh color and the crimson blush on the rind, the fruit is similar to the Thompson. Holds on the tree very well, but with some fading of color.[17] There are several named red varieties, Flame and Burgundy for example, planted in various areas, but for processing characteristics none are significantly different from the two major types listed above.

II. CULTURAL AND ENVIRONMENTAL CHARACTERISTICS

The grapefruit tree under optimal conditions is one of the largest citrus trees, requiring more space than any other citrus tree. It is vigorous, exceptionally heat tolerant, and only slightly less cold resistant than the sweet orange. Grapefruit trees can also tolerate considerable neglect, which in combination with the exceptional heat tolerance allows them to adapt to hot desert, humid semitropical or tropical environments. The environmental effects on the fruit size and quality characteristics are considerably more pronounced than can be observed in oranges.[17] Grapefruit, in general, are a little more susceptible to low temperature damage than oranges. Grapefruit trees sustain damage below 26°F, the fruit below 28°F. As a result of the greater sensitivity to cold, in addition to the high heat requirement for quality fruit, growing of grapefruit is confined to warmer areas of the countries of production[8,39] (Table 1).

As with the oranges, cultural practices have a significant influence on the quality of fruit produced.[9] In contrast with the oranges, the color development is only minimally influenced by the variation of day and night temperatures, particularly with the white varieties. However, the total heat exposure has quite a noticeable effect in the development of the pigment in the pink and red varieties.[30] In general, higher total heat exposure causes better color development, although the early and late maturing varieties show variations in this behavior. The color in the pink and red varieties is primarily due to the accumulation of lycopene, with some of the varieties accumulating β-carotene as the lycopene level

TABLE 3
Effect of Rootstocks on Selected Juice Characteristics

Rootstock	Juice yield (%)	TSS[a] (%)	TA[b] (%)	TSS/TA
	Marsh seedless grapefruit			
Sour orange	48	10.5	1.3	8.1
Rough lemon	47	9.4	1.4	7.4
	Ruby Red grapefruit			
Uvalde	58	9.6	1.25	7.7
Morton	42	9.4	1.23	7.6
Citrumelo F 80-9	51	9.1	1.12	8.1
Trifoliate orange	54	9.0	1.13	8.0
Smooth Flat Seville	50	9.0	1.10	8.2
Cleopatra mandarin	49	9.0	1.14	7.9
Troyer citrange	52	8.9	1.12	8.0
Rusk citrange	52	8.9	1.17	7.6
Citrumelo W-2	50	8.8	1.09	8.1
	Red grapefruit			
Trifoliate orange	49	9.5	1.12	8.5
Sour orange	50	9.5	1.16	8.2
Sweet orange	49	9.3	1.16	8.0
Rangpur lime	50	9.2	1.11	8.3
Cleopatra mandarin	49	9.1	1.15	7.9
Sweet lime	50	9.1	1.11	8.2
Rough lemon	51	8.2	1.06	7.7

[a] Total soluble solids.
[b] Titratable acidity.

Adapted from Reference 43.

decreases later in the season.[30] These varieties produce a salmon-pink colored juice as a result of the expression of both color components. Lycopene accumulation in the red (and some of the pink) varieties reaches a maximum about the first third of the maturation stage, with the color steadily declining towards the latter part of the season.[30] The influence of rootstocks on the development of color is not as marked as with oranges, with slight changes attributable to certain rootstocks on peel color.[43] Plant growth regulators, such as gibberellic acid (GA), have been utilized with grapefruit to delay color development in late season harvests to avoid the yellow-orange peel color, suggestive of overmature fruit.[15] In addition to GA, 2,4-dichlorophenoxyacetic acid has also been used to extend the harvesting season.[15] Certain bioregulators act by increasing or inducing the accumulation of lycopene in grapefruit, both in the peel and to some extent in the flesh of the fruit. These compounds, while interesting experimentally, are not utilized to any extent commercially, due to variations in response.[42]

As with oranges, higher total heat exposure also affects the accumulation of total soluble solids.[26] However, in the hot and humid tropics and semi-tropics the effect of variations in rainfall is superimposed on the cumulative effect of heat exposure, reducing the concentration of solids and acidity, while the fruit size increases.[13] In addition to climatic effects, rootstocks also influence the level of total soluble solids[43] (Table 3).

TABLE 4
Effect of Mineral Nutrition and Irrigation on Juice Quality

	Macronutrient					Micronutrient					
Quality factor	**N**	**P**	**K**	**Ca**	**Mg**	**Mn**	**Zn**	**Cu**	**Fe**	**B**	**Irrigation**
Juice content	+	0	−	0	0	0	0	0	0	0	+
Soluble solids	+	0	−	0	+	0	0	0	+	0	−
Titratable acid	+	−	+	0	0	0	0	0	0	0	−
SS/TA[a]	−	+	−	0	+	0	0	0	0	0	+
Juice color											
red	+	0	−	?	?	?	?	?	?	?	0
yellow	+	0	−	?	?	?	?	?	?	?	+

Note: +, increase; −, decrease; 0, no change; ?, no information.

[a] Soluble solids/titratable acidity.

Adapted from Reference 22.

Both fertilization and irrigation play important roles in fruit production and fruit quality.[22] Nitrogen affects the production and fruit quality more than any other nutrient. The effect of potassium is not as apparent. Other elements do not play a significant role in fruit quality, except in cases of severe deficiency, or in some cases, toxicity due to excessive application.[22] Table 4 summarizes the qualitative effects of the various nutrients and irrigation.

Various environmental stresses, including weather effects, pathogens, insects, and pre- and post-harvest mechanical damage can also affect the quality of the fruit. While several of these primarily influence the suitability of the fruit for the fresh market, others can cause sufficiently severe damage to make the fruit unsuitable for processing.[1] Similarly, nutritional deficiency or toxicity of applied materials can lower internal quality. Judicious use of growth regulators,[1,15] nutritional and pesticide sprays[1,22] may serve to alleviate these problems. Pruning is also necessary to produce healthy trees of manageable shape and size for efficient production. Mechanical injuries are problematic under severe weather conditions, primarily resulting in scarring, which may open a route for pathogens. Damage during harvest may be a problem only for fruit stored excessively before processing.

In grapefruit, as in oranges,[2] maturity is primarily indicated by leveling of the accumulation of soluble solids and the slow decrease of titratable acidity. Thus, the ratio of total soluble solids to titratable acidity (Brix/acid ratio) is used to legally define maturity in Florida,[30,31,32] Texas, Arizona, and California.[30] In some of these states additional standards exist involving juice content, peel color, total soluble solids, date of harvest, and geographic location within the region.[30,31,32] Several producing countries rely solely on size, color, and taste, although the majority base regulations on the Brix/acid ratio.[30] Table 5 shows representative values of internal quality factors in grapefruit grown in the U.S.. The harvesting season, in light of the above considerations, depends on the variety and the climate in which the particular variety is grown.

III. FACTORS INFLUENCING THE QUALITY OF THE CROP FOR PROCESSING

The desirable characteristics for the fresh market, such as good, uniform color and blemish-free skin, are less important for fruit intended for processing. Cosmetic rejects from packing houses are routinely sent to processing plants as long as they are wholesome and meet minimal standards established for maturity. These standards, discussed earlier, involve

TABLE 5
Composition of Grapefruit Juice—Average Values

Total soluble solids (TSS)[a], °Brix	7—12
Total sugars[a], g/100 ml	5.0—8.3
Reducing sugars[a], g/100 ml	2.3—5.3
Nonreducing sugars[a], g/100 ml	2.4—4.0
Titratable acidity (TA)[b], % by w.	0.4—2.64
Citric acid[a], g/100 ml	0.42—2.10
Malic acid[a], g/100 ml	0.03—0.23
Succinic acid[a], g/100 ml	0.05—0.09
Ratio (TSS/TA)[c]	7.05—12.80
Naringin[c], ppm (Davis test)	350—1509
Limonin[d], ppm (HPLC)	0.3—12.8
Protein (N × 6.25)[a], g/100 ml	0.4—0.5
Ash[b], mg/100 ml	218—414
Potassium[b], mg/100 ml	89—171
Magnesium[b], mg/100 ml	5—9
Calcium[b], mg/100 ml	5—12
Sodium[b], mg/100 ml	2—7

[a] Adapted from Reference 24.
[b] Adapted from Reference 8.
[c] Adapted from Reference 5.
[d] Adapted from Reference 4.

internal quality factors, such as required level of soluble solids, acidity, Brix/acid ratio, and juice content of fruit.[34]

A moderate amount of bitterness is a desirable characteristic of grapefruit flavor. Bitterness is caused by the presence of naringin, a flavanone glycoside, and limonin, a triterpenoid lactone. Excessive bitterness in the juice suggests either very early fruit or harsh processing conditions (high extraction and/or finishing pressures).[24] Levels of naringin are high in the early part of the season, slowly declining as the fruit matures. About 62% of the total naringin is found in the peel, 36% in the pulp and the rag and the remainder in the juice. The presence of high levels in the finished juice, particularly later in the season, is a good indicator of improper extraction or finishing conditions. The intensely bitter limonin does not exist in appreciable levels in most parts of the fruit. However, its precursor, limonoic acid A-ring lactone, is present in intact fruit or freshly extracted juice.[24] This precursor is converted to limonin in the presence of acid conditions on standing, or by heating under normal processing conditions. The appearance of limonin brings with it a delayed bitterness, which is masked in grapefruit juices by the presence of naringin. However, it does contribute to the total bitterness in grapefruit juice. Limonin is highest in early season fruit, gradually decreasing throughout the season.

Grapefruit scions are normally grafted on various rootstocks for a variety of reasons, primarily to reduce the period of juvenility associated with seedlings grown on their own roots. In addition, the selection of the appropriate rootstock for the specific soil conditions of the growing area is quite important. Effects of various rootstocks have been observed on juice yield, soluble solids, acidity, and bitterness.[43] Table 6 shows observations of rootstock effects on juice yield, soluble solids, and titratable acidity. As these factors are the basis of minimum standards for maturity of grade,[31,32] judicious selection of the appropriate rootstock with consideration given to suitability for a given soil type and disease resistance can be of great importance.

Fertilization and irrigation also play important roles in production of fruit for both fresh and processed purposes. Nitrogen and potassium are the two principal nutrients in citrus

TABLE 6
Effect of Rootstocks on Selected Juice Characteristics

Rootstock	Juice yield (%)	TSS[a] (%)	TA[b] (%)	TSS/TA
	Marsh seedless grapefruit			
Sour orange	48	10.5	1.3	8.1
Rough lemon	47	9.4	1.4	7.4
	Ruby Red grapefruit			
Uvalde	58	9.6	1.25	7.7
Morton	42	9.4	1.23	7.6
Citrumelo F 80-9	51	9.1	1.12	8.1
Trifoliate orange	54	9.0	1.13	8.0
Smooth Flat Seville	50	9.0	1.10	8.2
Cleopatra mandarin	49	9.0	1.14	7.9
Troyer citrange	52	8.9	1.12	8.0
Rusk citrange	52	8.9	1.17	7.6
Citrumelo W-2	50	8.8	1.09	8.1
	Red grapefruit			
Trifoliate orange	49	9.5	1.12	8.5
Sour orange	50	9.5	1.16	8.2
Sweet orange	49	9.3	1.16	8.0
Rangpur lime	50	9.2	1.11	8.3
Cleopatra mandarin	49	9.1	1.15	7.9
Sweet lime	50	9.1	1.11	8.2
Rough lemon	51	8.2	1.06	7.7

[a] Total soluble solids.
[b] Titratable acidity.

Adapted from Reference 43.

production. The other macronutrients and most of the micronutrients do not play significant roles except in case of severe deficiency.[22] In general nitrogen applications increase the juice content, soluble solids, and acidity. Excessive applications will lower fruit production. Potassium at high levels has a negative effect on juice quality, although its effect on increased fruit production per hectare leads to increased soluble solids per hectare. High rates of phosphorus cause lower acid content, thus increasing the Brix/acid ratio. When application is to soils low in phosphorus, increased fruit production is observed. Table 4 summarizes the observed effects of mineral nutrients on quality factors.[22]

Irrigation promotes tree growth and increases fruit production under limiting moisture conditions. The major role of irrigation is the enhancement of the effects of fertilization. Adequate irrigation will increase juice yield, soluble solids/acidity ratio, and fruit production. Excessive irrigation causes adverse effects in fruit quality[22] (Table 4).

Grapefruit, as oranges, are harvested primarily by manual labor. Mechanical harvesting equipment, either air blast or mechanical shakers, is still not used extensively. The tendency of the fruit to tightly adhere to branches is somewhat offset by the size and weight of the fruit, making it possible to remove the fruit with less force than required for oranges. However, the shaking can still cause some degree of peel injury, and possible tree damage.

Fruit harvested this way is not usually suitable for fresh use, and often, depending on the delay between harvest and delivery to the processing plants, is prone to microbial decay. Grapefruit, though, seems to be somewhat more resistant to decay than oranges. Research on abscission promoting chemicals to aid tree shakers to remove fruit more efficiently has shown that some of these are quite effective for grapefruit harvesting. With the aid of these abscission agents mechanical harvesting could become more practical, although the fruit probably will only be used for processing due to peel damage.[42]

IV. PROCESSING

The commercial processing of grapefruit began in Florida in 1912, when canned sections were first produced. Grapefruit juice was initially produced as single-strength canned juice, as a by-product of the sectionizing process.[24] Rapid expansion of grapefruit plantings, the relative ease of processing, and the storage stability of the canned single-strength grapefruit juice all contributed greatly to the leading role of grapefruit in citrus processing. This persisted until the development of the frozen concentrated orange juice.[23]

During the 1987 to 88 season more than half of all of the grapefruit produced by the major citrus growing countries was processed (Table 2). The U.S., the world's major producer, contributed about 75% of the total processed grapefruit products. About 88% of the total U.S. output of processed grapefruit was from Florida for the same period.

The primary product of grapefruit processing is juice, which can be frozen concentrated grapefruit juice (FCGJ), canned juice, and chilled juice. In addition, some grapefruit juice is marketed as blends with other citrus juices, or as grapefruit juice cocktail when blended with juices of other fruits. Some grapefruit is also utilized as chilled or canned sections, or as citrus salad, with other citrus fruit sections. Other products manufactured, minor in comparison to juice production, are dehydrated grapefruit juice, cold-pressed peel oil, dried peel and pulp for cattle and poultry feed, candied peel, and various specialty items.

A. HARVESTING, EXTRACTION, AND FINISHING

After harvesting, the fruit is delivered to processing plants, and in Florida, a sample of each load is inspected for wholesomeness and tested for maturity parameters by state inspectors. The sample is tested for minimum juice content (depending on the time of the season), color break, minimum required soluble solids, and minimum Brix/acid ratio.[30,31,32,34] A load not meeting the requirements is destroyed. The delivered fruit may be temporarily placed in storage bins. Fruit is then washed, sorted, with that intended for the fresh market receiving fungicide and wax treatment and the defective fruit removed. The fruit to be juiced, after washing and sorting for size, is sent by conveyors to mechanical extractors. In the U.S., either the FMC In-Line (FMC Corp.) or the Brown (AMC Corp.) juice extractors are used in most processing plants. Both types of extractors require fruit which is carefully sized.[6] The extractors and the extraction process are similar to that employed with oranges, with the exception of the larger size cups necessitated by the size of grapefruit.[9] Both types of extractors have limited capability of controlling the amount of peel oil in the extracted juice. Oil recovery from the peel is essentially identical to that of orange oil.[9,21,28]

The juice from both types of extractors is sent to finishers to remove pieces of peel, seeds, and excess pulp particles. Finishers are screening devices, either screw or paddle-type, and are used to adjust the level of pulp in the finished juice. They may be adjusted by varying the screen perforation size to set the nominal particle size of the finished juice, and by changing the pressure on the restrictor device of the peel and/or pulp discharge. The pressure applied determines the juice yield obtained and has an important bearing on the sensory quality of the juice produced. Grapefruit juice is expected to have some bitterness, characteristic of typical grapefruit flavor. Since the bitter components, naringin and limonin, are associated with the pulp, rag, and seeds, excessive extraction and finishing pressures can cause great increases of the levels of these compounds. Controlling the finishing pressure therefore is important in the production of high quality juice. Measurement of these two bitter components is routinely performed to monitor quality of the juices produced.[4,5,19,27]

B. JUICE PRODUCTION

The types of grapefruit juice produced commercially essentially fall in three basic categories (1) frozen concentrated grapefruit juice (FCGJ), (2) single-strength grapefruit juice, and (3) dehydrated grapefruit juice. Each category can be further subdivided into several variations of the type of the product, such as style (unsweetened or sweetened) and color (white/amber or pink/red).[37] Blends with other juices are also produced.

1. Canned Juice

Canned single-strength grapefruit juice is the oldest commercial juice product of the processed citrus juice industry. Due to the relatively high storage stability of the juice, considerable volume of single-strength grapefruit juice is packed in cans. Three types of canned single-strength grapefruit juice are produced commercially. The first type, made directly from freshly extracted, undiluted or unconcentrated juice of mature grapefruit, is heat sterilized and hot filled and sealed into tinned or enameled cans. Another type is made by reconstituting concentrate to standardized single-strength. This is the most common way that canned single-strength grapefruit juice is presently produced in Florida. In addition, a third type is produced in small quantities, where concentrated grapefruit juice, not more than 15% of the finished product's soluble solids, is added to the first type of product listed.[5] Prior to canning, the freshly extracted grapefruit juice is subjected to deaeration, deoiling, and heat treatment.[6,24] Deaeration serves primarily to reduce flavor deterioration, to prevent lowering of ascorbic acid levels due to dissolved oxygen, and to reduce frothing during the filling step. Deoiling is performed to control the peel oil level of freshly extracted and/or blended juices prior to packaging. It is done under vacuum with the application of mild heat to produce vapors which carry off about 90% of the volatile peel oil. The canned single-strength juices produced from concentrate are reconstituted to a standardized soluble solids level, flavored with peel oil, and canned following pasteurization.[10] The U.S. Standards for grapefruit juice specify an upper limit of 0.020 %(v/v) for peel oil content, measured by titration as d-limonene.[25,34] The heat treatment or pasteurization serves as the means to destroy microorganisms and to inactivate pectic enzymes, responsible for clarification of the juice. The high acidity of grapefruit juice permits slightly more efficient killing of the microorganisms associated with spoilage than in orange juice, and somewhat reduces the length of the exposure required to stabilize cloud.[6,24] The canning process has been thoroughly described in earlier publications.[6,8,24,39]

The natural level of acidity in grapefruit juice tends to be high, and the Brix/acid ratio too low to be acceptable to some consumers, although the grapefruit flavor seems to be desirable. To satisfy these consumers, some canned grapefruit juices are produced with added sugar or other sweetener. However, the starting juice must still meet maturity standards prescribed by law[31,32] for the standard unsweetened product. The added ingredients must be declared on the label.[31,32,37]

Canned juices are, in general, distributed as an ambient temperature product, subject to considerable storage abuse. The tin coating of the cans used in the canning process serves to stabilize the ascorbic acid level of the canned juice, as well as to aid in preserving the color of juices produced from pink and red grapefruit longer. The juices produced from the pigmented varieties are quite attractive initially, but have the tendency to deteriorate on standing, due to fading and the development of browning pigments. Nonenzymic browning is also a problem with white juices. With both juices, pre-canning abuse of the juice, temperature of storage, and oxygen are aggravating factors. The above factors cause accumulation of various compounds in the stored juices, resulting in deleterious flavor changes. Levels of diacetyl, oxygenated terpenes and furfural are determined to monitor microbial spoilage, oxidative deterioration and temperature abuse respectively.[24] Color changes, such as fading of the pink and red juices, and development of muddy, grayish-brown appearance are also indicative of temperature abuse. The flavor and color changes, described above, detract from the appeal of the product to consumers. Storage at the lowest practical temperature slows deterioration of flavor, appearance and ascorbic acid content.

2. Chilled Juice

The availability of home refrigeration, as well as the development of frozen concentrated citrus juices, caused a substantial shift in consumption towards chilled and concentrated grapefruit juices due to their higher quality. Chilled grapefruit juice is produced similarly to canned juice, up to the filling operation. In this process either sterilized glass or metal containers are filled with hot pasteurized juice, followed by rapid cooling, or sterilized coated paper, glass or plastic containers are filled with cold pasteurized juice under aseptic conditions. The product is stored under refrigerated conditions. The refrigerated storage permits the juice to maintain a high quality for longer, and eliminates the tinny flavor associated with canned juices.[24] Glass packages are preferred over cardboard cartons for chilled juices due to the increased storage life, flavor, and ascorbic acid retention. Chilled juice can be produced from fresh juice, although now it is virtually always made reconstituted from frozen concentrated grapefruit juice (FCGJ). Blending may be employed to achieve a more uniform product. Chilled juice reconstituted from FCGJ is pasteurized and peel oil is added for flavor.

3. Frozen Concentrated Grapefruit Juice

The development of frozen concentrated orange juice (FCOJ)[23] changed the production practices of citrus processors. The production of frozen concentrated grapefruit juice (FCGJ) is essentially the same as the FCOJ process.[6,8,9,24,39] The evaporators employed are primarily the high temperature/short time, falling film types. The most widely used are the thermally accelerated short time evaporators (TASTE),[11,12] which are multi-stage, multi-effect types, identical to those employed with FCOJ. As with FCOJ production, in the initial stages of evaporation, the volatile flavor and aroma components are collected. This product is an aqueous and oil mixture, which is subsequently separated into flavor fractions called aqueous essence or aroma and essence oil. After refining and standardization, these products are used to restore flavor to the concentrate before packaging. In recent developments, both the control of the evaporator heating and the juice flow is performed by microprocessor controlled equipment.[11]

FCGJ produced today is a highly standardized and regulated product, where numerous parameters have to meet certain federal and state requirements.[14] The FCGJ produced commercially is a product of 55 to 65° Brix, and is very low in flavor. It is normally pumped into bulk containers, usually plastic lined barrels, and is stored frozen. For distribution, the product is reconstituted to 38°Brix with either freshly extracted, pasteurized "cut-back" juice or with water and the addition of flavor fractions produced during evaporation. The product is packaged in 6, 12 or 32 oz. (177, 350 or 946 ml respectively) metal or composite cans and quick frozen at −40°C. This blend is distributed in frozen form through retail outlets.[24]

4. Dehydrated Grapefruit Juice

Dehydrated grapefruit juice can be produced by various methods, such as (1) vacuum 'puff' drying, (2) freeze drying, or (3) foam-mat drying. These processes all involve considerable difficulty and expense, thus they are not very widely used commercially. The quality of the product is not at par with the frozen concentrates. Only small amounts of dehydrated juice "crystals" are produced commercially, mainly by the 'puff' drying process.[6]

C. OTHER PROCESSED PRODUCTS

1. Grapefruit Sections

Grapefruit sections are normally produced by either manual sectionizers or by a largely automated system. With the manual method, the sectionizers either use the cold peel method, where the peel and the outer membrane is sliced off manually with a knife, or the hot peel method, where the peel is subjected to steam scalding or hot water immersion, which makes the peel puffy and easy to remove. The sections are separated with a knife from the peeled fruit, followed by packing into containers in juice or light syrup. The automatic peeling and sectionizing process still involves manual positioning of the fruit on the stand designed to hold the fruit. A series of blades score the fruit, followed by mechanical grasping fingers, which hold the peel while the fruit is raised to free the inner portion. The freed ball of the fruit is treated with a lye solution to remove the adhering albedo, followed by manual sectionizing and packing as with the completely manual method.[6] Control of quality of the product is in accordance with prescribed standards.[36,38] A newer, greatly improved method for producing high quality grapefruit sections, or sections from other citrus fruit, involves vacuum infusion of commercial pectic enzymes into scored grapefruit peel.[7] The peel of the treated fruit can be removed with ease, with the yield of the resulting sections increasing greatly. In contrast with manual or mechanical peeling, which averages about 60% yield of the edible portions of the fruit, the enzyme method permits nearly 100% utilization.[3] The product can be treated with calcium salts to improve firmness and coated with edible films to maintain section integrity and stored refrigerated in dry, fresh state, in low oxygen permeability packages. The quality of this product closely approximates fresh fruit, with the added convenience of having the peel already removed.[3]

The steps involved in the production and the products themselves are monitored for quality by state and federal inspectors located in the various processing plants.[10,25,34,40] In Florida, the Florida Citrus Code of 1949, governs the legal aspects of quality control, through which the Florida Citrus Commission (created by legislative action in 1935), now the Florida Department of Citrus, is empowered to develop strict regulations in the manufacture of processed citrus products. Quality regulations are monitored by a continuous inspection program of the Florida and U.S. Departments of Agriculture. This program covers all phases of the manufacturing process, from the fruit through equipment to warehousing.[10] The various grapefruit juice products are defined in the Standards of Identity under the Federal Food, Drug and Cosmetic Act.[36,37,38]

2. Flavor Fractions

The volatile components found in grapefruit juice are removed with the water vapor during the initial stages of evaporation. These volatile fractions contain most of the compounds responsible for grapefruit flavor. These flavor fractions, collected throughout the season from the different varieties of fruit processed, may be standardized by the producer or sent to flavor companies specializing in the refining and standardization of the aqueous essences and essence oils. Aqueous essence, which consists mainly of water and up to 15% ethanol, contains smaller amounts of aldehydes, ketones, esters, and terpenoid hydrocarbons contributing to the characteristic grapefruit flavor. Aqueous essences are blended and standardized on certain compounds deemed important to good flavor and aroma.[29] The standardized essences are used to restore flavor to FCGJ prior to packaging. The essence oils, containing mainly terpenoid compounds, are also refined and used for various flavoring purposes.

Grapefruit peel oils are an important by-product of the production of grapefruit juice.

Peel oil is usually collected as an aqueous slurry either simultaneously with the extraction of juice (FMC), prior to extraction (AMC) or by cold pressing the peel residues after the extraction, as described earlier.[9] The oil-water slurry is put through a conventional finisher, followed by centrifugation to remove particulate matter and separate the aqueous phase. The oil phase is known as cold pressed oil and is used as flavoring for juices and various products desiring grapefruit flavor. The cold pressed oils are often distilled to produce oil products reduced in terpene hydrocarbons. The distillation step, known in the industry as "folding", enriches the oil in oxygenated components and higher boiling sesquiterpenes, removing most of the D-limonene. The folded oils are used as flavoring materials for various products. The hydrocarbon fraction, mostly d-limonene, is used as a solvent or base material for various synthetic products.[21,28]

3. Water Extracted Soluble Solids ("Pulp-Wash")

The finisher pulp residues from juice extraction contain considerable amounts of soluble solids trapped in the particulate material. Most processors employ a series of finishers with water flowing through the finisher pulp in a countercurrent configuation, sometimes with pectic enzymes added. The trapped soluble material thus extracted is then concentrated in normal juice evaporators and used in beverages. This material, known as water extracted soluble solids or more commonly as "pulp-wash", cannot be added to any product labeled as grapefruit juice which is packaged in Florida. This extract is quite bitter, as the rag and the peel contain most of the limonin and the naringin from the fruit. The "pulp-wash" is concentrated with TASTE evaporators, and is used for flavoring beverages or producing grapefruit flavored beverage bases. Florida regulations require the addition of benzoate to these extracts to enable tracing the pressence of this "pulp-wash" in various products.[31,32]

4. Other Processed Grapefruit Products

The production of grapefruit juice results in considerable amounts of remaining waste materials in the form of peel and pulp residues. These residues are converted into various by-products, as described for oranges elsewhere.[9,20] Grapefruit are also utilized in numerous specialty products, including flavonoids which can be converted to dihydrochalcones, used as noncaloric sweeteners, pectin, candied peel, syrup, and wines.[20] Commercial importance of these products is minor compared with grapefruit juice, and the named products above. An excellent compilation of manufacturing of various by-products and specialty products is discussed in Reference 20.

REFERENCES

1. **Albrigo, L. G.,** Stress factors affecting external and internal quality of citrus fruit. in *Factors Affecting Fruit Quality, Proc. 1988 Citrus Short Course,* Lake Alfred, FL. Ferguson, J.J. and Wardowski, W. F., Eds., 1988, 69.
2. **Bain, J. M.,** Morphological, anatomical and physiological changes in the developing fruit of the Valencia orange, *Citrus sinensis* (L.) Osbeck, *Aust. J. Bot.,* 6, 1, 1958.
3. **Baker, R. A. and Bruemmer, J. H.,** Quality and stability of enzymically peeled and sectioned citrus fruit. In *Quality Factors of Fruits and Vegetables: Chemistry and Technology.* Jen, J. J., Ed., American Chemical Society, Washington, D.C. 1989.
4. **Barros, S. M., Davis, J. E., Dougherty, M. H., and Griffiths, J. T.,** Inter-relationships of °Brix, Brix-acid ratio, naringin and limonin and their effect on flavor of commercial canned, single-strength grapefruit juice. *Proc. Fla. State Hortic. Soc.,* 96, 316, 1983.
5. **Barros, S. M., Dougherty, M. H., Moore, E. L., and Carter, R. D.,** Commercial canned grapefruit juices produced in Florida, 1973-74 through 1982-83 seasons, *Proc. Fla. State Hortic. Soc.,* 97, 92, 1984.

6. **Berry, R. E. and Veldhuis, M. K.,** Processing of oranges, grapefruit and tangerines, in *Citrus Science and Technology,* Vol. 2, Nagy, S., Shaw, P. E., and Veldhuis, M. K., Eds., AVI Publishing, Westport, CT. 1977, chap 4.
7. **Bruemmer, J. H.,** Method of preparing citrus fruit sections with fresh fruit flavor and appearance, U.S. Patent 4,284,651, Aug. 18, 1981.
8. **Burdick E. M.,** Grapefruit juice, in *Fruit and Vegetable Juice Processing Technology,* Tressler, D. K., and Joslyn, M. A., Eds., AVI Publishing Co., Westport, CT. 1961. chap 26.
9. **Buslig, B. S.,** *Orange,* In this volume, 1990.
10. **Carter, R. D.,** Florida grapefruit juice from concentrate: Reconstitution, packaging and distribution, Technical Manual, Florida Department of Citrus, Lakeland, FL. 1983.
11. **Carter, R. D.,** Heating steam control in a TASTE evaporator using a microprocessor controller, *Proc. Int. Soc. Citriculture,* 1988, 1717.
12. **Carter, R. D.,** Some recent advances in the citrus processing industry in Florida, *Proc. Int. Soc. Citriculture,* 1988, 1697.
13. **Carter, R. D., Buslig, B.S., and Attaway, J. A.,** The effect of precipitation on maturity parameters of Florida citrus, *Proc. Fla. State Hortic. Soc.,* 84, 92, 1971.
14. **Chadwell, K. C.,** Citrus processing regulations: Florida standards, *The Citrus Industry,* 69(12), 1988.
15. **Davies, F. S. and Jackson, L. K.,** Plant growth regulators and fruit quality in citrus, in *Factors Affecting Fruit Quality, Proc. 1988 Citrus Short Course,* Lake Alfred, FL. Ferguson, J. J., and Wardowski, W. F., Eds. 1988. 49.
16. Florida Department of Agriculture and Consumer Services, Citrus Summary, 1987-88, Florida Agricultural Statistics Service, Orlando, FL. 1989.
17. **Hodgson, R. W.,** Horticultural varieties of citrus, in *The Citrus Industry,* Vol. 1, Reuther, W., Webber, H. J., and Batchelor, L. D., Eds. University of California, Division of Agricultural Sciences, 1967, chap 4.
18. **Hume, H. H., Ed.,** *Citrus Fruits,* Macmillan, New York, 1957.
19. Idetek, Inc. Bitterdetek™ limonin test kit, Enzyme immunoassay test for the detection of limonin in citrus juice, Idetek, Inc., San Bruno, CA, 1988. U.S. Patent No. 4,305,923.
20. **Kesterson, J. W. and Braddock, R. J.,** By-products and specialty products of Florida citrus, *Bull. 784.* Agric. Experiment Stations, Inst. Food and Agric. Sciences, University of Florida, Gainesville, FL, 1976.
21. **Kesterson, J. W., Hendrickson, R., and Baddock, R. J.,** Florida citrus oils, Bull. 749. Agric. Experiment Stations, Inst. Food and Agric. Sciences, University of Florida, Gainesville, FL. 1971.
22. **Koo, R. C. J.,** Irrigation and fertilization effects on fruit quality, in *Factors Affecting Fruit Quality, Proc. 1988 Citrus Short Course,* Lake Alfred, FL, Ferguson, J. J., and Wardowski, W. F., Eds. 1988, 35.
23. **MacDowell, L. G., Moore, E. L., and Atkins, C. D.,** Method of preparing full-flavored fruit juice concentrates, U.S. Patent 2,453,109, Nov. 9, 1948.
24. **Nagy, S. and Shaw, P. E.,** Processing of grapefruit, in *Fruit and Vegetable Juice Processing Technology,* 3rd ed., Nelson, P. E., and Tressler, D. K., Eds., AVI Publishing, Westport, CT, 1980, chap 3.
25. **Redd, J. B., Hendrix, C. M., and Hendrix, D. L.,** Quality control manual for citrus processing plants, Intercit, Inc. Safety Harbor, FL, 1986.
26. **Reuther, W.,** Climate and fruit quality, in *Factors Affecting Fruit Quality, Proc. 1988 Citrus Short Course,* Lake Alfred, FL, Ferguson, J. J., and Wardowski, W. F., Eds., 1988, 9.
27. **Rouseff, R. L., Barros, S. M., Dougherty, M. H., and Martin, S. F.,** A survey of quality factors found in Florida canned singe-strength grapefruit juice from the 1977-78, 1978-79 and 1979-80 seasons, *Proc. Fla. State Hortic. Soc.,* 93.286, 1980.
28. **Shaw, P. E.,** Citrus essential oils, *Perfumer and Flavorist,* 3(6), 35, 1979.
29. **Shaw, P. E. and Wilson, C. W., III.,** Importance of selected volatile components to natural orange, grapefruit, tangerine and mandarin flavors, in *Citrus Nutrition and Quality,* ACS Symposium Series, No. 143., Nagy, S., and Attaway, J. A., Eds., American Chemical Society, Washington, D.C. 1980, 167.
30. **Sinclair, W. B.,** *The Grapefruit,* University of California Press, 1972.
31. State of Florida, *Florida Citrus Fruit Laws,* Chapter 601, Florida Statutes, State of Florida, Department of Citrus, 1980 ed. and subsequent revisions.
32. State of Florida, *Official Rules Affecting the Florida Citrus Industry,* Pursuant to Chapter 601, Florida Statutes, State of Florida, Department of Citrus, Effective January 1, 1975 and subsequent revisions.
33. **Swingle, W. T.,** The Botany of Citrus and its Wild Relatives (rev. by P. C. Reece), in *The Citrus Industry,* Vol. 1., Reuther, W., Webber, H. J., and Batchelor, L. D., Eds., University of California, Division of Agricultural Sciences, 1967, chap 3.
34. **Ting, S. V. and Rouseff, R. L.,** *Citrus Fruits and their Products: Analysis and Technology,* Marcel Dekker, New York, NY, 1986.
35. U.S. Department of Agriculture, Horticultural Products Review. Circular Series, F Hort. 7-85, 1-86, 1-87, 12-88, Foreign Agricultural Service, 1985, 1986, 1987, 1988.

36. U.S. Department of Agriculture, U.S. Standards for Grades of Canned Grapefruit, Effective October 25, 1973, Agricultural Marketing Service, Fruit and Vegetable Division, Processed Products Standardization and Inspection Branch., incl. in Code of Federal Regulation, Title 7, part 52, rev. Jan. 1, 1989.
37. U.S. Department of Agriculture, United State Standards for Grades of Grapefruit Juice, Effective September 12, 1983, Agricultural Marketing Service, Fruit and Vegetable Division, Processed Products Standardization and Inspection Branch., incl. in Code of Federal Regulation, Title 7, part 52, rev. Jan. 1, 1989.
38. U.S. Department of Agriculture. U.S. Standards for Grades of Canned Grapefruit and Orange for Salad. Effective January 11, 1984. Agricultural Marketing Service. Fruit and Vegetable Division. Processed Products Standardization and Inspection Branch., incl. in Code of Federal Regulations, Title 7, part 52, rev. Jan. 1, 1989.
39. **Veldhuis, M. K.,** Grapefruit juice, in *Fruit and Vegetable Juice Processing Technology*, 2nd ed., Tressler, D. K., and Joslyn, M. A., Eds., AVI Publishing, Westport, CT. 1971, chap 3.
40. **Wardowski, W. F., Soule, J., Grierson, W., and Westbrook, G.,** Florida citrus quality tests, *Bull. 188*. Florida Coop. Extension Service, Inst. Food and Agric. Sciences, University of Florida, Gainesville, FL. 1979.
41. **Webber, H. J.,** History and Development of the Citrus Industry, in *The Citrus Industry*, Vol. 1., Reuther, W., Webber, H. J., and Batchelor, L. D., Eds., University of California, Division of Agricultural Sciences, 1967, chap 1.
42. **Wilson, W. C.,** Plant growth regulator uses on citrus, in *Bioregulators: Chemistry and Uses*, Ory, R. L., and Rittig, F. R., Eds., American Chemical Society, Washington, D.C. 1984.
43. **Wutscher, H. K.,** Rootstock effects on fruit quality, in *Factors Affecting Fruit Quality, Proc. 1988 Citrus Short Course*, Lake Alfred, FL. Ferguson, J. J., and Wardowski, W. F., Eds., 1988, 24.

Chapter 4

THE APPLE

D. V. Fisher and J. A. Kitson

TABLE OF CONTENTS

I. INTRODUCTION

A. HISTORICAL DEVELOPMENT OF NORTH AMERICAN CULTIVARS

The original home of the common European apple *Malus sylvestris* is not definitely known. This species is indigenous to the region south of the Caucasus from the Iranian province, Ghilan, on the Caspian to Trebizond on the Black Sea, and from prehistoric times it is certain to have existed in Europe, both wild and cultivated over an area extending from the Caspian Sea to the Atlantic Ocean, except in the extreme north.

Native crabapples of several species are indigenous to North America, but the European apple was introduced by the early European settlers both in eastern Canada and the U.S. before the year 1600. Few trees were grafted until the early 1800s and until that time most orchards were of seedling origin. In fact, it is from such orchards that most of our modern varieties originated.

Starting in 1848, the American Pomological Society[1] assumed the job of selecting and recommending promising cultivars of all fruits. In 1905 S.A. Beach published the historic 2 volumes *The Apples of New York*[2] in which over 2,000 named cultivars and seedlings were described in great detail. Since 1944 new apple, and other cultivars, as introduced, have been listed and described under *New Germ Plasm Releases* in *HortScience* published by the American Society for Horticultural Science.

B. CHARACTERISTICS OF MAIN APPLE CULTIVARS

1. Delicious

The Delicious apple originated on a pioneer general farm in Peru, Iowa, and first fruited in 1879. The originator was Mr. Jesse Hiatt who had planted a seedling orchard. One tree was out of the row, and twice he cut it down. However, it sprouted again, and because he had a natural sympathy for anything that would stand adversity he looked at it and said "If you must live, you may". The tree flourished, produced up to a barrel of apples, and he named it Hawkeye, the state emblem of Iowa. Purchased by Stark Bro's Nurseries, it later was named Delicious.

Today, Delicious (and mainly its sports) is the most widely and heavily produced cultivar in the world. Its rise to fame can be attributed to certain very good cultural and fruit characteristics: perhaps even more so to the advertising by Stark Bro's Nurseries who bought, named, and introduced it in 1896. Starks introduced an improved color sport, Starking, from Lewis Mood of New Jersey in 1921, followed by the spur-type strain, Starkrimson, from Bisbee Orchards in Oregon in 1957 and later by additional high color and spur strains such as Ultra Red. Other nurseries throughout the U.S. and Canada have introduced numerous and excellent strains of both standard and spur types and the count at present in these is well over 200.

The trees of standard Delicious are vigorous, spreading, somewhat slow to commence bearing, but once in production yield heavily and annually. The cultivar is self-sterile, and needs ample provision for cross pollination. The fruit is medium to large, firm with tough skin which resists bruising, is picked mid-season and has a pleasing delicate aroma and flavor eaten at its prime, but quickly becomes over-ripe and mealy on exposure to warm temperatures. The variety does not suffer from mildew, but scalds badly if picked immature. Delicious is not a good processing apple due to low acidity. The cultivar is expected to dominate the fresh fruit market at least until the turn of the century. Present production is 40% of the North American apple crop.

The trees of spur-type Delicious generally produce very highly colored fruit with pronounced calyx lobes, bear very young and are inclined to biennial bearing unless well thinned. Trees have an upright growth habit which requires limb spreaders to widen branch angles. Most Delicious planted today are of spur strains because of compact tree growth and early and heavy bearing.

2. Golden Delicious

The second most widely planted and produced apple in the world is Golden Delicious. Like Delicious, it originated as a chance seedling on the hill farm of Anderson Mullins in Clay County of West Virginia. Supposedly it is a seedling of Grimes Golden crossed with Golden Reinette. Recognizing its superior qualities, Mullins sent fruit samples to Stark Bro's Nurseries. Paul Stark, Sr. visited the orchard and owner in 1914 and was so impressed he bought the tree for $5000.00 and erected a steel guard fence around it. By 1917 the variety was introduced by Stark Bro's Nurseries and given the name Golden Delicious to capitalize on the success of the Delicious cultivar. It was rapidly accepted as an important new cultivar and holds second place today in world production. It is widely grown in the U.S., Canada, France, Holland, Italy, South Africa, Australia, New Zealand, and many other locales.

The tree is moderately spreading, early bearing, and annually productive if well grown, and the fruit well thinned. Poor cultural practices can lead to biennial bearing. The fruit is greenish yellow to yellow in color and for fresh fruit purposes and long storage is picked when color changes from green to greenish yellow. Golden Delicious apples are thin skinned, bruise easily, subject to russeting, and often difficult to pick as the stem does not separate easily from the spur. On the plus side, the apple has a beautiful golden finish, particularly when grown in dry climates, has cream-colored flesh, excellent quality resulting from a good blend of sugar and acid, and stores well, especially in low oxygen controlled atmosphere (CA) storage. With modern storage techniques, it is now marketed from September to August. Golden Delicious, in addition, is excellent for processing as juice, apple sauce, or pie filler. Many people consider Golden Delicious the choicest, general purpose cultivar.

Unfortunately, this cultivar is currently in the doldrums because of overproduction and many thousands of acres have been removed in most heavy producing countries. When grown in damp climates, skin russeting detracts from economic value (but not quality). A Golden Delicious seedling selection, Smoothee, and French sport, Lysgolden, possess Golden Delicious qualities with greatly reduced russeting tendencies.

3. McIntosh

For many years, McIntosh was second only to Delicious in North American production. However, in the last 20 years it has been overtaken by Golden Delicious and slipped into third position with production of around 15 million bushels. Numerous surveys in Canada have continued to rate it as the most popular cultivar, a reputation enhanced by its extended market season associated with CA storage.

McIntosh originated as a seedling of unknown (possibly Fameuse) parentage on the farm of John McIntosh, Dundela, Dundas County, Ontario, about the year 1810. It was the remaining tree from a group of seedlings planted about 1800 and by about 1820 the fruit quality of this new apple was so outstanding that seedlings from the tree were grown and sold. However, starting about 1835, John McIntosh commenced selling grafted, true-to-name trees of his new cultivar. Trees were sold in Ontario, Vermont, and New York. The original tree was believed to produce blush type apples, but subsequent populations contained trees which had reverted to striped color and for many years orchards contained whole trees or part trees of blush or striped apples. A number of improved color strains have been introduced, the Hamilton strain from Ontario and the Summerland strain from B.C. being among the best. In 1912, a monument was erected in Dundela, Ontario in honor of the original tree.

The cultivar gradually grew in reputation, but because of competition from Fameuse, Baldwin and other favorites of that era, and also because of McIntosh's susceptibility to apple scab, really extensive plantings did not take place until around the turn of the century. From that time on its success was assured because of excellent quality for eating, home cooking and processing, and tree characters of good yield and hardiness. McIntosh is the most widely grown apple in Canada.

Similar to the Delicious story, the whole course of McIntosh history was changed by the discovery in 1962 and 1967, in British Columbia[3] and subsequently elsewhere, of "spur" strains of McIntosh. These strains produce early and heavy bearing trees of compact growth, great structural strength, and good blushed fruit color characteristics. The important strains all arose as sports of Summerland Red McIntosh. The best known strain is MacSpur, and spur strains of McIntosh comprise 90% of new plantings in North America. Unfortunately because of some reversions in spur type McIntosh to standard growth habit and lack of identification of reverted trees used to produce wood for propagation, some orchards contain significant numbers of standard McIntosh trees purchased as "spurs". This situation will be resolved as nurserymen become more aware of the importance of proper selection of propagating wood.

4. Rome Beauty

This cultivar, the fourth most important apple grown in America, has a current production, mostly in the middle latitudes of the eastern U.S. of 14 million bushels. Its originator, Joel Gillett of Proctorville, Rome Township, Laurence County, Ohio, gave the seedling to his son, Alanson, in 1816. The tree produced fine apples and was first propagated about 1828, and named Rome Beauty after the Township. History states that the original tree planted on the bank of the Ohio River was washed away by a flood in 1860. H. N. Gillett was largely responsible for propagation of the new variety and brought it to the attention of the Ohio Convention of Fruit Growers in 1848. (Writers note: Ohio, New York, Indiana, and Virginia were major apple states of that era.)

Rome beauty comes in different strains. The old original strain produced striped, often poorly colored apples. Then came the numerous red sport strains, some good and some subject to reversions, and ultimately the high color 'spur' compact tree strains. All produce large, late maturing fruit with fair quality and medium to long storage life. Traditionally, because of size and somewhat coarse texture, Rome has been prized as an apple for baking purposes. Latterly, because of the striking color of the new standard and spur strains (Law and Lawspur) for, example, Romes are often found in supermarkets competing for shelf space with Delicious and other red cultivars. The cooler the location in which Rome fruit will mature, the better the fruit color.

The tree of Rome is early to come into bearing and heavy bearing. It is only of moderate size, bears annually and forms fruit buds on tips of shoots (tip bearer). It blooms late so is not a good pollenizer unless for other late flowering varieties. Being self-fertile it requires no pollenizers. For the foreseeable future, good strains of Rome Beauty seem assured of a place in North American apple growing.

5. Jonathan

Jonathan is an 'old timer' which still retains its popularity as the fifth most important apple grown in the U.S. This cultivar originated as a seedling on the farm of Mr. Philip Rick of Woodstock, Ulster County, New York. It was first described by Judge J. Bull of Albany, New York, in 1826 by which time it appears to have been in propagation under several different names by a number of nurseries.

While grown for many years all over the U.S. and Canada, principal production is now located in the U.S., midwest (Michigan, Illinois, Ohio) and in Pennsylvania. Its popularity stems from its excellent fruit quality, the result of a fine blend of sugar, acid, and flavor. This cultivar is also in high demand for processing. Jonathan keeps fairly well in regular cold storage and responds particularly well to CA storage. The original strain produces striped fruit, but many excellent high color and several 'spur' strains are available.

This cultivar, however, has its problems. The tree type tends to be a bit willowy, and the tree is very susceptible to fire blight and powdery mildew (The latter also causes skin russeting.) The fruit tends to be small and particularly, if large, the fruit is susceptible to Jonathan flesh breakdown. Jonathan breakdown is accentuated by overmature harvesting. Calcium foliar sprays during the growing period have greatly reduced incidence of flesh breakdown. The other storage problem with Jonathan is Jonathan spot, a black, disfiguring lenticel spotting associated with over-long cold storage; CA storage markedly reduces this problem.

6. York Imperial

York Imperial dates back to the early 1800s where it was found on the farm of a Mr. Johnson, adjoining what was then the borough of York, Pa. Recognizing its value, Mr. Jonathan Jessop commenced propagating it before 1830 under the name Johnson's Fine Winter. The name later was changed, about 1853, to York Imperial by nurseryman, Charles Downing. York Imperial became widely distributed throughout Pennsylvania, Maryland, West Virginia and Virginia where it became the leading cultivar. Approximately 8 million bushels are still produced in this general area.

In the early 1900s, York, gained popularity due to an active export business to Britain, mainly because of its good keeping quality. When export restrictions were imposed, emphasis was placed on domestic outlets. However, due to indifferent color it competed poorly with better colored red varieties, and gradually became regarded as a processing apple, the trend being accentuated by demand during World War II.

York Imperial is best characterized by its lop-sided shape. However, it has a creamy colored flesh, small core, and gives a high yield of processed product. The cultivar matures late, requires a long growing season, is firm, and bruise resistant at harvest, high yielding, and stores well. York Imperial is subject to cork spot, a problem somewhat reduced by boron sprays at bloom time. It is fairly tolerant of apple scab.

When well pruned, thinned and fertilized this cultivar is a fairly regular cropper, but can become biennial bearing if neglected. The tree is large, upright, and a vigorous grower. Because of its firmness, many York Imperial orchards are mechanically harvested for processing.

Despite the introduction of red strains, York Imperial will never become an important dessert apple because of its shape and stiff competition from better known 'reds'.

7. Stayman Winesap

This old cultivar, more generally referred to as 'Stayman' after its originator, Dr. J. Stayman of Leavenworth, occurred as a seedling of Winesap and first fruited in 1875. However, it was not cataloged as a cultivar until 1894 to 95. Stayman, and more particularly its excellent red strains, is still extensively grown in the east central U.S. The tree is vigorous, productive, but inclined to be frost tender and unsuited to northern locations where fruit does not always fully mature. Being a triploid, it is useless as a pollinator. The fruit is of good size, crisp, juicy, and suitable for both fresh use and processing. It retains its popularity because of the fine appearance of its red sports and its general purpose qualities. Annual U.S. production is 5 million bushels.

8. Newtown

Until the recent appearance in North American markets of Granny Smith, Newtown (Yellow Newtown, Albemarle Pippin) had the distinction of being the only acceptable green dessert apple on this continent. The production of 4.0 million bushels is limited mostly to California, Washington, Oregon, and British Columbia, and the further north it is grown the better is its quality and cold or CA storage life.

The exact history of Newtown is not precisely known, but the cultivar is believed to have originated as a seedling sometime between 1700 and 1750, on the property of Gershom

Moore in Newtown, Long Island, N.Y. Originally grown extensively in Virginia, it enjoyed, for many years, an excellent export market to Great Britain.

The tree is vigorous, spreading, cold tender, and susceptible to perennial canker, and the fruit on light crop trees is inclined to have poor storage life and to develop bitter pit. This cultivar is slow to commence bearing and because of its alternate bearing tendency, produces greenish fruit in "on" years and large, yellowish, off-shapen fruit in "off" years. Newtown does not require a pollenizer.

The crop is picked the second or third week of October and can be stored until April or May. The flesh is very firm, juicy, of excellent dessert quality, but because of rapid browning after cutting is unsuited for processing.

9. Cortland

This cultivar[6] resulted from a cross between Ben Davis and McIntosh made at the New York State Agricultural Experiment Station in 1898. Introduced in 1915, it has the distinction of being the first apple resulting from a planned breeding project in North America to achieve and still retain some commercial importance. Its only other counterparts, resulting from breeding programs (and introduced much later), to achieve any degree of commercial importance are Spartan, Idared and Empire. Present production is about 3.5 million bushels.

The fruit of Cortland is of good size with a fair degree of red striped color, with white flesh, good flavor and texture, and stands handling well. It is picked about two weeks after McIntosh, stores well, and is good for eating or processing. At least one red strain is available. The tree is hardy, productive, and annual bearing. It has a slender, willowy growth habit and is very susceptible to mildew. It is grown mostly in New York and New England states and in Eastern Canada.

10. Rhode Island Greening

This ancient cultivar, still an important processing apple in New York state, has an annual production of about 3.5 million bushels. It originated as a seedling, probably close to 300 years ago on the property of a Mr. Green in the vicinity of Newport, Rhode Island. It was mentioned as a popular cultivar in Coxe's *Pomology* published in 1817. The tree is large, has tendency to biennial bearing, is slow to come into bearing, and apple-scab susceptible.

The fruit is green, sometimes slightly blushed, of good size, and excellent quality for eating or processing. It is picked mid to late season, and if harvested immature is very susceptible to scald. The future of Rhode Island Greening is uncertain especially since to the knowledge of the writer, no spur strains have so far been reported.

11. Winesap

For many years Winesap was a most important cultivar in the Eastern States and later in the Pacific Northwest. Winesap retained its popularity for a great many years owing to its excellent long storage life. However, the advent of CA storage, resulting in a longer marketing period for preferred cultivars such as Delicious and Golden Delicious, has resulted in greatly reduced demand. Present production in North America is about 3.5 million bushels.

The originator, date, and place of origin of Winesap is not known. Undoubtedly it was discovered before 1800 because Coxe in 1817 noted that Winesap was becoming "the most favorite cider fruit in West Jersey". This was in an era when production of hard apple cider was an enormous business.

The tree of Winesap is vigorous, reasonable hardy, and annually productive. The standard (old) strain of Winesap produced a striped red apple often deficient in color. Numerous red color sports have been reported, perhaps the most widely planted being the Mosebar. A number of 'spur' types have been discovered but have not achieved any importance perhaps because their color is no better than the old Winesap. Winesap has a distinctive, fair to good flavor, perhaps a bit high in tannin, and because of a high acid content is in demand for blending in juice processing. Present prices and demand for Winesap suggest this cultivar may gradually disappear.

12. Northerm Spy

Northern Spy, once an important cultivar in many areas of eastern Canada and U.S. now has a production of only 2.5 million bushels, mainly in New York, Ontario, and Michigan. This apple, famous for its quality, has an interesting history. It resulted from seedlings planted about 1800 at East Bloomfield, New York, by Heman Chapin. The original tree died before bearing fruit, but threw up vigorous root suckers which were planted on the neighboring orchard of his brother-in-law, Roswell Humphrey, where, in time, these fruited and established the reputation of the apple. A bronze plaque was erected in 1964 near the location of the original tree in recognition of its origin and importance.

As a cultivar, Norther Spy is notoriously slow in coming into bearing (unless grown on dwarfing roots), has a strong tendency to biennial bearing, and is very susceptible to bitter pit. It is late in coming into blossom but overlaps sufficiently with many other varieties to become pollinated. Northern Spy is essentially streaked with bright red color, but many blushed sports are available. The keeping quality of these sports is considered by some to be inferior to the parent types.

The Northern Spy tree is upright in early years, but with training and fruiting assumes a more spreading form. The cultivar continues growing late into the fall, the fruit is picked very late, and the tree is considered somewhat winter-tender. The apple is of exceptional quality for fresh eating and is well accepted for processing. Processors pay a premium price for this cultivar. Spy has a long storage life and a high vitamin C content. Because of poor color fresh market grades are low, and the future for Northern Spy is not considered bright.

It has been used in many breeding programs and of the many named seedlings it has parented, only Spigold, Sandow, and Wayne have achieved any importance. Perhaps its most important contribution to pomology has been the resistance displayed by its seedlings to root woolly aphids. It is one parent of the English woolly aphid resistant rootstock series M106, 104, and 111, and the U.S.D.A. rootstock Spy 227.

13. Idared

This cultivar,[6] the progeny of Jonathan × Wagenar, was originated by Dr. Leif Verner at the Idaho Agricultural Experiment Station. It was selected in 1935 and named and introduced in 1942. Idared has been planted world-wide, but greatest production is centered in New York State. Current U.S. production is about 2.5 million bushels.

The tree of Idared is early to come into bearing, productive, only medium in size, and moderately hardy. It is somewhat susceptible to fireblight, similar to its Jonathan parent.

The fruit of Idared is large, firm, late maturing (2 weeks after Delicious) with a good red skin color and exceptional flesh firmness which makes it a late keeper. The fruit quality is fair to good, flesh texture fine, and core area small. Similar to its Jonathan parent, in some years Idared develops Jonathan spot. This cultivar will probably retain some importance where it grown well and a receptive local market is assured.

14. Granny Smith

Because of its relatively recent world appearance and great popularity on North American markets, many people look on this cultivar as "new". Far form it! Granny Smith apple was named after Mrs. Marie Ann Smith, an English immigrant of 1839 who settled near Sydney, Australia. Mr. and Mrs. Smith developed a small orchard and nearby was a neglected small apple tree which a neighbor had noticed produced a few green apples of good eating quality.

Mrs. Smith said the seedling arose from some rotten Tasmanian apples that remained in some gin cases obtained from the Sydney market, and which she threw out. The neighbor, Mr. Small, and a member of the Smith family propagated and planted a fairly large block of the new cultivar. By the turn of the century, it became evident that here was a superior new apple, popular with consumers, and which was easy to handle and productive for the grower.

Gradually, Granny Smith has become a leading cultivar in Australia, New Zealand, South Africa, Argentina, and most recently, the northwestern U.S.[3]

The Granny Smith produces a leggy tree of upright habit unless branches are spread, the leaves hang on the trees very late in the fall, the cultivar is susceptible to mildew, and tree hardiness has not yet been established in colder areas.

The trees are productive, require pollenizers, and the fruit is pale to light green at harvest. There are also strains which show faint red striping. Granny Smith is available in both 'spur' and standard strains. A long, warm growing season is required and the fruit should be harvested a month or more after Delicious, probably 180 d after bloom. This cultivar is susceptible to bitter pit, and unless picked at good maturity develops severe apple scald.

While current U.S. production is only about 2 million bushels, some forecasts indicate that in 20 years, Granny Smith might displace Goldent Delicious as the No. 2 apple in North America. There is no doubt that with the present volume of production, a strong, profitable North American market exists for this cultivar.

15. Spartan

Spartan originated as a cross between McIntosh and Newtown made at Agriculture Canada Research Station, Summerland, B.C., Canada.[4] The cross was made in 1926 by the late Dr. R. C. Palmer and introduced in 1936 to lengthen the McIntosh marketing season. It has been heavily planted in British Columbia and in some northern European countries, notably Denmark. In nonirrigated orchards in Eastern Canada and U.S., Spartan fruit is reported to be small, but this appears largely to be the result of inadequate pruning and thinning, and too early picking. The tree is vigorous, early to come into bearing, and when well managed, produces annual crops. The fruit is of fine texture, crisp, juicy, and matures just before Delicious. It keeps well in cold storage until February and retains remarkable firmness and quality in CA storage. In some orchards, in some years, fruit flesh breakdown has been a serious problem in storage but this largely has been eliminated by pre-storage calcium dips or pre-harvest foliar calcium sprays.

C. CULTIVAR SELECTION

The elimination or retention of apple cultivars and the development of new kinds is a ruthless and on-going procedure. Of the cultivars grown and recommended 50 or 80 years ago, very few still remain. For example in the 1900 era, Ben Davis, Winesap, Baldwin, Northern Spy, Greening, Russets, etc. held sway. Many other cultivars such as McIntosh, Rome, Newton, Jonathan and Delicious were grown but not on nearly as important a scale. Newer cultivars such as Golden Delicious, Cortland, Idared, Empire, Granny Smith, Mutsu, Jonagold, and Spartan were at that time mostly little known or not yet introduced.

Elimination of older cultivars has been the result of poor tree and fruit characteristics such as lack of hardiness, difficulties in fruit handling, disease susceptibility, unsuitable color or loss of consumer appeal. Many newer cultivars, while lacking the flavor and overall quality of older kinds are more foolproof for the grower and distributor and often have multiple commercial uses.

D. DEVELOPMENT OF IMPROVED CULTIVARS

The development of high color sports in many cultivars has been their salvation. Still more important has been the discovery and popularization of compact tree growth 'spur sports' which now are available in all of our mainline cultivars, and 90% of all kinds grown commercially. These have resulted in early producing, high density orchards.

The most exciting development in apple breeding, which in the next 50 years will have a far reaching effect on cultivar selection, is the breeding of apples with complete resistance to apple scab *Venturea inequalis*, the most serious apple disease. Resistance has been bred into standard apples of the *Malus sylvestris* type by a dominant resistance gene obtained from the small crabapple, *Malus floribunda*. Major programs are in progress all over the world but the initial work published in 1953 by Hough, Shay, and Dayton[5] in the U.S. set the pattern. Active U.S. programs are now located in New Jersey, Illinois, Indiana, New York, and in Canada at St-Jean, Quebec. Now cultivars such as Prima, Priscilla, Sir Prize, Liberty, Freedom, Redfree, and Jonafree[6] have been available commercially for some time. Other selections soon will be released. While fruit quality, size, and appearance of the new cultivars is excellent, they tend to have a short storage life and their ultimate commercial importance will hinge, in large measure, on further breeding to incorporate long, cold, or CA storage life.

E. PRODUCTION OF MAIN APPLE CULTIVARS

Subject to annual fluctuations the current production of apples is listed by cultivar as follows:

Annual Production in 40 lb (18.2 kg) Units

Cultivar	U.S. (millions)	Canada (millions)
Delicious	82.0	5.5
Golden Delicious	32.0	1.5
McIntosh	16.0	15.0
Rome Beauty	14.0	
Jonathan	9.0	
York Imperial	8.0	
Stayman Winesap	5.0	
Newtown	4.0	
Cortland	3.5	
Rhode Island Greening	3.5	
Winesap	3.5	
Northern Spy	2.5	1.0
Idared	2.5	
Granny Smith	2.0	
Spartan	—	1.5
Miscellaneous	12.5	1.5
Total	200.0	26.0

II. STORAGE AND QUALITY OF FRESH APPLES

Apples are available almost worldwide all year, thanks to highly developed storage systems. First to be developed were regular cold storage facilities where fruit is usually held at temperatures of $-1°$ to $+1°C$ and 85 to 90% relative humidity. Under these conditions, many cultivars have storage life of 2 to 5 months and some such as Newtown and Winesap may keep up to 8 months.[7]

Apart from cultivar differences, the storage life of apples is also greatly influenced by handling immediately following harvest. If fruit is not cooled rapidly, the increased respiration can cause accelerated ripening and possible increases in breakdown and losses from rot developing in storage.

An increasing quantity of apples is being held for fresh market beyond the normal period for cold stored fruit by use of CA storage. In a typical CA apple storage, oxygen content of storage air is reduced from a normal 21% to about 2.5%. Carbon dioxide is raised from 0.03% in normal atmosphere to 1.5 to as high as 5.0% for storage of certain cultivars.[7]

The apple's respiration is greatly reduced by holding in a low oxygen, high carbon dioxide atmosphere and the onset of senescence is delayed. This allows some cultivars to be marketed up to ten months after harvest. Use of modified storage atmospheres may also allow specific cultivars such as McIntosh to be held at higher temperatures, up to 3.3°C. Storage at this temperature reduces the core browning which is a problem in McIntosh held at − 1° to 0°C.

A number of systems have been devised for controlling CA storage atmospheres. The first developed and simplest systems use the fruit's own respiration to reduce the oxygen content. Respiration products, particularly carbon dioxide are partially removed in various types of wet or dry scrubbers to reach the desired final CO_2 level. Probably the simplest system, known as the dry lime scrubber was developed by Eaves[8] and has been widely used in Canada and in a number of other countries. In this system the storage atmosphere is circulated through a chamber filled with sacks of hydrated lime. This reacts with CO_2 to form calcium carbonate. By adjusting the circulation rate of the CA room atmosphere through the scrubber, it is possible to maintain the desired CO_2 level.

A number of atmosphere generators for CA storages are commercially available.[9,10,11] These systems have the ability to adjust the CA storage atmosphere usually within 2 to 3 d compared with 10 to 15 d in storages using natural fruit respiration and scrubbers. Rapid reduction of oxygen content in CA storage atmospheres has been shown to improve significantly flesh firmness and acidity retention in CA stored apples from a wide range of production systems.[12] Also, if it is necessary to open a CA storage for repairs or to check the quality of the stored fruit then it is desirable to be able to re-establish the low oxygen atmosphere again as rapidly as possible. In these instances, the system of using the fruit's own respiration is too slow.

Some CA atmosphere generators, the Gen-O-Fresh[9] system for example, use oxygen in the CA room atmosphere to support combustion of a hydrocarbon fuel such as propane. This combustion rapidly reduces the oxygen content of the air and at the same time generates an excess of carbon dioxide, some of which is removed by a scrubber.

A recently developed and patented CA atmosphere generator, Oxydrain[10], produces nitrogen by thermally cracking ammonia into nitrogen and hydrogen and then reacting the gas mixture with oxygen according to the formula:

$$2N_2 + 6H_2 + 3O_2 \rightarrow 2N_2 + 6H_2O$$

The nitrogen and water vapor mixture is cooled and much of the water vapor is condensed before the gas is admitted to the storage chambers. Nitrogen generating equipment does not produce carbon dioxide during its operation and hence requires scrubbers only to remove the CO_2 produced by the fruits' natural respiration.

Among the newest CA developments is a nitrogen generator which uses molecular sieves to remove preferentially the oxygen from compressed air leaving nitrogen remaining. In the Gastec[11] system two adsorbers operate alternately on a 60 s cycle. While one side of the adsorber system is producing nitrogen the other side is going through a 60 s decompression and regeneration cycle. Using this system the room air is displaced by nitrogen and the oxygen level is thus rapidly reduced.

Fruit is often loaded into CA storages in 450 kg bulk bins which are frequently not pregraded. After storage periods of from 6 to 10 months, the CA chambers are opened and the apples sorted and packed. Usually, if the fruit was carefully selected from a good orchard, the percentage of sort-outs for size, shape, and color will be small, however, there are always some apples which fail to meet fresh market standards and must be converted into processed products. Apples from CA storage are generally satisfactory for processing, provided they are used within 3 to 4 d. Long holding periods lead to excessive softening and make juice extraction or preparation of peeled apple products difficult.

During any form of storage, apples tend to lose acidity and pectin, with the change in pectin content being associated with a loss of firmness or crispness. Texture change in apples is frequently followed in commercial practice by use of a pressure tester which in its simplest form is a spring scale fitted with a plunger, which is forced into the cut surface of a fruit.[13] Along with a decrease in acidity during storage there has been a general worldwide trend to grow lower acid apple cultivars such as Red and Golden Delicious. An exception to this trend has been the expanding production of the green and relatively high acid cultivar Granny Smith. However, with the exception of Granny Smith this has meant that fruit with low levels of acidity is generally available and is probably well accepted by a majority of consumers. Average acidity of Delicious was reported by Strachan et. al.[14] as 0.27% with a range from 0.20 to 0.58%. For Golden Delicious the average was 0.41% and the range from 0.32 to 0.51%. By these standards McIntosh should be rated as a high acid apple with an acidity range of 0.33 to 0.74% and an average of 0.54%.

Apples for fresh market are usually sorted carefully to remove fruit that may, from its location on the tree, not mature properly. Thus many reported analytical data for apples based on fresh market quality fruit may not be representative of the fruit available for processing. As a general rule processing grade fruit, particularly that available early in the season, will tend on the average to be lower in soluble solids than fruit of the same cultivar packed for sale as fresh fruit.

As an example, fresh market quality Delicious were reported by Strachan et. al.[14] to have soluble solids averaging 13.2% with a range of 12.3 to 13.6%. This is somewhat above the seasonal range of 11.8 to 12.8% found[15] for commercially made Delicious juice from the same growing area.

Immediately after harvest many apples show a small increase in sugar content as starch present in the tissue is converted to sugar. Following this rise sugar content tends to decline because of conversion of sugars to carbon dioxide due to respiration. If apples become significantly desiccated in storage, however, the loss of water may offset the respiration losses and sugar content may actually increase.

Apples are a good source of dietary fiber with the principal components being cellulose and pectin. Peeling the fruit reduces the fiber content from 1.0% to 0.6%.[16] The fiber content of processed peeled apple products such as sauce remains essentially the same as that of fresh apples unless it is diluted by addition of sugar, water or other ingredients.

Among mineral constituents, apples are a good source of potassium, 1100 ppm; and low in sodium, 10 ppm; calcium, 60 ppm; phosphorous, 100 ppm; and iron, 3 ppm.[16] Fresh apples are also a reasonably good source of ascorbic acid. A range of 22 to 110 ppm was found for British Columbia cultivars,[14] while up to 220 ppm has been reported in older U.S. cultivars.[17] During processing into juice, ascorbic acid is largely destroyed by oxidation so, unless fortified, apple juice is not a significant source of this vitamin.

III. FACTORS INFLUENCING THE QUALITY OF APPLES FOR PROCESSING

Most apples grown commercially are intended for sale on the more profitable fresh market. Because of this the fruit available for processing is not necessarily of the optimum size, shape, texture or flavor for processed products.

These are exceptions, as in the case of some of the special cultivars such as Tremlett's Bitter, grown specifically to produce the hard ciders popular in England. These apples may have a high level of tannin to impart a desired astringent flavor to the finished product.

In general, apples intended for peeled apple products, should have a spherical or spheroidal shape with a relatively smooth surface. A good shape along with a minimum diameter of about $2^1/_2$ in will result in mimimum peeling loss and a minimum amount of trimming.

For products such as applesauce, color and flavor preferences differ. In some parts of North America where McIntosh has been traditionally popular, the whitish green color, smooth texture and relatively high acidity of this apple is preferred in applesauce, and it may be sold under the cultivar name. In other areas the less acid, more grainy texture, and yellow color of Golden Delicious are preferred.

For canned apple slices and pie fillings, preferences vary depending upon what is available in any particular region where apples are processed. In the U.K. the full flavored Bramley Seeding is highly rated for use in pie slices in spite of its somewhat irregular shape. In Australia and New Zealand the cultivar Granny Smith is rapidly gaining in popularity as a pie slice apple and will tend in future to replace other regional varieties. In California, Newtown Pippin has been a popular peeling apple, while in Washington and British Columbia Golden Delicious has become a major cultivar for slices and sauce. With further extensive plantings of Granny Smith and Golden Delicious in many parts of the world, the importance of these cultivars in peeled apple products will continue to rise.

Good quality applesauce and canned or frozen slices require that the fruit, of whatever cultivar used, be fully mature but with a firm texture. Over mature fruit becomes difficult to peel and usually yields soft textured slices or sauce with an excessively smooth paste-like appearance. In many instances processing apples are sorted from orchard run fruit with the best colored apples from the top and outside of the tree being selected for fresh market. If trees are not well pruned and trained to produce a relatively uniform quality crop, a portion of the apples sorted out for processing will comprise immature fruit with high starch content, high chlorophyll levels, and otherwise poor color. Use of such fruit in sauce and slices will result in generally poor quality processed product.

With the exception of fruit size, the same quality factors required for peeling apples are also required for those used for juice. Fruit with high soluble solids content and low to medium range of acidity (0.3 to 0.4%) seems to be most highly accepted by consumers. This is particularly true in areas where low acid cultivars such as Delicious are popular as fresh fruit. Government grade standards in the U.S. set minimum soluble solids standards and acidity ranges as follows:[18]

U.S. Grade A	Soluble solids minimum—11.0%
	Acidity (as malic) 0.24—0.67%
U.S. Grade B	Soluble solids minimum—10.5%
	Acidity (as malic) 0.19—0.77%

Canadian regulations for similar grades require a slightly higher level of acidity:[19]

Canada Fancy	Soluble solids minimum—11.5%
	Acidity (as malic) 0.35—0.70%
Canada Choice	Soluble solids minimum—10.5%
	Acidity (as malic) 0.30—0.80%

Immature apples are particularly undesirable for juice manufacture because the starch present causes difficulty with clarification and may also cause post-process haze formation in previously clarified juice or concentrate.

Overmature fruit on the other hand is difficult to handle in most juice extraction systems, yields less juice, and frequently, juice with quality defects such as excessively low acidity. Now that CA storage is becoming widely used and fruit is sorted aften ten months or more in storage, some CA stored apples are available for juice making up until the new crop is harvested each year. Apples from CA storage may have a firm texture and provide a good yield of juice, however, they are frequently almost totally lacking in aroma volatiles. Aroma may be regenerated by holding the fruit at room temperature for up to a week after it is removed from storage, however, by that time the texture is frequently too soft to allow satisfactory juice extraction.

In apple products with reduced water content, such as low moisture dehydrated apples or apple juice concentrates, the soluble solids content of the fruit used has a significant influence on the yield of finished product. A 1000-kg lot of 11% soluble solids apple juice, for example, will yield 152.7 kg of 72% concentrate.

A similar quantity of 12% soluble solids, juice will yield 173.6 kg of the same product.

Because apples are usually not grown specifically for dehydration or concentrate manufacture, there has been no major breeding program aimed at producing high soluble solids fruit. The economics of producing such fruit for processing, however, become worth considering as rising energy prices increase the cost of water removal by evaporation.

IV. UTILIZATION OF APPLES FOR PROCESSING

About a third of the worlds commercial apple crop is converted into processed apple products. These fall into two main categories, those in which the apples are first peeled and cored, and those in which the whole apples are converted into liquid apple products.

Because of the economics of peeling operations, the peeling process is usually restricted to apples of $2^1/_2$ inch or larger diameter, preferably with a spherical shape. Peeled and cored apples are converted into a wide range of dehydrated, canned, and frozen apple products.

Dehydrated or reduced moisture content products made from apples comprise two major groups. The first group consists of dried apple rings, slices, and occasionally other cuts with 22 to 24% moisture content and usually a high content of sulfur dioxide to preserve color and to prevent spoilage. These products are known in the trade as dried or evaporated apples. They may be used directly in apple pies, turnovers, and other baked goods or further dried to become products in the second group. The second group consists of low moisture dehydrated apple products usually containing less than 4% moisture. These include: low moisture apple pieces in a number of shapes, either in expanded (puffed) or regular form and various forms of apple powder or dehydrated apple sauce.

Apples are canned most commonly in North America in the form of apple sauce, either sweetened or unsweetened. A second major canned product processed in many parts of the world consists of solid-pack pie slices, usually packed in institutional size cans. Of lesser commercial importance are other canned apple products including canned apple pie filling, canned dyed spiced apple rings, canned baked apples, and canned apple pieces in syrup.

Peeled apple slices for pies are the most important frozen apple product. Most of these are packed in poly bags or plastic pails containing up to 30 lb, but these is also a small volume of retail size, 2 to 5 lb poly bags of this product.

Frozen apple turnovers, consisting of a pie filling mix in a special pastry that is fried in deep fat before serving, are a mjaor institutional apple product used in fast food outlets. Frozen apple pies are a more significant product for home use by consumers, although several processors make frozen apple turnovers for home use.

Specialty apple products that are frozen include apple dumplings and baked apples. There has also been a small commercial production of frozen applesauce but although the quality of this product is high, it has not become popular probably because of the inconvenience and time required to thaw before serving.

The major volume of processed apples are made into liquid products. These consist of clarified or unclarified apple juices, juice concentrates of various soluble solids contents, sweet or unfermented cider, and fermented ciders. Some fermented ciders are distilled to make calvados or apple brandy.

Most apple juice sold on world markets is a clarified product packed in glass or metal containers. There is now, however, a rapid growth of aseptically packed apple juice in containers formed in the final filling operation. These containers usually consist of various laminates of paperboard, plastics, and aluminum.

Apple juice concentrates are manufactured in many parts of the world including eastern and western Europe, China, Japan, India, South Africa, Argentina, Australia, New Zealand, U.S., and Canada. Most of this production is sold for remanufacture into reconstituted apple juice and is packaged and shipped in plastic or metal barrels, larger tank type cargo containers or large size bag-in-box packages. Apple aroma or essence is collected and concentrated seperately during the juice concentration process. This is usually sold along with the concentrate to reprocessors.

Nonfermented or sweet cider is made in the same way as apple juice but is usually not pasteurized. Much of it is treated with chemical preservatives, although there is a significant market for a raw product which is sold refrigerated and has a short shelf life.

Fermented cider is particularly popular in Europe and the U.K. where it is a traditional beverage. There is smaller scale production in other parts of the world with significant volumes being produced in Australia and Canada.

Among specialty juice products made wholly or partially from apples are carbonated apple juice and apple juice blends with other juices such as pear, apricot, cranberry, and grape. A specialty juice, distinct from cloudy or "natural" applejuice, is liquid apple or apple nectar which contains a higher proportion of pulp than other unfiltered products.

A. PROCESSING TECHNIQUES FOR APPLE PRODUCTS

1. Bulk Handling

Apples to be made into higher value peeled apple products are usually moved to the processing plant from cold storage in bulk bins containing 450 kg or more. In many parts of North America it is also common practice to handle apples intended for juice and concentrate in the same type of bins. Some juice grade apples in North America and a major portion of this product in Europe, the U.K., and Argentina are handled in bulk trucks which dump into bulk receiving tanks at the processing plants. From bulk storage tanks, holding up to 1000t or more of fruit, apples are moved with water into the juice plant. Fruit in bulk bins is either unloaded by a bin tipper which discharges fruit directly into a washer and inspection belt or by a system which submerges the bin in water and allows the fruit to float out into the washer.

2. Washing — Sorting

Apples for processing are washed in water to remove any field soil that may have been accumulated during harvest and post harvest handling. This may be a particular problem if the fruit has been picked up from the ground as windfalls. Any water soluble pesticide residues may also be removed by washing. In older systems apple washers have a fluming or water immersion section followed by sprays. Sprays are frequently applied to apples on a roller belt which rotates the fruit to expose all sides to the spray action. This system does not necessarily remove much of the soft or rotten portions of the fruit and thus must be followed by inspection and hand sorting. New systems of combined brush and abrasive roll spray washers are being introduced rapidly by apple juice processors. This equipment will remove most soft apple tissue and greatly reduce the need for sorting and trimming of fruit coming into the processing line.

3. Peeling, Coring, Slicing, Sectoring, and Trimming

In some developing countries apples for canned slices and other whole apple products are peeled, cored, and cut by hand. This is a highly labor intensive process that does not usually yield as good a quality product as mechanized peelers used in most parts of the developed world.

Various models of hand loaded and hand powered mechanical peelers have been available for more than eighty years. Later improvements in manually fed machines involved motor drives and multiple heads which increased output per operator to 60 apples per minute. The past 20 years have seen the introduction of a mechanically fed fully automatic mechanical peeler and a chemical peeler which eliminates the cost of manual feeding. Both machines reduce greatly the 30 to 40% peeling and coring losses common to older design mechanical peelers. The chemical peeler claims to have the lowest peel loss, but has the disadvantage of reducing the by-product value of the peel and core waste produced. The fully automatic mechanical peeler removes somewhat more edible apple tissue during peeling, but its waste has more value for by-products such as juice for vinegar and apple pulp for jam.

Depending on the final use of the peeled apples, they may be trimmed either before or after cutting into sectors, rings, or other shaped pieces. Apples intended for sauce are requently cut into sectors and only lightly trimmed to remove discolored spots. Most pieces of skin and carpel tissue are removed in the sauce finisher or during final inspection of the cooked sauce before filling. Apples being made into canned or frozen slices or into pie filling are normally trimmed after peeling, but before cutting into pieces. This reduces the number of units requiring trimming if peeling is not complete.

For dehydration, apples are frequently cut into rings, 4 to 5 mm thick or into 6 to 7 mm cubes or other pieces of relatively uniform thickness.

4. Sulfite Treatment

Most apple pieces destined for freezing or dehydration are treated with sulfur dioxide (SO_2) either in solution or as a gas. In frozen apples SO_2 is used as an enzyme inhibitor to prevent browning induced by polyphenol oxidase naturally present in the fruit. It is usual practice to immerse apple slices for about 5 min in a 0.2% sulfur dioxide content solution. This will usually result in a residual SO_2 content of about 125 ppm in the frozen product. Because SO_2 penetration into slices ceases once they are frozen, it is usual to hold treated packaged slices for eight h or more before freezing. A simple spot test with 1% catechol solution is used to indicate if SO_2 penetration is complete before freezing.

Traditional evaporated or dried apples with 22 to 24% moisture content may contain up to 0.25% SO_2.[19] This high level serves not only as an antioxidant, but also as a preservative in this relatively high moisture content product. The high concentration is achieved by holding prepared apple pieces in an atmosphere of fumes from burning sulfur or SO_2 gas for up to 4 h.

For production of low moisture apple products with 5% or less water content, apple pieces are usually dipped in sprayed with SO_2 solution but the content in the finished product is well below 100 ppm.

5. Dehydration

Some older processing plants dry apples on slatted trays which are loaded onto carts and passed through a tunnel drier. In North America, at least, most drier trays are now made of moulded plastic to minimize maintenance costs and to maintain good sanitation condition. In spite of mechanical tray loading, unloading, and stacking and unstacking systems, tray driers are generally being replaced with continuous belt driers. In these large and expensive pieces of equipment, apple pieces are deposited on a perforated metal belt and conveyed through various zones where they are subjected to heated air under varying temperature and velocity conditions. During the early stages of drying, until the product weight is reduced by half, the rapid evaporation of water allows the product to remain cool in spite of high drying air temperature. During later stages, product temperature tends to approach that of the drying air, and to prevent heat damage due to carmelization, the drying air temperature must be reduced. While it is possible to dry apple products slowly to low moisture at atmospheric pressure, there are many advantages in finish drying from 15 to 20% down to 4 to 5% moisture under vacuum. The temperature of the process may be reduced with consequently less carmelization and the apple pieces may undergo a certain amount of expansion yielding a desirable shape and texture for a snack food. Vacuum driers are usually batch type units with product to be dried held on trays.

It is also possible to produce low moisture apple snacks by freeze drying, however, the quality advantages are not generally considered sufficient to warrant the higher process cost.

Dehydrated apple sauce is not a large volume commercial product, however, it has been manufactured in various countries. The simplest process is to crush low moisture dry apple pieces to desired particle size. Another relatively simple and low cost process is to dry prepared unsweetened applesauce on a drum drier.[20] This produces dry applesauce flakes which reconstitutes readily and has acceptable quality although it tends to be low in apple aromas. More costly processes that yield dry sauces with better aroma retention include freeze drying and foam mat drying. In the latter, process[21] sauce is whipped into a foam which is spread on a perforated belt. Craters are blown into the foam by high velocity air passing through the holes in the belt and the thus perforated foam is dried in a drying tunnel.

Apple sauce or pulp from peels and cores may also be dried on shallow trays to form what is called a fruit leather. Leathers contain 10 to 15% moisture and are a sold as a substitute for candy bars. Vacuum frying in oil at 105° to 110° C is a novel method of dehydrating food products. When applied to thinly cut apple pieces, it yields a product similar to potato chips. Because the porous cell structure of an apple tends to absorb an excessive amount of fat, it is usual practice in this process to saturate the apple slices before frying with a solution of maltodextrin.

6. Canning

Apart from applesauce, the majority of canned apples are in the form of institutional size cans of "solid pack" slices. An important step in preparing apple pieces for canning is to remove the oxygen and other gases trapped in the intercellular spaces of the tissue. If this oxygen is not removed it will cause rapid can corrosion and product failure.

The most commonly used process for gas removal is to hold the prepared apple pieces under vacuum for 5 to 10 min and then to blanch the evacuated slices in hot water or steam. This not only removes the tissue gases, but also saturates the apple pieces with water, gives them a translucent appearance, and prevents piece breakdown during later baking in pies or other apple desserts. Vacuum treated and blanched slices are tightly filled into cans with 4 to 5% water added. Since heat transfer in large cans of solid pack apples is slow, the product is usually filled as close to 82°C as possible. The sealed cans are given a short process in boiling water to ensure commercial sterility.

7. Pie Fillings

Apples for use in canned pie filling are prepared in the same way as for canned solid pack. After vacuum blanching, they are mixed with a sugar-starch slurry, possibly with other stabilizers and spices added. The mix is cooked in an agitating cooker, either of the batch or continuous type, and hot filled into cans. Large size institutional packs of this product may have sodium benzoate or potassium sorbate preservative added. In this instance, the product may be packed hot in plastic pails and air cooled as rapidly as possible.

Fillings for frozen apple pies and turnovers are made in similar fashion, however, the fresh fruit pieces are not necessarily vacuum treated. Since some of these frozen products are made from previously frozen or dehydrated apples, they may contain SO_2. Thickener systems for frozen pies usually involve modified starches that are stable to freezing and thawing and also will not "boil out" during final heating of the product.

8. Specialty Products from Peeled Apples

A number of specialty products may be made from whole apples. These include canned or frozen "baked" apples and frozen apple dumplings. The "baked" apple products are made by coring and, in some cases, partially peeling large high quality apples. The prepared fruit is vacuum treated to remove entrapped gases, then blanched in hot syrup. The intercellular spaces become filled with syrup and the enzymes are inactivated during this process. Apples pretreated in this manner may either be filled into cans with syrup and heat processed until a soft texture similar to that of a regular baked apple is achieved, or they may be frozen with a suitable filling in the core cavity. Apple dumplings are made from peeled cored whole apples wrapped in pastry and frozen. Another well established specialty apple product is apple butter. This is particularly popular in the United States and in parts of Western Europe. Traditionally apple butter was made from finely sieved apple pulp and boiled cider concentrated to at least 43% soluble solids. Boiled cider refers to an apple juice concentrate made by evaporating juice in an open kettle. The carmelized flavor and color of this ingredient impart much of the characteristic flavor and color the finished product. Optional ingredients may include various forms of white or brown sugar, invert syrup, glucose, fructose, honey or other sources of sugar. Spices and vinegar or other acidulants may also be added. Apple pulp is also used as an ingredient in low cost blended fruit jams. The product used for this purpose is frequently made by extracting pulp from apple peels and cores and is usually preserved by addition of up to 2000 ppm of sulfur dioxide.

High sugar content products made from apples also include apple jelly. Most processors find it convenient to make this from clarified, depectinized apple juice concentrate, using added pectin and sugar, glucose and/or high fructose corn syrup. The mixture is usually concentrated to approximately 68% soluble solids and the pH is adjusted if necessary to about 3.2.

9. Juice Extraction

The major steps in making single strength apple juice include milling or grinding washed and sorted fruit, pressing the milled pomace, clarifying, filtering, pasteurizing and packaging the juice. In making juice concentrate aroma is stripped from the unclarified juice along with approximately 10% of the volume of the juice. This dilute aroma solution is then concentrated, in a fractionating column operated at atmospheric pressure, by a factor of 15 to 20 times. This yields 150 to 200-fold aroma concentrate. The stripped juice is carefully depectinized, filtered, and then concentrated to 72% soluble solids. The concentration step is usually carried out in multi-effect evaporators operated under reduced pressure.

Apples for juice are ground in two types of mills. One type, the grater or rasp mill, grinds the apples with a rotating drum fitted with rows of serrated teeth. This type of unit is more commonly used in Europe. In North America apple juice processors more commonly use hammer mills in which apples are chopped through a screen surrounding blades fixed to a rotating shaft. Both milling systems have advantages depending on the maturity of fruit being processed.

Juice is extracted from milled pomace by a wide variety of pressing systems. Until recently, the most widely used has been the rack and cloth press. Here pomace is enclosed in loosely woven press cloths to form layers or cheeses each up to 10 cm thick. Cheeses are separated with wooden racks built in the form of a lattice which allow the juice to drain during pressing. Layers of cheeses and racks are built up to form a load on a tray that is usually fitted with wheels. The loaded tray is placed in a hydraulic press and pressure is applied to extract the juice.

Pressing is complete when juice no longer flows from the cheeses. The racks and cloths are unloaded and the pressed pomace is shaken from the cloths and sent to the pomace disposal system. Rack and cloth presses are probably the most efficient conventional system of extracting juice from milled apples with yields of 72% or more from optimum quality fruit. They have the major disadvantages of being labor intensive, difficult to clean, and to maintain. Various automatic batch and continuous systems have been introduced to replace rack and cloth presses. Those in common use include piston presses or pack presses, continuous belt presses, and continuous screw presses.

The most common piston press, the Bucher Guyer consists of a large rotable cylinder with a hydraulically operated piston. The piston is connected to the opposite end of the cylinder with a number of longtitudinally grooved, flexible reinforced rubber rods. These grooved rods are each fitted with a woven sleeve of synthetic cloth similar to fine press cloth. During press operation the sleeves allow juice from the milled apples to pass into the grooves and thus into the juice collecting system. Milled apple pulp is pumped into the press with the piston in the extended position. As the cylinder cavity fills, the piston is pushed back to its retracted position and the grooved rods stretched straight. During this phase of the press cycle, easily extracted "free run" juice is collected. The cylinder may be rotated to distribute the pomace load evenly then the piston is compressed to complete the juice extraction process. The cylinder walls retract to allow automatic discharge of the press cake and the cycle repeats automatically.

In continuous belt presses, apple pulp is either deposited on the lower of a pair of porous cloth conveyor belts or in the nip of a single belt that is folded to form a vertically oriented trough. In both types of presses the pomace enclosed in moving cloth conveyor is compressed between a series of rollers to extract the juice. Pomace removed from the end of the press is frequently pressed again in a second continuous belt press. If juice is to be made into concentrate, water may be added to the press cake between the two stages to increase the efficiency of the extraction.

Continuous screw presses are a popular juice extraction system in North America. Two basic types are the vertical type Jones press and the horizontal type Reitz press. Both types consist of a screw conveyor insider a screen cylinder. Pulp is retained in the cylinder by a hydraulically pressured end cone. Most apple cultivars when adequately mature for juice making are difficult to press in screw presses without some form of press aid. The most common press aid is purified wood cellulose which is mixed with apples in the mill at a rate of 1.0 to 1.5% of the weight of the fruit. The higher additive rate is required with more mature fruit. Other press aids include course fibers of shredded alder* and rice hulls.

It is common practice to use three or four screw presses in series to obtain the maximum possible juice yield. The first stage in such a system will usually be a large unit which removes most the free-run juice or approximately half of the total available. Following presses are progressively smaller in size and yield less of the total volume of juice.

A radically different method of obtaining juice from apples is the so-called diffusion extraction process which is an adaptation of the process used to extract sucrose from sugar beets. This process is based upon counter current extraction of thinly sliced fruit with warm water. The process, which was developed first in Europe and used first on a large scale in South Africa, was introduced commercially in the U.S. in 1982. A major advantage is the high yield of 90% or more of the weight of fruit compared with 72 to 75% from hydraulic rack and cloth presses and less from continuous belt or screw presses. One disadvantage is that the juice is diluted and must be concentrated to single strength concentration of 11 to 12% soluble solids from an initial concentration of 6 to 9%. Diffusion extraction may also

* Silvacel, a product of Weyerhauser Inc., Longview, Wa.

yield a juice with higher concentrations of tannin and other pigments. This may be an advantage in manufacture of dry hard cider, but might not be so desirable in apple juice or concentrate.

Pressed juice is usually screened on vibratory screens of 100 mesh per inch size to remove larger particles of apple tissue. Solids from the screens are returned to the incoming milled pulp for extraction in the juice presses. Juice containing suspended solids and colloidal substances such as pectin is pumped to clarification tanks. The main clarification processes consist of treatment with pectic enzymes or pectinases and the addition of gelatin which in combination with naturally occuring or added tannins flocculates colloidal substances in the juice. Some processors combine both of these clarification systems. Others may also use Bentonite clay suspensions as one component of the clarification process.

Once the colloidal material in juice is flocculated, a major portion of the floc may be removed by centrifugation. Both vertical and horizontal continuous centrifuges are used for this purpose. Partially clarified juice from the centrifuge is polish filtered through a precoat pressure filter. The filter plates are precoated with diatomaceous earth and additional diatomaceous earth is added to the juice prior to filtration. In some juice plants further filtration through a paper plate-and-frame filter will be used to achieve an extra-high clarity product.

Clarified juice is either packaged and sold as single strength juice or made into concentrate which is later reconstituted into single strength juice.

The most usual process for packaging single strength apple juice is to hot fill into glass or metal containers at 88°C or higher temperatures and to hold the containers briefly before cooling to 37°C or less. Recently aseptic packaging of apple juice has been introduced and is rapidly increasing in commercial importance. In the aseptic process currently used, juice is sterilized and cooled to room temperature in a high-temperature-short-time heat exchanger system. At the same time the packaging material, usually a multilayer laminate of paper, plastics, and aluminum foil is cold sterilized with hydrogen peroxide solution. The sterile juice and sterile packaging material are brought together in a sterile form-fill-seal system. The cool sterile packaged juice is then cartoned and moved to storage and distribution.

A relatively small volume of unfiltered or ''natural'' apple juices are packed, usually in metal or glass containers. In many of these products the colloidal systems are stabilized with heat treatment soon after pressing. This may also be combined with addition of ascorbic acid which, if added during or immediately after juice extraction, prevents enzymatic browning.

Apple juice blends with other fruit juices are commercially significant products in a number of countries. Juices used in blends include cranberry, grape, orange, lime, apricot, and pear.

10. Juice Concentrates

A large volume of apple juice is converted into concentrate with a standardized concentration of 72% soluble solids (72° Brix). This is a relatively stable product that is packaged and shipped in barrels or larger bulk containers on a worldwide basis. Storing and shipping concentrate at cool temperatures is desirable because high temperatures and the concentration of sugars and amino acids in the product tend to accelerate nonenzymatic or Maillard reaction browning.

Concentrate is made from clarified juice which has been, as far as possible, completely depectinized. One step in the juice making process differs, however, from that for single strength juice. This involves stripping the volatile aroma before clarification with pectic enzymes. If this step is attempted on juice that has been treated first with pectin methyl esterase (PME), which is a typical component of the mixed enzyme systems used for juice clarification, the methanol content of the aroma solution will be unacceptably high. Similarly if there has been any fermentation of the raw juice before aroma recovery, the volatiles obtained will be high in ethanol. The usual procedure is to distill off and condense 10% of the total juice volume. This volume of distillate is usually recognized to contain all of the commercially recoverable aroma. Dilute solutions of apple volatiles are then concentrated in packed distillation columns operated at atmospheric pressure. The final aroma solution obtained is 150 to 200 times stronger in volatiles than the original juice.

The stripped juice is clarified and filtered as previously described and then concentrated to 72° Brix (or 72% soluble solids) in a multistage vacuum concentrator. A typical system will use three stages all operating at progressively higher vacuums and progressively lower temperatures. Heating between the evaporation stages is usually accomplished by means of tubular or plate heat exchangers. Many concentrators employ vapor recompression as a means of reducing energy use. A well designed concentrator subjects the product to minimum heat damage by using high efficiency heat exchangers with short residence time.

Concentrate is cooled to at least ambient temperature before packaging. The most usual containers are 202 l (45 U.S. gal) polyethelene or coated steel barrels, however, concentrate is also packed in bulk aseptic containers containing up to 8 t. Apple aroma concentrate is normally sold along with the juice concentrate, but in separate packages. The two components are usually remixed during reconstitution with water to make single strength juice. Good quality of reconstituted juice is very dependant upon minimum heat damage during concentration and upon minimum damage during storage. Cool storage of concentrate is ideal. High temperatures accelerate both Maillard browning, involving reducing sugars and amino acids, and carmelization of the concentrated sugar solutions. These reactions result in dark colored reconstituted juice, frequently with carmelized flavor.

Processors using blends of concentrates to make reconstituted juice have an advantage in being able to adjust soluble solids : acid ratios to the range desired. This is provided they are able to obtain the desired quantities of high and low acid concentrates for blending. With the current large world wide supply of the low acid cultivar Delicious, it is frequently a problem for processors to obtain sufficient higher acid content concentrates.

REFERENCES

1. American Pomological Society, *North American apples* : Varieties, Rootstocks, Outlook, Mich. State University Press, E. Lansing, 197, 1970.
2. **Beach, S. A., Booth, N. O., and Taylor, O. M.,** *The Apples of New York,* State of New York, Department of Agriculture, 1905, Vol 1.
3. **Ballard, J. K.,** Granny Smith, an important apple for the Pacific Northwest, Extn. Bull. 0814, Wash. State University, College of Agriculture, Pullman, 1981.
4. **Lapins, K. O. and Fisher, D. V.,** Fruit tree cultivars in British Columbia, I, apples and pears, *Publ. SP53,* Agriculture Canada, Summerland, 1972.
5. **Hough, L. F., Shay, J. R., and Dayton, D. F.,** Apple scab resistance of *Malus Floridbunda, Proc. Am. Soc. Hortic. Sci,* 63, 341, 1953.
6. **Brooks, R. M. and Olmo, H. P.,** *Register of New Fruit and Nut Varieties,* 2nd ed. Univ of California Press, Berkely and Los Angeles, 1972, 7.
7. **Porrit, S. W.,** Commercial storage of Fruits and Vegetables, Publication 1532, Agriculture Canada, Ottawa, 1974.
8. **Eaves, C. A.,** A dry scrubber for CA apple storages, *Trans. Am. Soc. Agric. Eng.,* 2, 127, 1959.
9. "Gen-O-Fresh", Samif TransFresh Corp. P.O. Box 1906, Salinas CA 93902.
10. "Smit Oxydrain", Holec gas generators B. V., 6500 AB Nijmegen, Holland.
11. "Gastec", Gastec California Company, Divn. of Gastec Corp., 15245 Charter Oak Blvd, Salinas, CA 93907.
12. **Lau, O. L.,** Storage responses of four apple cultivars to a "rapid CA" procedure in commercial controlled atmosphere facilities, *J. Am. Soc. Hortic. Sci.,* 108, 4, 530, 1983.
13. **Haller, H.,** Fruit pressure testers and their commercial application, *Circ. 627,* U.S. Dept. of Agric. 1941.

14. **Strachan, C. C., Moyls, A. W., Atkinson, F. E., and Britton, J. E.,** Chemical composition of British Columbia tree fruits, *Publ. 862,* Agriculture Canada, Ottawa, 1961.
15. **Reid, C.,** SunRype Products Ltd., Personal communication 1981.
16. **Watt, B. K. and Merrill, A. L.,** Composition of Foods, *Agric. Handbo. No. 8,* U.S. Dept. of Agriculture, Washington, D.C., 1963, 190.
17. **Smock, R. M. and Neubert, A. M.,** *Apples and Apple Products,* Interscience, New York, 1950, 106.
18. **Anon.,** *The Almanac of the Canning, Freezing and Preserving Industries,* E. E. Judge and Sons Inc., Westminster, MD. 1983, 333.
19. **Anon.,** Canada Agricultural Products Standards Act, Processed Fruit and Vegetable Regulations, Agriculture Canada, Ottawa 1979.
20. **Lazar, M. E. and Morgan, A. I. Jr.,** *Instant Applesauce Food Technol.,* 20(4) 179, 1966.
21. **Morgan, A. I. Jr., Graham, R. P., Ginnett, L. F., and Williams, G. S.,** Recent improvements in foam mat drying, *Food Technol.,* 15, 37, 1961.

Chapter 5

THE PEAR

R. E. Wrolstad, P. B. Lombard, and D. G. Richardson

TABLE OF CONTENTS

I. INTRODUCTION

A. HISTORY OF COMMERCIAL VARIETIES

There are about 22 primary species in the genus *Pyrus*, all of which originated in either Asia or Europe.[1] The pear has been cultivated in China for at least 3000 years.[2] The first species to be domesticated, no doubt, was *P. pyrifolia* (Burm.) Nakai because the wild type is edible without selection. Later, the hardy northern Chinese type *P. ussuriensis* Maxim probably became cultivated after selection from the wild type which is small, has "stone" cells, and is very astringent. Natural hybridization between these two wild species likely occurred in China to produce the modern "ussuri" cultivars in northern China. They are complex hybrids of *P. ussuriensis* and *P. pyrifolia* and are hardier than the *P. pyrifolia* cultivars but less hardy than the wild *P. ussuriensis*. The Asian cultivated pears are crisp in texture in contrast to the melting texture of European pears and will ripen on the tree like apples. Various terms have been used for Asian pears, such as Nashi in Japan, New Zealand, and France, salad pear, pear-apple (misnomer), and Oriental pear in the U.S.. However, the term Asian pear is used in the U.S. trade.

In other parts of the world, cultivated pears have been derived from *P. communis* L. Cultivation of pears in western civilization was first mentioned by Homer nearly 3000 years ago.[3] However, Theophrastus some 600 years later described in "Enquiry into Plants" the cultivation of pear in detail. He noted that pears grown on grafted trees did not produce fruit of the "degenerate kind" as from seedling trees. He spoke of other cultural considerations such as the necessity of "cross pollination" that had been recognized and that pear trees are apt to shed their fruit before ripening. Therefore, pear culture was nearly as well established in Greece in 300 B.C. as now in the 20th Century.

P. Communis var. pyraster and/or *P. communis* var. *caucasica* were probably the ancestors of the common pear of Europe, but "French" cultivars are probably complex hybrids of these two. The snow pear *P. nivalis,* has been used for making pear cider in Europe.[4] A review of the development of the cultivated pear in Europe has been made by Hedrick.[3]

The *pyrus* genera is not endogenous to North America and consequently many of the early pear varieties were brought from Europe. All of the major pear cultivars of North America originated as chance seedlings rather than from breeding programs. All are more than 100 years old and some are nearly 200 years old. Bartlett (Williams), Anjou, Bosc, Hardy, Comice, and Winter Nelis originated in Europe between about 1795 and 1850.[5] Seckel, Keiffer, and Clapp Favorite originated in the U.S. between 1800 and 1860. All are *P. communis* except Keiffer which is a hybrid of *P. communis* × *P. pyrifolia.*

II. AGRONOMIC CHARACTERISTICS

A. CLIMATIC REQUIREMENTS

The best climate for growing quality pears is one that is warm to hot during the summer and has enough frost-free days to mature the fruit. The length of the growing season required varies from 100 to 200 d, depending upon the variety. Where irrigation is not available, 900 mm or more of annual precipitation is needed. Wet and humid conditions during spring and early summer are undesirable because they cause disease, such as pear scab and fire blight and cause fruit russeting.

Domestic pears are generally more hardy to low winter temperatures than peaches but less hardy than apples. A few cultivars of *P. ussuriensis* from northern China are as hardy as most apples. Most commercial cultivars require 1000 to 1500 h between 0.0 and 7.5°C for chilling. But a few varieties such as Packham's Triumph, Florida Home, and Eldorado require less than 800 chilling hours for cropping and therefore can be grown in latitudes less than 35°, in areas such as the southern U.S., Brazil, Mexico, Australia, and South Africa.

B. DISTRIBUTION

World distribution of pears is primarily between the 35° and 45° latitudes both North and South. Europe ranks first with about 5,000,000 tons, with Italy, U.S.S.R., Spain, France, and Germany as the major production areas.[1] Bartlett, Passe Crassane, Abate Fetel, Conference, Bosc, Jules Guyot, and Comice are the principal cultivars.

Asian production is estimated at 2,500,000 tons with principal pear regions in China and Japan.[1] The principal Asian pear cultivars are Ya Li, Tzik, Chioubaili, Shiyue Huali, Pinguoli, Tsu Li, and Jingbaili in China and Nijisseiki (20th Century), Chojuro, Kosui, Hosui, Kikisui, Shinseiki (New Century), and Shinsui in Japan.

North American pear production is about 824,000 tons with 91% in the U.S. Four cultivars, Bartlett, Anjou, Bosc, and Comice make up 98% of the production. Cultivated pears in the U.S. are grown primarily in the three western coastal states, Washington, Oregon, and California and constitutes about 97% of the U.S. production. The primary reason for west coast production is due to a disease, fire blight, *Erwinia amylovora* (Burr., Winsl. et al.) making pear culture difficult in the eastern U.S. except for resistant cultivars.

Other world production regions are Turkey with 260,000 t, South Africa with 128,000 t, Argentina with 154,000 t, and Australia at 137,000 t.

C. CULTURE REQUIREMENTS

Pear culture is similar to apple requiring annual dormant pruning to increase light exposure, to improve accessibility for spraying and harvesting, and to reduce the crop load for increased fruit size and quality. However, such pests as pear psylla *Psylla pyricola* (Forster), fire blight and blossom blast *Pseudomonas syringae* (van Hall) readily infest pears specifically plus codling moth, *Laspeyresia pomonella* (Linn.), San Jose scale, *Quadraspidiotus pernicious* (Comstock), and pear scab, *Venturia* species. Pear trees are generally more susceptible to spider mites, *Tetranychus urticae* (Koch) and *Panonychus ulmi* (Koch) than apple.[6] Pear psylla requires several dormant and summer insecticides for control to prevent a russeting while fire blight and blossom blast requires additional pruning to remove the infection plus spring sprays of streptomycin, terramycin, or copper dust. Also, pears bloom several weeks in advance to apples during a more susceptible frost period. Frost of less than -2 to -5°C prior to bloom through post-bloom can either reduce the crop or cause considerable frost ringing (epidermal injury), Therefore, pear production costs generally exceed the farm gate return because of increased pesticide costs and lower yields.[6] Pesticide costs are from 15 to 20% of the cash costs for a pear orchard while frost protection amounts to about 14% of the cash costs in the northwestern U.S.

The fleshy part of the pear fruit is nonovarian, fused calyx, and receptacle tissue (cortex) around 5 ovaries each with a maximum of 2 seeds, but at times the fruit may develop without fertilization. However, while only fertilized ovules develop embryos (seeds), seed number is related to fruit size in terms of weight or diameter at least up to 5 to 6.[7] Therefore, conditions which favor fertilization such as climate and cross-pollination can improve fruit size by the stimuli of seeds. The period of cell division of the young developing fruit usually occurs for 7 to 9 weeks after anthesis. Then sometime during the cell division, cell enlargement begins and proceeds at a rapid rate through the season until pear harvest. Concurrent with cell enlargement, air spaces increase gradually until harvest. Vacuoles form in the cells and increase in size as cells enlarge, and ultimately they occupy most of the cell space. Fruit specific gravity of developing pears decreases gradually from 1.10 to nearly 1.04 g/cc at harvest. Pear fruit density is considerably greater than apple because of greater soluble solid content and less intercellular air space. Combined fruit growth including cell division, cell enlargement, and air space results in a general sigmoidal curve when fruit volume as weight is plotted as a function of time. Unlike drupe fruits which have a double sigmoid curve, the pear is harvested mature but unripe prior to a reduced rate of fruit growth (the upper part of the S curve). Therefore, delaying of harvest can increase tonnage and fruit size by 1% on a daily rate.

Maximum fruit growth rate of pear is achieved with a leaf to fruit ratio from 20 to 40 leaves per fruit.[8] Fruit size is generally less of a factor in production costs compared to apples because pear cultivars, particularly Anjou and Comice, set less fruit than most apple cultivars. Also, alternate bearing is less of a problem with most pears unlike many apple cultivars. However, dormant pruning, warm post-bloom temperatures, and some fruit thinning, and sufficient soil moisture during the rapid fruit growth period prior to harvest is generally important for fruit sizing in pears. However, regions with low chilling will produce fruit with larger cells (and fewer cells) requiring special efforts to enlarge the fruit.

Like most fruit trees, pears are budded/grafted on specific rootstocks to produce trees true to the variety wanted, to tolerate certain soil and disease conditions, and to control tree size, production, and fruit quality. Principal rootstocks used are seedling stocks of *P. communis,* and rooted cuttings of common quince, *Cydonia oblonga* L. for the European pear and seedlings of *P. pyrifolia, P. calleryana, P. betulaefolia,* and *P. ussuriensis* for the Asian pear. *P. calleryana* and *P. betulaefolia* root stocks are used on a limited basis on small fruited cultivars such as Seckel, Bartlett, and on Asian pears to increase fruit size. Use of *P. pyrifolia* and *P. ussuriensis* stocks can cause a physiological disorder, "black end'' on the fruit of both Asian and European pears. Because these two species when used as rootstocks on European pears, are highly susceptible to pear decline, a mycoplasma disease spread by the pear psylla, they are seldom used in North America.

Rootstocks can also affect quality such as the fruit acidity and soluble solids, texture, maturity, storage ability, and fruit finish. However, differences are subtle in most cases and little research has been conducted in this area. But we have noted that the rootstock, Old Home, will produce Anjou pears free of the condition of ''cork spot'' because of the ability to extract greater amounts of calcium from the soil.[9]

III. PRODUCTION AND MATURATION OF THE PRINCIPAL PEAR CULTIVARS

Pear cultivars are described by the condition or color of the fruit epidermis such as clear skin or russeted; green, yellow, blushed, or red pears. Russeted or red skinned mutants of some green and yellow fruited cultivars have been selected for production, i.e., Red Bartlett, Red Anjou, Red Comice, and Red Clapp Favorite. Generally, fruit of russet mutants are of higher quality in terms of fruit size, soluble solids, and texture than those of clear skin or green, yellow, or red fruits of the same cultivar, particularly Bartlett. However, red fruited mutants at present are favored in the fresh market based on the return to the grower. Also, Red Bartlett, a mutant of green Bartlett, has been used for breeding several new red fruit cultivars such as Reimer Red, Canal, and Cascade. However, red skin pears are not processed because the red pigment stains the flesh dark brown. Otherwise, russeted and all other green, yellow, and blushed European pears are processed for juice or puree. Generally, only Bartlett and Winter Nelis pears are used for canned halves or cocktail salad pieces.

European pears are noted for their melting flesh when ripened, commonly referred to as a buttery texture. However, ''stone'' cells, which are brachysclereids or sclerenchyma cells, are randomized throughout the flesh of various cultivars, i.e., Bosc and Anjou, but they are usually concentrated around the core and the inner cortex of the pear. These ''stone'' cells can not be softened by processing and, in fact, they are more noticeable in processed pears. The virus disease stony pit, which is commonly found in trees of certain varieties such as Anjou and Bosc, can produce pockets of stone cells in the flesh and the fruit must be culled because of noticeable exterior pits on the fruit surface. However, Bartlett trees are symptomless carriers of the virus and its fruit is free of stony pit.

Bartlett has been the preferred cultivar for canning worldwide for: (1) its melting texture without stone cells; (2) its flesh remains light colored and does not turn pink as other cultivars are apt to; and (3) its flavor components predominate after processing. However, fruit of other cultivars is processed for canned halves by local custom: Keiffer, Bosc, Seckel, Packham's Triumph, Anjou, Magness, Clapp Favorite, and Harrow Delight. Bartlett, Comice, Anjou, and Bosc fruit are used in pies; Bosc and Comice fruit are baked; and Seckel fruit are pickled in a sweet brine. Sliced or fruit halves of Bartlett and Bosc can be dried successfully with or without skins.

For juice, European pears are usually pressed green, firm, and mature, but unripe.[10] However, the juice has no distinctive flavor or aroma. But we have found that if the pears are allowed to ripen at room temperature for 4 to 5 d after storage, pressing of the partially softened pear with a pressing aide, pectinase, will produce excellent juice for concentrate purposes or pear cider (perry).

Asian pears are ready to eat from the tree, unlike European pears, and are sweet but subacid, crisp, juicy, and have a distinctive flavor recognized by those familiar with them. Asian pears do not change texture after picking or storage such as Bartlett, etc. They are often called pear apple (misnomer) because they are crisp and juicy like apples. They are also called salad pears (U.S.), Nashi (Japanese for pear), Oriental, Chinese, or Japanese pears (Nihonnashi). Asian pear is the preferred trade name. There are three types: (1) yellow skin and round shaped, generally of Japanese origin; (2) brown skin (russet) and round shaped, also of Japanese origin; and (3) green with or without russet and pyriform (pear) shaped, generally of Chinese origin. Fruit maturity is best determined by soluble solids (12 to 13%), by a change in skin color from green to yellow for clear skin fruit or from a greenish brown to yellowish brown for russeted fruit, and/or by tasting the flesh. Most Asian pears are thin skinned and bruise easily if mishandled, particularly at harvest and during packing operations. Stems are cut close to the base to prevent puncture and fruits should be hand packed with gloves and sorted without moving belts.

Production of Asian pears is predominantly in California, but there are new production areas in Oregon and Washington. Although the Asian people in North America are the principal consumers, many non-Asians have discovered them recently and demanded their availability. Since the fruit is traditionally eaten fresh there has been very little processing of the fruit. However, whole, sliced, or diced Asian pears can be processed or fried without losing their firm texture, unlike European pears. Several processors are interested in such characteristics to supplement other processed foods in fiber content. No juice products have been made from Asian pears but they are much easier to juice than European pears.

Production, utilization, harvest, and storage characteristics of the principal pear cultivars are given in Tables 1 and 2. Also, a brief description of each cultivar is given for the same cultivars.

A. ANJOU (BEURRE D'ANJOU), RED ANJOU

Anjou (Beurre d'Anjou) or Red Anjou is grown in cooler regions of the northwestern U.S. because of their adaptation and long storage life, and is, therefore, referred to as a winter pear. Lack of fruit set is common and the tree requires heavy pruning, sunlight, and high nitrogen applications, which are more than 2 to 4 times as much as that required for Bartlett. It requires cross-pollination for adequate set and fruit size. Fruit finish trees and production of Anjou are susceptible to insect damage and pests, and in particular, mites must be carefully monitored.

Anjou matures about 145 d after bloom and is principally utilized fresh. It requires 50 to 60 d in storage at -1°C to ripen properly unless picked one month later. It can be cold stored about 7 months at -1°C in air and 9 months in controlled atmosphere storage (2% O_2 and 1% CO_2).

B. BARTLETT (WILLIAMS' BON CHRETIEN), RED BARTLETT, RUSSET BARTLETT

Bartlett is the principal processed pear and is used also for fresh consumption because of its high quality. Trees are productive under a wide range of latitudes and conditions. Bartlett trees are productive in warmer areas and climates where its seedless fruit can be produced without cross pollination (stimulative parthenocarpy). The seedless fruit is less pyriform and more ideal for processing equipment. Bartlett also produces late or "rat tail" bloom about a month after the normal bloom period increasing the danger of fire blight infection but also producing misshapen fruit. Bartlett can be either thinned to reduce the crop level in June or double harvested to obtain adequate fruit size of 6.35 cm diameter. A preharvest spray of NAA is recommended to prevent preharvest fruit drop.

Premature ripening of the cultivar is common in cooler climates making it difficult for processors to uniformly ripen the fruit. Bartlett matures early, about 115 to 135 d after bloom. The fruit stores well at -1°C for 4 months and is still able to ripen afterward although it is considered a summer pear. Storage prior to ripening is not necessary for Bartlett.

TABLE 1
Production Characteristics of Principal European and Asian Pear Cultivars

Cultivar	Thinning requirements	Commercial cropping level (t/ha)	Market sizes (fruit) Diameter ranges (mm)	Weight (g)
European				
Anjou also Red Anjou	Usually none	20—60	54—80	180
Bartlett also Red Bartlett	Usually	20—80	57—70	165
Russet Bartlett	Usually none	20—80	61—73	175
Bosc	None	30—80	51—76	145
Clapp Favorite also Red Clapp, Starkrimson	None	20—40	54—80	160
Comice also Red Comice	Usually none	10—60	61—83	220
Hardy	None	20—80	57—80	160
Harrow Delight	Usually	—	—	165
Harvest Queen	Usually	—	—	165
Highland	None	—	—	180
Kieffer	None	20—0	57—80	190
Magness	Usually	—	—	165
Nelis	Usually none	20—80	51—70	150
Packham's Triumph	None	20—60	57—80	190
Seckel	Considerable	20—40	45—64	110
Asian				
Hosui	Usually	20—30	70—90	350
New Century (Shinseiki)	Considerable	25—35	60—80	270
Tsu Li	None	20—30	65—85	300
Twentieth Century (Nijisseiki)	Considerable	25—35	60—80	250
Ya Li	Usually none	25—35	55—73	200

C. BOSC (BEURRE BOSC, KAISER)

Bosc is a widely grown dessert winter pear throughout southern Europe, the southern hemisphere, and the western U.S. The trees are very productive and can set seedless fruit but they generally require pollinizers for increased fruit size. Fruit thinning is not generally necessary. Bosc pears mature 150 to 160 d after bloom. Fruit is heavily russeted and the fruit remains firm when ripe. Bosc store up to 6 months at -1°C. They ripen better after 20 d of storage.

D. CLAPP FAVORITE, RED CLAPP FAVORITE (STARKRIMSON)

Clapp Favorite and Red Clapp Favorite are grown in Canada and the U.S. but not extensively. Fruit matures before Bartlett and fruit size is somewhat small although the trees crop poorly. The fruit can be attractive when mature, particularly the red fruited mutation. The skin can bruise easily when near ripe and this can make the fruit less attractive. The fruit has poor shelf life and can be stored for only 2 months at -1°C.

E. COMICE (DOYENNE DU COMICE), RED COMICE

Comice is considered the best eating pear because of its juicy, buttery texture and well balanced flavor. It is grown throughout Europe primarily, but also in western Oregon and California. Comice trees require close attention to crosspollination, pruning, nutrition, irrigation, and frost control to produce only a moderate crop of very large fruit. They are seldom hand thinned, but large sized fruit is required for catalog sales. The fruit finish is susceptible to russeting from powdery mildew, spider mites, and wind marking. Comice fruit are harvested about 140 to 155 d after bloom and require 45 d of chilling before ripening. Comice are generally stored up to 5 months at -1°C.

TABLE 2
Utilization and Harvest Characteristics of Principal European and Asian Pear Cultivars

Cultivar	Utilization	Growing season length (d)	Harvest firmness (N[a])	Comments
European				
Anjou also Red Anjou	Fresh, juice	135—140	77—68	Ca fruit excellent quality
Bartlett	Fresh, juice	115—125	77—99	Low solids
	Canned	120—130	65—82	
Red Bartlett	Fresh	120—130	77—99	Low solids
Russet Bartlett	Fresh	120—130	77—99	High solids
Bosc	Fresh, juice	145—155	72—82	Good handler and shipper
Clapp Favorite	Fresh	105—110	68—77	
Comice	Fresh, gift	140—145	45—63	Prime eating, poor handler
Hardy	Process, fresh	125—135	59—68	Processed in fruit cocktail
Harrow Delight	Fresh, process	110—125	—	Home use
Harvest Queen	Fresh, process	115—130	—	Similar to Bartlett
Highland	Fresh	110—125	59—68	Excellent quality
Magness	Fresh	130—140	—	Excellent quality
Nelis	Fresh, process	160—170	55—64	Pollinizer
Packham's Triumph	Fresh, process	160—170	63—82	Excellent quality
Seckel	Fresh, pickled	140—145	76—67	High quality, small fruit
Asian				
Hosui	Fresh	135—145	58—63	Russet skin
New Century (Shinseiki)	Fresh	120—130	53—58	Susc. to skin abrasion
Tsu Li	Fresh	165—175	58—62	Handles well
	Fresh	140—155	53—58	Susc. to skin abrasion
Ya Li Twentieth Century (Nijisseiki)	Fresh	165—170	53—58	Core breakdown if stored improperly

[a] Newtons × 9.8 = kilograms firmness.

F. HARDY

Hardy was originally grown in the Santa Clara Valley for export markets. Recently it has been used for cocktail fruit mixes. Processors prefer Bartlett and, therefore, Hardy production is declining rapidly. The fruit has an attractive russet which gives the appearance of a greenish brown pear at harvest and it becomes light brown when ripe. The fruit ripens quickly on the tree after reaching maturity. The fruit sizes well without thinning.

G. HARROW DELIGHT

Harrow Delight, another fire blight resistant cultivar from Agriculture Canada, matures about 2 weeks before Bartlett. The oval shaped fruit is green with a blush over 20% of the surface without russet. The flesh is melting in texture, juicy, and has some ''stone'' cells. Flavor is good but distinctly different from Bartlett. The fruit can be processed for canning and puree.

H. HARVEST QUEEN

Harvest Queen, a second fire blight resistant cultivar from Canada, matures a week before Bartlett. The fruit is oval shaped with yellow, russet free skin when ripe. The flavor and quality is equal and almost identical to Bartlett with equal canning quality. The main defect is the small fruit size, which is 10% less than Bartlett.

I. HIGHLAND

Highland was released from Cornell University in 1974 and is of interest in southern Europe because of its high quality. The fruit is moderately sized, pear shaped, and the skin is yellow and smooth but with a light russet. Fruit quality is excellent and rated higher than Bosc and Anjou. It matures about a month after Bartlett and keeps in storage for 5 months.

J. HOSUI

Housi is a very large round russeted (bronze) pear with an aromatic flavor. Although it is a relatively new pear cultivar from Japan it is most popular in Japan and in the U.S., since it is very juicy and has excellent flavor, size, and appearance.

K. KEIFFER

Keiffer is widely planted in the southern and lower midwestern states since it is relatively resistant to fire blight. The fruit is large, resembling a Bartlett in color and shape but with a longer calyx end (opposite to the stem). Flavor is good compared to fruit from other Asian × French pear hybrids and, in fact, the flavor is similar to Bartlett but with less melting flesh. Fruit matures after Bartlett in late September. Fruit is usually consumed fresh but it can be canned.

L. MAGNESS

Magness, a fire blight resistant cultivar from the U.S.D.A., matures about 2 weeks after Bartlett. The fruit is lightly covered with russet, medium sized, and is generally oval shaped like Anjou. The fruit flesh is soft, very juicy, and free of "stone" cells. In prime conditions its fruit quality rates very high.

M. NELIS (WINTER NELIS)

Nelis has been the principal pollinizer for Bartlett in California. It was used as a winter pear in southern Oregon for long storage and late spring shipment. The fruit is heavily russeted, small, and unattractive. Eating quality is less than Bartlett but has high quality for a late winter pear. It has been used as canned pears in California.

N. NEW CENTURY (SHINSEIKI)

New Century also has a light green to yellow green skin and round shape but is less attractive and has less flesh quality than 20th century. Fruit size is larger and requires less thinning than 20th century. It is harvested about Bartlett season and can be kept in storage for 4 to 5 months.

O. PACKMAN'S TRIUMPH

Packham's Triumph is considered to be a southern hemisphere pear since little production occurs in Europe or North America. Limited production occurs in Oregon where the fruit is primarily sold in Canada because of poor consumer acceptance in the U.S. Winter chilling requirement of Packham's is less than most other commercial varieties and, therefore, the cultivar can be grown successfully in subtropical regions. Fruit quality is rated among the highest as a late winter pear because of its fine melting texture and juiciness and without internal breakdown after late storage. However, the appearance of the fruit is poor due to the bumpy and non-uniform shape of the fruit plus the light russeted green skin. The mature fruit is quite firm and handles well. Fruit thinning is generally not necessary since fruit load is generally low. Heavy nitrogen applications are used in Packham's orchards in the southern hemisphere to increase fruit size. Southern production regions market most Packham's fruit soon after harvest but Packham's can be stored 8 months at −1°C.

P. SECKEL

Seckel pears are generally small and, therefore, trees must be thinned heavily to produce marketable sizes with fruit diameters of 4.75 to 6.35 cm. The fruit matures after Bartlett, when a few ripe Seckels are found on the tree. Seckels are picked at optimum firmness of 67 to 75 N. The fruit is used for fresh and sweet pickling. They are very sweet, 16 to 20% soluble solids, and have a characteristic cinnamon flavor. They are somewhat gritty from "stone" cells, and subacid. Fruit desiccation can occur if not stored in high humidity and plastic carton wraps. Seckels soften quickly and are not stored for more than 4 months. They are primarily grown in Oregon, Washington, New York, Michigan, and Pennsylvania. Seckel can be stored up to 4 months.

Q. TSU LI

Tsu Li, another Chinese cultivar, has a large, somewhat non-uniform but pyriform shaped fruit with a crisp, juicy, and excellent flavored flesh. The fruit is generally green with some russet. It has been rated as the best flavored Chinese pear cultivar. It seldom needs thinning because of light crop load and large fruit. The fruit matures during Ya Li season and can be stored at 0°C for 6 to 7 months.

R. TWENTIETH CENTURY (NIJISSEIKI)

Twentieth Century is the principal cultivar imported from Japan or grown in the U.S. This attractive round fruit with light yellow colored skin is medium-sized with excellent flavor and is the most popular of Asian pears. This cultivar is more difficult to size than others and so must be heavily thinned. Harvest of 20th century occurs about one month after Bartlett. It can be stored about 6 months at −1°C but it is usually marketed soon after harvest.

S. YA LI (DUCK PEAR)

Ya Li is a Chinese cultivar of moderately sized pyriform shaped fruit with light green skin. The fruit is crisp and somewhat juicy with good to excellent flesh quality. Fruit thinning can increase fruit size when crop load is heavy. Harvest is usually about 2 months after Bartlett. Storage requirements involve a slow rate of cooling with a beginning temperature of 7°C but ending at 0°C over a 3-week period in order to prevent internal breakdown. Common storage at 5°C can maintain Ya Li fruit for several months without breakdown. Therefore, cold storage may be of only limited advantage.

IV. QUALITY AND STORAGE OF RAW PEARS

Quality is the sum of all of the positive attributes which are deemed to be important for a specific intended use for a fresh fruit or vegetable.[11,12] This certainly applies to pears since they may be used fresh, canned, frozen, dried, or as juice, pureed, etc. Taking only the fresh pear quality attributes for a moment, there are some very dominant characteristics. First of all, the consumer seeks a pear fruit with the color, shape, and size which is considered typical of the variety. Further, the condition of the fruit is important, it should be free of mold, cuts, bruises, extraneous matter (soil, etc.), insects, or insect wounds. The grades and standards of most importing and exporting nations emphasize these surface visible characters and add tolerances for major (usually 1 to 2%) and minor (usually 5 to 10%) defects or deviances from the standard. There may also be requirements for internal quality to meet grade standards. Often these specify absence of certain types of internal disorders or pathologic conditions, but may also set minimum % soluble solids or minimum % soluble solids to titratable acids ratios, e.g., in the case of citrus fruits or grapes. It may also be desirable to have the pear in an advanced stage of ripeness which whould allow for immediate consumption. These are quality considerations for the consumer or buyer at the point of sale, and beyond these, other quality aspects are often not very influential. At the point of consumption, however, the internal aspects of quality rise to importance. Besides the surface color and gloss, internal color, aroma, flavor, texture, juiciness and sugar/acid balance, and very mild astringency, all interact to compose quality. Lest the ending precede the rest of the story, we need to recognize that many preharvest factors may interact to ultimately produce high quality fruit for fresh consumption or processing.

A. PREHARVEST FACTORS

Ultimately, pear fruit quality depends upon the variety, how well it is adapted to a particular growing region, the rootstocks on which it is grown,[9,13] the prevailing climate for that growing season, relative crop load, irrigation,[11,14,15] and mineral nutrition provided by the soil or by added fertilizers. Each have their impact and interaction on fruit quality.[12]

1. Growing Season

Climate variations regionally and from year to year impact pear quality in several ways. First of all, a rapid progression from cold winter and cold early spring to a sudden sustained warm-up leads to a rather synchronous bloom condition. This tends to carry through during fruit development to give very uniform fruit-to-fruit maturity at harvest. Contrast this with common mild winter and spring conditions where bloom is strung out over time and this persists throughout fruit development and it can become difficult to judge when a crop should be harvested. It may even be necessary to selectively harvest the more mature fruits first in order to try to manage these different maturities for orderly storage and marketing strategies. Cooler climates result in less photosynthate reaching the fruit, and even though fruit size might not suffer, internal sugar concentrations will be lower, acids will persist at higher levels and the fruits usually will not mature as quickly. Interestingly, the best combination for climate on fruit quality is warm, sunny days, and clear, cold nights. This allows the maximum movement of sugars to move to the fruits. However, if the nights are too cool (more than 36 h at less than 7°C) during the last few weeks of maturation, the problem of premature ripening can be induced in Bartlett pears.[16]

2. Maturity

The decision as to when to harvest a given pear variety is based on its intended use. Pear fruits for processing are generally harvested at a later stage of maturity than that for fresh market.[17] For fresh market there is even a compromise based upon research and observation about the ability of the pears to undergo prolonged storage and yet ripen with good texture and flavor without internal or external senescent-type disorders, and with enough resistance to storage pathogens.[18,19] In this regard, pears are a bit unusual in that they are harvested in a mature, but as yet unripened stage. They have reached a developmental stage which allows them to later ripen upon exposure to ethylene (either internally generated, or externally supplied) and carry out all of the normal ripening functions. In contrast to many other types of fruits, most pears left to ripen on the tree develop very poor mealy texture and often unpleasant flavors. Thus it is critical to have accurate maturity indices as to when is the best stage of maturation for harvest which optimizes storage potential, ability to ripen following storage, and yield. Almost everywhere that pears are grown commercially, days from full bloom (DFFB) are used as a rough indicator of when harvest might begin, but this is fine-tuned with a flesh firmness test to more accurately determine best firmness ranges for a given variety, location, and marketing plan. Table 2 lists the common commercial pear varieties and recommended firmness values in Newtons (N) for peeled pears punctured by an 8-mm tip. Firmness in Newtons (N) is converted to kilograms by multiplying N × 0.1016, or to pounds by multiplying N × 0.2235. Hotter growing regions may start harvest at firmness values as much as 0.5 to 1.0 kg higher. While % soluble solids (Brix) or % titratable acidity or starch conversion might be of some use, in reality they have little predictive value in the case of pears even though they may be genuine components of fruit quality once the fruit is ripened. Crop load variation and climatic variance in the last few weeks of fruit development can have dramatic effects on soluble solids that do not reflect accurately the progressive physiological status of the fruits. Thus, flesh firmness is the best single indicator and predictor of harvest maturity. As a general rule, Bartlett and other summer pear varieties lose about 0.125 kg firmness per d. Winter pear varieties lose about 0.062 kg per d.

Asian pears are an exception in that firmness values are not as useful for maturity indices. Rather, taste tests are often used since these pears remain quite crisp and do not undergo typical melting flesh type of softening. Taste tests of Asian pears attempt to detect loss of astringency, increase in flavor volatiles, and sweetness.

Asian pears *(Pyrus serotina)* have some distinct characteristics which differ from the *Pyrus communis* varieties. They do not undergo the typical rapid softening as they approach the edible stage.[20] Thus, flesh firmness tests are not very useful. While DFFB is fairly useful, taste tests and skin color changes are more valuable as maturity (ripeness) indicators. Still another difference relates to the susceptibility of some varieties to what appears to be a form of chilling injury. That is, they exhibit internal breakdown disorders if stored at temperatures near 0°C, but not at temperatures 5°C and above. This chilling sensitivity seems to be variable from season to season, and is not well characterized at this time.

B. HARVEST AND PACKING

Besides the preharvest and maturational considerations, there are many harvesting and handling factors which can influence pear quality.[21-23] Avoiding mechanical injuries to the fruit must rank near the top in reducing losses due to bruising, scuffing, cuts, and the entries for pathogenic fungi that ensues. Instructions to picking crews to avoid mechanical damage from bumping fruit bags on ladder rungs, making sure that pickers have trimmed fingernails, and that fork-lift operators gently lift and set down palletized loads of fruit are examples of the careful management practices used by field supervisors to reduce damage.

Other than avoiding mechanical injuries, harvesting crews generally do not do any sorting in the field. Sorting, sizing, and grading usually occur at a centralized packing facility. Pears are moved from the orchard in wooden bins containing about 500 kg. Upon arrival at the packing shed, they are either pre-cooled in the bins and held in storage until they can later be sorted and packed, or they are unloaded for movement through the packing operation. There are two common methods of unloading pears from the bins, direct dumping and flotation dumping. Flotation dumping has become the method of choice as it usually results in less mechanical injury to the fruits. Constant monitoring and fungicidal treatment (active chlorine or oxidants) to maintain clean water in the dump tanks is necessary to prevent pathogen inoculum buildup. Early removal of any fruits showing rot is a necessary sanitation step that helps to control storage rots and keeps equipment surfaces from potentially spreading fungal pathogens. Conveyor belts, channeling bars, cornering structures, rollers, and other surfaces which contact the fruits should be inspected frequently for rough spots, or possible needs for padding in order to avoid injuries to the pears being carried through the packing house. Points where fruits change elevation, or go around a corner are particularly troublesome. While there are still many sizing devices in use which are of the divergent roller type, newer packing houses predominantly have installed the individualized cup sizers linked to computer tallying devices. While little data on comparative fruit damage is available, weight sizing is probably an improvement over divergent roller sizers, particularly for nonspherically shaped fruits such as pears.

C. GRADES AND STANDARDS, QUALITY CONTROL

U.S. Department of Agriculture (USDA) Food Safety and Quality Service (FSQS) has the responsibility of establishing, altering, and implementing grade standards. These are published in the Federal Register.[24,25] The European Economic Commission (EEC), under the division of Organization for Economic Cooperation and Development (OECD), has developed similar international standards for fresh fruits and vegetables. Basically, standards create a definition of terms for communications regarding quality between buyer and seller.

The USDA Standards for Summer and Fall Pears is quite similar to that of Winter Pears. "U.S. No. 1" grade specifies "that the pears be of one variety, mature but not over-ripe, carefully hand-picked, clean, fairly well-formed, free from decay, internal breakdown, scald, freezing injury, worm holes, black-end, and from damage caused by hard-end, bruises, broken skin, russeting, limbrubs, hail, scars, drought spot, sunburn, sprayburn, stings or other insect injury, disease, or mechanical or other means".

"U.S. No. 2" differs only in allowing for more misshapen pears. Tolerances are allowed up to 10% provided that not more than 5% are seriously damaged by insects, nor more than 1% by decay or internal breakdown.

Quality control is accomplished via trained sorting and grading crews who select the fruits to conform to the standards applicable to the intended market. Records are kept on the pack-out quality of each lot of pears so that growers can be paid equitably for their quality and sizes. Inspection of the packed pears is done randomly by state or federal inspectors to assure that standards are met. Inspections may also be done at destination to assess condition as well as grade specifications. Inspection records are legal documents in the event claims are not settled via normal procedures.

D. STORAGE

The most commonly recommended storage temperature for all varieties of pears is −1°C, with the possible exception of Ya Li Asian pear, which may be truly chilling injury susceptible. Ya Li apparently is often slowly cooled in stages in caves in China and this slow cooling may allow it to avoid symptoms of chill injury.

Pears are quite sensitive to storage temperature. Porritt[26] has shown that a storage temperature only 1°C warmer can result in a 10% reduction in storage life. Similarly, delays of only a few hours in cooling the summer pears, e.g., Bartlett, can drastically shorten storage life and leads to rapid chlorophyll and firmness loss and ethylene accumulation which could adversely affect other fruits. While the winter pear types do not show the extreme temperature sensitivity of Bartlett, they are nonetheless sensitive enough to take precautions to get them cold quickly. Storage operators try to maintain storage humidity above 90% in order to reduce problems with stem-end shrivel of the fruits, even if they have been packed in boxes with pin-hole polyethylene film liners. Since pears are so sensitive to even minor increases in temperature, it is critical to stack pallets and space them in storage to achieve uniform cold air movement in contact with at least one surface of each box. This means 10 cm between pallets, and 20 cm away from the storage walls. The development of rapid forced-air cooling (pressure cooling) by Mitchell, et al.[27] has been a significant advancement in improving pear storage operations. Equally important is multiple location temperature measurements within cold storage chambers, including pear fruit core temperature monitoring to know where there may exist slightly warmer or colder temperatures so that the stacking pattern or air flow direction may be corrected. This is best accomplished by thermocouples or thermistors attached to a recording device. Pears held under less than optimal storage temperatures risk quality losses in color, flavor, and especially in texture, where mealiness and dry, non-juicy ripening may result.

Controlled atmospheres have been very important to the market developments of pears all over the world. There are numerous references to CA storage conditions for pears.[28-40] Controlled atmosphere conditions for most commercial pears, including Asian pears, have been summarized by Richardson.[28] Generally, pears benefit at oxygen less than 2 to 3%, and can be taken safely down to 0.5% oxygen, provided that CO_2 is removed below 0.05% by some method. Pears under low O_2 conditions are very sensitive to even low levels of CO_2, and will develop CO_2 injury symptoms without some scrubbing method.

CA storage can be expected to extend the useful storage life by 25 to 35%. CA helps to conserve organic acids,[38] reduces ethylene production,[41,42] retains fruit firmness and chlorophyll, and even can retard certain types of spoilage organisms.

E. PHYSIOLOGICAL DISORDERS

There are several common physiological disorders of pears, including russeting, superficial and senescent scald, corkspot, hard-end, premature ripening, core flush, core breakdown, and some controlled atmosphere related disorders from low O_2 or high CO_2. Russeting is related to high humidity conditions in the orchard during the early stages of fruit development. Bosc, Hardy, and Nelis are fairly easily russeted and, in fact, mild russet is considered a desirable identifying feature for some varieties. Certain spray materials, such as copper, may induce russeting. Scald is also more pronounced under cooler, moister climates and is further influenced by harvest maturity. Less mature pears are more susceptible to superficial scald, whereas overmature pears are more likely to develop senescent scald. Optimally mature pears are the least sensitive to either type. Preventive treatments utilize anti-oxidant scald inhibitors such as ethoxyquin (a tocopherol derivative) which must be applied within the first 5 to 6 weeks after harvest to be effective. Essentially, all varieties of pears, including the Asian types, have some susceptibility to scald and the longer the fruit is held in storage, the greater the incidence and severity of scald, even for ethoxyquin-treated fruits. Mellinthin et al.[37] and Chen et al.[39] have shown the usefulness of low O_2 controlled atmosphere storage conditions for prevention of both types of scald on Anjou pears which had not been treated with ethoxyquin.

Cork-spot is a localized fruit calcium-deficiency disorder.[43-50] This disorder is fully analogous to bitter pit in apples. It is characterized by yellowed, slightly sunken spots externally, which when peeled show initial tan-yellow, somewhat diffuse spots that rapidly oxidize to dark brown spots. These spots may be only a few mm under the skin or, more typically, they run up to a few cm into the flesh. The Anjou variety is the most susceptible, although cork-spot has been observed in Bartlett, Bosc, Eldorado, and Packham's Triumph. Originally termed "drought spot", is has been known that hot, dry, droughty conditions are significant as part of a complex of factors leading to this disorder. Pre-harvest calcium sprays on the developing fruits have been effective in reducing the incidence and the severity of this disorder,[43,46,47,50] although complete control is rarely accomplished. Richardson and Al-Ani[44] established that Anjou pears at harvest should have greater than 7 mg calcium per 100 g fresh fruit weight as a critical minimum to avoid cork spot. They further established that as the N:Ca ratio increased above 10:1, the critical minimum Ca needs also increased. Another closely related disorder is "green-stain" or "alfalfa-greening" of Anjou pears which appears to be from excessive nitrogen or at least from a high N:Ca ratio.[51] Green stain is a failure to lose chlorophyll during maturation and ripening, thus leaving the pears with dark green blotches against a yellowing background. Often this is accompanied by small sunken pits in the lower two-thirds of the fruit, and is known as "freckle-pit". An outgrowth of studies on low calcium effects on pears was the discovery that beyond the benefits of avoiding the Ca-deficiency cull problems, there were added benefits to increasing fruit calcium which have helped to reduce pathogen susceptibility[52,53] not only in Anjou, but in Bosc and possibly other varieties as well. Further, Gerasopoulos[54] has shown the effects of higher fruit Ca on prolonging the chilling requirement for ripening to occur in Anjou, on slightly reduced respiration rate, and on delayed ethylene synthesis. High Ca pears hold up several months longer in storage and still are capable of ripening to good texture and flavor, whereas fruits with lower Ca had already lost the capacity to ripen with quality.[50]

F. POSTHARVEST PATHOGENS

Pears are attacked in storage mainly by Botrytis, Penicillium, Mucor, Pezicula, and Phialophora fungal pathogens. Additional, but relatively minor pathogens include Cladosporium, Venturia, Phytophthora, and Coprinus.

While excellent postharvest decay control has been effected by the use of Benomyl, recent controversies (1989) relating to potential mammalian health threats have resulted in this substance being voluntarily withdrawn from postharvest use by the manufacturers. Thiabendazole (TBZ), a closely related compound, has been pressed into service in the wake of the benomyl withdrawal. Newer fungicides in development are showing some promise.[55] Bacterial problems on pear fruits are insignificant. There are some virus problems, the most serious of which is stony pit. This virus causes fairly deep pits to occur in Bosc, Seckel, Packham's Triumph, and to a lesser extent in Anjou pears. It rarely, if ever, affects Comice or Bartlett. Somehow the virus stimulates the lignification process and quite large stone cells develop. Some of these may be several mm in size. The only remedy for this virus problem is excision of the affected limb in the hope of eliminating the source, and in removal of the fruits as culls.

G. APPEARANCE

Just as in the case of all other fresh fruits, appearance is what initially attracts a consumer to purchase pears. Thus it is important that they be uniformly shaped and typical of the variety. While a glossy, waxy surface may help in apple attractiveness, this is limited in the case of pears because the addition of wax may interfere with the ripening process by limiting gas exchange. Nonetheless, the natural wax can be carefully brushed to improve the glossiness of pears, especially Anjou. Waxes and other coatings adapted to the special permeability needs of pears have been under development in the past few years and look promising.[56]

H. RIPENING

Essentially, all pears are harvested in a mature, but unripe stage of development. This allows them to be stored until they are either marketed fresh or processed. In both cases, to achieve optimum eating quality, they must be ripened. Fresh pears are usually marketed in the unripened or only partially ripened state, the fruit ripening is left to the consumer or retailer to accomplish. Processors will ripen pears to an appropriate stage of ripeness which will differ depending upon intended use. Upon removal from cold storage temperatures, pears are best ripened at 20°C.[16,57,58] Normally, Bartlett pears will ripen to good dessert quality in four days, Bosc and Comice in 4 to 5 d, and Anjou and Packham's Triumph in 7 to 10 d.

Winter pears have an interesting trait, as they require a period of time in −1 to 0°C cold storage in order to develop the capacity to ripen with good quality.[59] This cold or chilling requirement time duration depends upon the variety. Bartlett can ripen without this chilling, whereas Bosc requires about 15 d. Comice and Eldorado need 45 d, and Anjou and Packham's Triumph need 50 to 60 d.[54,60] Interestingly, pear fruit calcium concentration may relate to the chilling requirement, as fruits low in Ca have a shorter chilling requirement and high Ca pears have a longer chilling requirement.[49]

Pears are classified as climacteric-type fruits and thus are typically harvested at the preclimacteric respiratory minimum stage to achieve longest storage potential. Winter pears produce very little (<0.05 μl/l) internal ethylene at normal harvest maturity.[54,59-62] Ethylene synthesis capability is reached very slowly during the time in cold storage when the chilling requirement is being satisfied. After chilling is accomplished, autocatalytic ethylene (ranging from 1 to 30 μl/l) is capable of driving the full set of ripening responses including chlorophyll loss, rapid softening to a juicy, buttery texture,[61] some reduction of organic acids, and development of characteristic flavor volatiles, and loss of astringency flavors by polymerization of monomeric polyphenolic compounds such as chlorogenic acid.[61] Responses to ethylene are strong in that immature pears or mature pears not having satisfied the chilling requirement can be driven to ripen by ethylene treatments.[61-63]

Several enzymes are activated or synthesized by ethylene. Cell wall degrading enzymes, polygalacturonase, pectin esterase, and cellulase cause substantial hydrolysis of pear cell walls which results in texture proceeding from crisp and juicy to buttery and melting.[61,65-68] The dominant feature of this is the dissolution of the middle lamella and the depolymerization of polygalacturonic acid chains to soluble pectin, i.e., galacturonic acid. Studies[69] have shown that pears possess pectinase inhibitors, but the extent to which they regulate softening has not been clarified. Protein synthesis is required for ripening,[60,70] and malic enzyme increases dramatically.[71]

There is an important shift in the ethylene biosynthesis pathway via *S*-adenosylmethionine wherein metabolic flux through ACC is accelerated and the flux through the polyamine pathway is decreased.[72] Winter pears commonly accumulate ACC during cold storage which results in a rapid conversion to ethylene upon warming which triggers the complex of ripening events.

When pears have been held in cold storage beyond a certain time, they lose the ability to ripen properly. Attempts to ripen them result in slightly softer, but rubbery and tough textured fruit, flavor is virtually absent, and the pears are insipid, lacking both acids and sweet flavors. Ethylene synthesis is still possible, but production is lower than normal. Fruits held long in storage may undergo a flattened and prolonged climacteric respiratory increase without several other features of ripening.

Some market research has shown that preconditioned or partially ripened pears (down to 40 N firmness) can be safely sent through market channels with more enthusiastic consumer acceptance, but longer storage at destination is much curtailed.

I. FLAVOR AND AROMA

Flavor is a complex interaction of many organic substances, Compared to many other fruits, pears have a very subtle flavor profile which makes them especially attractive to beverage industries who use pear juices and concentrates as a natural sweetness base that can be easily modified by stronger flavors. As public interest increases demand for unprocessed sugars, the natural fruit juice bases such as pears will become more prominent in the future. While we seek to achieve gross estimates of quality based upon sugars (% Brix) and acids (titratable acidity), in reality there are so many contributors to flavor that it is unlikely that we will ever be able to totally define pear flavor in biochemical quantitative terms. Still, there are some major substances of high importance. Much of the sweetness comes from the simple sugars, mainly fructose and glucose. Generally, sucrose is quite low. Sorbitol is a significant contributor in some varieties. The organic acids also significantly impact flavor and the two most prominent in pears are malate and citrate. Characteristically, these acids decrease over time in storage, and their losses lead to losses in flavor upon ripening. A significant benefit of CA storage is that it helps to conserve the organic acids.[38] Pears held too long in cold storage have an insipid flavor as well as very poor texture, and may even have off flavors which may be associated with abnormal Kreb's cycle metabolism that accumulates succinic acid. Volatile components also contribute to flavor and aroma in pear quality.[77] Creveling and Jennings,[74] and more recently Russell et al.[75] and Quamme,[76] have shown the volatile profiles related to pear flavor, particularly the methyl-, ethyl-, propyl-, and butyl-esters of 2-*cis*,4-*trans*decadienoic acid quite characteristic of Bartlett pear flavor.

Pears have a very mild flavor profile, which makes pear juices and concentrates desirable to beverage industries as a natural juice base which can be flavored with minimal impact from the juice base itself.

Another important change in flavors upon ripening is the loss of astringency as the unpolymerized polyphenolics, such as chlorogenic acid, become polymerized or further metabolized.[41,77] Most of the substances are located near the peel.

J. TEXTURE

Upon ripening, there is a dramatic change in texture of *Pyrus communis* varieties. Within a period of 4 to 10 d at 20°C, the firmness decreases from 65 N to 10 N, and the mouth feel of the texture passes from a firm, crisp, somewhat dry fruit to a juicy, melting, almost buttery texture.[35] Numerous studies on the cell wall biochemistry have shown substantial degradation of the middle-lamellae pectinaceous substances during this process.[65-68] The degradation of pectin also occurs in the case of Asian pears, but the extent of degradation is not nearly as complete.[19] While similar trends have been observed in apples, it appears to be much more complete in the case of pears. Although ethylene is capable of dramatically stimulating the softening mechanisms, recently we have observed that under certain conditions this process can slowly occur without production of endogenous ethylene.[54] Perhaps this should not be so surprising, since many non-climacteric fruits, e.g., strawberry, soften dramatically without ethylene. Similarly, prior to the climacteric, there is considerable softening during maturation and this is clearly the case for many pear varieties. While there is some starch storage in pears, the shift from starch to free sugars does not play a significant role in texture and likely makes only a small contribution to sweetness. Mention should be made, however, of the role that stone cells play in the easily perceptible gritty texture of many pear varieties. These tiny stone cells are lignified, very hard, and coarse, and usually highly localized near the peel, surrounding the core, and adjacent to vascular bundles.

V. FACTORS INFLUENCING THE QUALITY OF PEARS FOR PROCESSING

Improved postharvest technology has effectively extended the season for fresh fruit and allowed for a greater proportion of the crop to be sold for the fresh market. Processing the fruit into a variety of products—canned pears, clarified juice and juice concentrate, single-strength and concentrated purees, dried fruit and fruit leathers, wine and brandy, jams and confections—allows for complete utilization of the crop and subsequent economic benefits for both the grower and the processor. Fruit having certain quality defects for premium fresh market quality such as being either too large or too small in size, misshapen, having surface blemishes, etc., can be used to produce high-quality processed products. It should be emphasized that regulatory standards such as those promulgated by Codex Alimentarius, the U.S. Code of Federal Regulations and USDA grades and standards will not permit the use of rotted and microbially spoiled fruit and fruit contaminated with pesticides and agricultural chemicals.[78-80] Principal quality factors of concern to the consumer and the processor will be the product's color and appearance, its flavor, and its textural and nutritional attributes.

A. COLOR AND APPEARANCE

1. Bleaching of Canned Pears

The USDA grading system for canned pears stipulates that Grade A pears shall have a "light yellow-white, greenish-white, or beige-white color". A pink tint, tan or beige color, "chalky" or "dead-white" appearance results in down-grading to Grade B or substandard.[80] Canning in unlacquered cans allows for chemical reduction of phenolic compounds and subsequent bleaching of the fruit. Reaction between tinplate and H^+ from the fruit acids results in the formation of metal ions (Sn^{++} and Fe^{++}) and H^2 gas. Phenolic compounds serve as H^2 acceptors and are chemically reduced, resulting in a bleaching of the fruit and a driving of the reaction to completion. These reactions result in a change in flavor as well; pears canned in enameled cans or retort pouches have taste qualities which are perceived differently. The metal ions generated in the detinning reaction can play a role in haze and sediment formation by complexing with phenolics, proteins, and/or pectic substances. The extent of detinning (which can lead to pinholing) is dependent on the quality of tinplate as well as length of storage time.

2. Pinking of Canned Pears

High concentrations of leucoanthocyanins have been implicated as the cause of pink discoloration which confronts processors of canned pears from time to time.[81] Chandler and Clegg[82-84] identified the pigment in discolored pears to be an insoluble tin-cyanidin complex, cyanidin being generated from leucoanthocyanins. Reducing agents such as sulfur dioxide and stannous ions were effective in preventing discoloration in canned pear puree. Pinking can be more of a problem with lye peeling as less of the upper dermal layers of fruit tissue is removed than in mechanical peeling, and this portion of the fruit along with the peel contains higher concentrations of leucoanthocyanins. Czerkaskyj[85] found a relationship between skin blush on fresh fruit and pink discoloration after processing. Red skin color was indicative of higher leucoanthocyanin content in the fruit. Only blushed fruit was prone to pinking and the blushed half of any single fruit was more susceptible than the unblushed half. Extending the cooking time from 10 to 20 minutes did not increase the incidence or severity of pinking. The "red-blush" pigments of Bon Chretien pears have been identified as the 3-galactosides of cyanidin and peonidin.[86] The increasing popularity of the red-skinned pears for fresh market is likely to result in pink coloration of processed products such as juices and purees with the possibility of extracting the peel anthocyanins.

3. Enzymic Browning

Brown discoloration in canned pears occurs if time delays preclude thermal inactivation of polyphenoloxidase.[87] The browning reaction is very pronounced in pureeing of fruit and in the crushing and milling operations of juice processing where the antive enzyme reacts with molecular oxygen and the natural substrates—chlorogenic acid, leucoanthocyanins, (+)-catechin, (−)-epicatechin, *p*-coumaryl quinic acids and caffeic acids.[88] Tate et al.[89] studied the characteristics of Bartlett pear polyphenol oxidase; Halim and Montgomery[90] reported polyphenoloxidase from Anjou pears to have a pH optimum of 7.0, be relatively heat stable, and particularly sensitive to inhibition by L-cysteine and diethyldithiocarbamate. Wissemann[91] purified the enzyme and determined the isolectric points and molecular weights of the isozymes. Montgomery and Petropakis[92] could produce a light colored pear juice by adding 300 ppm of ascorbic acid during milling with subsequent heat inactivation at 90°C. In a later study, the same authors[93] reported that boiling of ascorbic acid treated pear juice for 1 min was required for complete inactivation of polyphenol oxidase. Skorikova and Lyashenko[94] recommended 75 to 85°C for maximum retention of polyphenolics in pear juice, higher temperatures causing chemical degradation of the phenolics and lower temperatures enzymic oxidation; leucoanthocyanins were the least stable polyphenol fraction. While cysteine is not an approved additive for juice manufacture in the U.S., Montgomery[95] found that initial browning of pear juice concentrates could be eliminated by treatment with 2 m*M* cysteine.

4. Nonenzymic Browning

A major quality defect of pear juice concentrate is its brown color which progressively increases during storage. While enzymic browning is the major cause of browning during processing, Cornwell and Wrolstad[96] demonstrated that nonenzymic browning predominates during storage. Removal of amino acids by treatment of pear juice with cation exchange resin could effectively stop browning. Treatment with anion exchange or mixed bed resins will remove nonvolatile acids as well. Overripe Bartlett pear juice contains lower concentrations of free amino acids than normal fruit[97] and presumably would exhibit reduced rates of browning during storage. Some commercial pectinase preparations contain proteases which can hydrolyze proteins, increase the amino acid concentration, and aggravate nonenzymic browning problems. Akhavan and Wrolstad[98] reported a 6% loss of sugars during accelerated storage (16 weeks at 36°C) of pear juice concentrate; Beveridge and Harrison[99] detected a 20% loss of amino acids but no loss of carbohydrates when pear juice concentrate was heated at 80°C for 4 h. The latter authors[100] determined the zero order rate constants for browning of pear juice concentrate during storage and concluded that the reaction could be stopped by storage at −13°C. Ascorbic acid after being oxidized to dehydroascorbic acid will participate in browning reactions itself, hence its effectiveness in delaying enzymic browning is confounded by its participation in nonenzymic browning reactions.[101]

5. Haze and Sediment Formation

In addition to minimizing brown color, it is desirable that pear juice and pear juice concentrate be free of haze and sediment for most beverage applications. Potential contributors to haze include proteins, phenolics, metal ions, pectin, starch, and other polysaccharides. Common commercial practices are to hydrolyze the pectic substances by treatment with pectolytic enzymes, remove starch by centrifugation or other physical means, remove solubilized starch with commercial amylase preparations, remove proteins with bentonite fining, and polyphenolics with gelatin fining. Chang[102] examined the sediment in ripe Bartlett pear juice which had been depectinized and had its pH adjusted to 3.5 with citric acid; the precipitated colloid was characterized as being 35% protein, 34% phenolic, 23% lipid, 8% sugars, and 1% minerals. Recently, Babsky et al.[103] characterized the polysaccharide which was responsible for development of a thermolabile haze in clarified pear juice concentrate during storage as being an arabinan. It is believed that commercial enzyme preparations in combination with efficient mechanical presses facilitated the extraction of this cell wall constituent which is insoluble at reduced temperatures and higher concentrations. Commercial pectinase preparations possessing arabinase activity can hydrolyze the solublized arabinan, preventing the problem of haze development.

B. TEXTURE

Better texture and flavor is achieved in both fresh and canned Bartlett pears if they are ripened after picking. Common harvest practice is to pick when they are full size but still green and hard (17 to 19 lb. firmness test). Claypool et al.[104] in a two year California study investigated optimum conditions for postharvest handling of fresh pears for canning. They recommended ripening at 20°C and approximately 2 lb pressure for "more natural" canned pear flavor. Current commercial practice is to ripen Bartletts' to 3 to 4 lb pressure to achieve optimum texture of the canned product. Solubilization of pectic substances accounts for the pronounced textural changes during ripening. During the later stages, hydrolysis of the pectin occurs. Deesterification of pectin also takes place during ripening, Bartlett pears decreasing from 95% esterification for hard, green pears to 64% for soft, ripe fruit.[105] Juice for pear concentrate is traditionally pressed from hard unripe fruit, the mechanical expression of juice being easier than from soft ripe fruit. Use of press aids such as cellulose and rice hulls and treatment with pectolytic enzymes enables the recovery of juice from ripe fruit and the utilization of peels and cores from cannery wastes. Thus in the U.S. both "hard" and "soft" pear juice concentrate is commercially available, the essential compositional differences being due to the maturity differences of the fruit. Early season, unripe fruit contains higher concentrations of starch and treatment of juice with amylase enzymes is often required to hydrolyze solubilized starch and prevent development of haze.

The characteristic gritty or granular texture of both fresh and canned pears is due to lignified stone cells. Ranadive and Haard[106] investigated peroxidase activity and its role in pear stone cell development. They concluded that mineral nutrition (Ca^{++}) delimited peroxidase localization and lignin deposition in pear fruit. In juice processing, the cells normally remain in the presscake or are removed in screening, filtering, or centrifuging unit operations. Càsimir[107] reported that milling of pear puree reduced the average particle size of stone cells from a range of 160 to 260 μm to 20 to 50 μm, resulting in a smooth-textured product. Yamane[108] reported on procedures for isolating stone cells from Japanese pears and described their chemical and physical characteristics.[109]

O'Brien and Singh[110] studied the effect of bruising on the quality of dried pears. Bruised portions suffered from collapse of structure with a leathery texture resulting.

C. FLAVOR

1. Sweetness

Pears are relatively high in sugar content, even when harvested at the green and hard, but full-sized stage of maturity. In their study of changes in sugars and acids during the ripening of Bartlett pears, Akhavan and Wrolstad[98] found green, hard pears to contain 12.0 g sugars per 100 g which increased to a maximum of 13.5% and then decreased to 12.4% at the fully ripe stage. Juice is easier to express from the green hard fruit and processors often prefer to press at that stage of maturity. There is no standard of identity for pear juice in the U.S. however it is common commercial practice to use 12.0° Brix to represent single strength pear juice. The German RSK values[111] which are widely applied as typical compositional indices for single strength juice report a range of 11.18 to 13.54 and mean of 12.13° Brix. The USDA jelly standard[79] for single strength pear juice is 15.4° Brix, which is considered by most juice processors to be unrealistically high. Wrolstad and Shallenberger[112] in their literature compilation of free sugars and sorbitol content of fruits found relatively good agreement among workers for the values reported for pears. The sugar profile is characterized by its high fructose content, high fructose:glucose ratio, and high sorbitol content. The high sweetness of fructose will influence the taste quality of processed pear products, particularly when they are consumed at chilled temperatures as the sweetness of fructose is more pronounced at lower temperatures. Sorbitol values ranged from 1.21 to 2.80 g/100 g. This compares with a range of 1.10 to 2.64 g sorbitol/100 ml juice reported by Weiss and Saemann.[113] As sorbitol is non-fermentative, sorbitol will provide a residual sweetness in pear wines that have been fermented to dryness.

2. Acidity

Pears are relatively low in titratable acidity, a range of 0.31 to 0.62 and mean of 0.44 g/100 ml as malic being given for pear juice in the RSK values.[111] Fernandez—Flores et al.[114] reported total acidity values of 0.53 g/100 ml and Akhavan and Wrolstad[98] found values of 0.21 g/100 ml. The total acidity varies considerably with ripening, Akhavan and Wrolstad[98] reporting a range of 4.4 to 6.0 meq/100 g fruit during the ripening of Bartlett pears. The low level of acidity results in a high sugar:acid ratio which contributes to the sweet taste quality of pear products. Czerkaskyj[85] found that addition of 0.1% w/w/citric or malic acid significantly improved the flavor of canned pears as rated by a taste panel.

Malic is considered the principal acid in pears. However, citric can predominate depending on the maturity.[98]

3. Astringency and Bitterness

The leucoanthocyanins which are responsible for the pink discoloration in canned pears may also possess taste properties which need to be given attention. Lea and Timberlake[115] established that polymeric procyanidins were major contributors to the astringency and bitterness of cider apples. The procyanidins are present in larger concentration in the peel and the upper dermal layers of fruit tissue. The concentration of these compounds will be greatly reduced in clarified juices and wines as a result of their polymerization through oxidative and enzymic reactions and subsequent removal in fining and filtration.

4. Aroma

Jennings and his co-workers have studied the volatiles of Bartlett pear in a series of classical papers.[116 to 123] The lengthy list of compounds which have been identified is dominated by esters and short-chain alcohols. The methyl, ethyl, propyl, and butyl esters of *trans*-2,*cis*-4-decadienoic acid are of particular interest as they possess Bartlett pear "character impact" and correlate well with sensory evaluation of pear essence. Russell et al.[124] compared the flavor quality of Bartlett, Magness, and Kieffer pears and found that Bartlett, which had the highest flavor panel scores, contained the largest number and greatest concentrations of flavor constituents. Bartlett contained much greater quantities of the decadienoate esters. These unsaturated esters reach a maximum shortly after the climacteric point when the fruit possess optimum flavor. Jennings and Tressl[122] studied the formation of volatiles in ripening Bartlett pears and found that the major esters (acetates and decenoates) and α-farnesene were produced at rates varying in a cyclic matter. They proposed that some control mechanism alternatively activated (or inhibited) the biosynthetic route for the terpene and the acetate esters. The decadienoate esters are believed to be produced through a β-oxidation pathway. Strandzhev's[125] finding that blanching of pears resulted in a 2.5 times lower content of aroma compounds substantiates the enzymic biogenesis of aroma compounds. Russell[124] demonstrated synthesis of flavor volatiles when Bartlett pear puree was incubated with C_{18}-unsaturated fatty acids.

The decadienoate esters are high boiling and their unsaturation renders them susceptible to oxidation. These factors probably account for some of the difficulties in producing high quality essence. Pear juice concentrate, which is marketed with the essence separate or added back, is often used as a neutral base for blended juices, juice drinks, and "pop" wines. For these applications, a neutral product not containing the characteristics essence is often preferred. Hydroxymethylfurfural forms in pear juice concentrate during storage[96] and it may mask or interfere with other flavor compounds.

D. NUTRITIONAL QUALITY

The nutritional rational for fruit consumption gives particular attention to ascorbic acid, potassium, and dietary fiber content. Fresh pears are not a very good source for ascorbic acid (4.0 mg ascorbic/100 g).[126] Considerable losses occur with canning (75%) and other processing procedures. Pear juice concentrate contains essentially no ascorbic acid of nutritional significance. Addition of ascorbic acid in the processing of pear juice has been tested.[92] However, the major objective for its addition was for its ability to inhibit enzymic browning, rather than for nutritional fortification.

Pears are a relatively good source of potassium (125 mg/100 g of fresh fruit.[126] Individuals with concerns of hypertension are advised to reduce their sodium intake and consume foods high in potassium in order to achieve an improved Na/K ratio. Use of ion-exchange technology in juice processing may alter the Na/K ratio depending on the materials used to charge the resin bed. By using cation exchange resins that are potassium charged, it should be possible to produce ion-exchanged pear juice that maintains a desirable potassium content.

While pears contain higher concentrations of amino acids than apples and many other fruits, the protein-amino acid content is too low for them to be considered a significant protein source from a nutritional standpoint. Amino acids are in fact detrimental to quality as they participate in nonenzymic browning and Strecker degradation reactions leading to brown color and off-flavor formation. Soluble proteins are also detrimental in juice, concentrates, and wines as they can combine with phenolics and/or metal ions to form haze and sediment. Bentonite and other fining agents are used to remove proteins from solution and prevent such reactions in clarified beverages.

More attention has been given in recent years to the benefits of non-nutritive fiber in foods and its possible role in reduced incidence of colon cancer and diverticulitus. High fiber diets may also result in lowered levels of blood cholesterol and have a preventive role in atherosclerosis. Holloway and Grieg[127] included pears in their investigation of the water holding capacity of various fruits and vegetables. Pears had a water holding capacity of 16.4 g water per g dry matter and a 2.2% yield of hemicellulose from freeze-dried fruit. Similar studies will likely be done in the future to determine the relative effectiveness of fiber from different sources as well as the effectiveness of the different fractions in preventing or treating various disorders. Fiber is removed from the fruit in the processing of clarified fruit juice and the presscake provides a fiber source for use as a food ingredient.

V. PROCESSING

A. CANNING

Canning of pear fruit into a variety of different pack-styles—pear halves, slices, chunks, dices, purees—has been the principal processing method used to preserve the fruit. Bartlett is the most common variety used for canning because of its shape as well as for texture, color, and flavor considerations. The unit operations involved in canning pears are thoroughly reviewed by Luh and Kean.[128] Cruess[129] gives an earlier description of this long-established processing operation.

Mechanical peeling, coring, and stemming generates processing waste accounting for 30 to 35% of the original weight.[128] Lye peeling removes less of the upper dermal tissue, reducing the quantity of waste. As the upper dermal tissue contains higher concentrations of leucoanthocyanins, pinking of canned fruit peeled by this procedure may be more of a problem. Dry caustic peeling conserves on water and results in a two-thirds reduction in BOD and COD as compared to conventional chemical peeling.[130] A high vacuum flame sterilization process (HVFS)[131] has been developed for sliced pears which incorporates blanching and deaeration unit operations into the canning process. The HVFS process does not use added water or syrup and the process requires a minimum of 7.5 min for biological stability. After 18 months of storage, the HVFS fruit possessed texture and flavor more like fresh fruit as compared to the conventional process. Use of retort pouches offers potential savings in packaging and shipping costs, however the absence of tin in the packaging material will prevent the catalytic reduction of pigments. The fruit will not have the bleached appearance which is preferred by many consumers. Beuchat and Heaton[132] were able to shorten the process time for canned pears through the addition of potassium sorbate and sodium benzoate. This advantage was offset by the development of some adverse changes in sensory quality; the preservative-containing samples were still considered acceptable.

Consumer interest in fresh fruit and the availability of high quality pears for much of the year have been contributing factors to the decline in consumption of canned pears in the U.S. This has led to an increased demand for winter varieties such as Anjou, Bosc, and Comice and a decline in Bartletts, which are the preferred variety for canning.

B. JUICE AND JUICE CONCENTRATE

Processing pears into juice and juice concentrate has become a major means of utilizing fruit which contain defects or are misshapen, too large, or too small, making them inappropriate for the fresh market. Schobinger and Mueller[133] evaluated three pear cultivars as to harvesting characteristics, juice yield, composition, and taste. Juice can also be recovered from cores and peels, minimizing processing waste and providing a product which can be used as a replacement for sugar syrup in canning. The neutral flavor and attractive pricing has made pear juice concentrate a good base for juice drinks, soft drinks, syrups, and fruit wines. The sweet, neutral flavor of pear juice blends well with more intensely colored and flavored fruits such as red raspberries and Concord grape. Spayd et al.[134] reported on the development of polymeric color and haze and the loss of anthocyanin pigment in pear juice blended with Concord grape, red and black raspberry, and Bing cherry juice.

For these product applications it is desirable to have a concentrate which is free of haze and has minimal brown pigmentation. These quality factors were discussed in the previous section. Low temperature storage is a practical method of reducing the browning which occurs during storage and treatment of the juice with cation exchange resin to remove amino acids is an effective means of lowering the browning potential.[96] Montgomery[95] demonstrated that cysteine is an effective inhibitor of browning in pear juice concentrate, but it is not approved for use in the U.S. at present. Process modification to improve the color quality of pear concentrate will need to consider inactivation of polyphenoloxidase and/or removal of the pigments by treatment with charcoal or various resins, and in addition, reduction in the concentration of amino acids to retard the Maillard reaction.

Processing pears by the DDS-diffusion technique results in increased yields as compared to conventional pressing methods.[135] Enzymatic liquefaction through the combined action of pectinase and cellulase can also improve yields. This procedure is particularly useful in pressing ripe and overripe fruit. Schobinger et al.[136] found that juices of enzymically liquefied pulps had higher concentrations of solids (more galacturonic and quinic acid, glucose, polyphenols, methanol and minerals) and also received lower sensory assessments. The solubilization of araban through such procedures and its subsequent precipitation was discussed in the previous section. Kirk et al.[137] used ultrafiltration to clarify pear juice. This technology offers an alternate means of removing enzymes as well as proteins, polysaccharides, and polyphenolics. It may offer a practical means of removing color and haze from a low quality concentrate; it also may be efficient to use it in combination with traditional fining procedures.

Concentration to 68° Brix or higher is necessary to obtain microbial stability. The low water activity at the high concentration levels results in greater rates of browning, however. Kaupert et al.[138] used a combination of heat and irradiation to totally inactivate the microbial population of pear juice concentrate. This would enable one to concentrate to a lower °Brix with subsequent improvement in color and other quality attributes.

C. WINES, CIDERS, AND OTHER ALCOHOLIC BEVERAGES

Perry, a cider-like fermented beverage, is made from pears or blends of pears and apple juice. In Great Britain not more than 25% of Perry can be from apple juice and the minimum alcohol content is 4% by volume. Special varieties of pears of high tannin content are grown in Great Britain, France, Switzerland, and Germany and are considered the "true" Perry pears which may be blended with apples and dessert pears in the manufacture of this product. Beech and Carr[139] have thoroughly reviewed the standards and production statistics as well as the historical, compositional, and technological aspects of Perry production.

A certain amount of pear wine is produced in the U.S. and Europe from dessert pears which tend to be low in acid, sweet, and consumed as a dessert wine. Waste peels and cores have been utilized for the production of vodka and pear brandy,[140] the former product being a highly rectified distillate practically devoid of fruit flavor. Pear juice concentrate in the U.S. has often been used as the fruit source in the production of "pop" wines, the principal flavoring materials coming from other sources. If the distinctive pear aroma compounds could be retained in fermentation or captured in a high-quality essence, it should be possible to produce a liqueur, brandy, or other alcoholic beverages which would capitalize on the distinctive pear flavor.

Tanner[141] recommended heating the mash to reduce the methanol content of spirits manufactured from pears. Heat inactivation of the native pectinesterase enzymes reduced a lowering of the methanol content by 40 to 90%. Brandies produced from pear peels cores, and stems produced twice as much methanol as that produced from the fruit flesh.[142]

D. PEAR PUREE

The bland sweet flavor, smooth texture, and easy digestibility of canned pear puree has made it a highly acceptable item for use in infant foods. Luh[143] has discussed the processing conditions for manufacture of a high quality puree. Casimir[107] compared purees made from unpeeled and uncored pears. The puree had a gritty texture from the stone cells which had an average particle size of 160 to 260 μm; a smooth texture could be obtained by putting the puree through a colloid mill which reduced the size of the stone cells to 20 to 50 μm. Leonard et al.[144] made pear puree from whole Bartlett pears and obtained a 77% increase in yield as compared to puree manufacture from peeled and cored fruit. Return of captured essence gave a puree which had more acceptable flavor.

Concentrated pear puree has become a significant item of commerce because of the economic advantages in packaging, shipping, and storing a concentrated product. A method of manufacturing aseptically concentrated pear puree was described by Luh and Sioud.[145] A usual practice is to concentrate the puree to 28% total solids and then pack aseptically in 55-gal drums. A refreshing nectar can be prepared from the product through addition of water, sweetener, acid, and optional flavoring. Other applications of the product is as an ingredient in fruit bars and fruit leather analogues. While the concentrated puree is processed to be microbiologically stable, undesriable flavor and color changes will occur if it is not stored and handled properly. Luh and Sioud[145] examined the quality changes in canned concentrated pear puree during storage and observed no distinct color changes in puree stored 371 d at 0° and 20°C. Formation of hydroxymethylfurfural was detected after 180 d storage at 30° and 37°C and tin corrosion and hydrogen gas formation was also noted at high temperature storage.

E. DRIED PEAR COMMODITIES

Sun-drying of ripe pear fruit has traditionally been done by exposing the peeled, cored, and halved fruit to the fumes of burning sulfur.[146] Bartletts are the principal variety used. Tunnel and conveyer dryers both shorten the drying time and enable closer control of moisture content. Demand for dried pears has increased in recent years with the popularity of dried fruit as a snack by itself and its use in cereals and confections. McBean[147] experimented with a number of pre-drying modifications including steam peeling, steam blanching, infusion of citric acid, and flavoring with passion fruit juice and ginger essence. There is considerable interest in lowering or eliminating the SO_2 content of the product. SO_2 is very effective in inhibiting enzymic browning. Dehydrated pears have been prepared where a combination of 1% ascorbic acid and 0.25% malic acid as a 3 min dip was used as an alternative to bisulfite.[148] While the color of the dehydrated product was not equal to that of the sulfured product, the flavor was judged to be superior. A line of dried pear solids or flakes has been developed where pear puree is combined with modified starch and drum dried in 50 s.[149] The short drying time gives a product with good nutrient, flavor, and color retention. Osmotic drying is yet another processing technology which has been applied to pears.[150] Moisture is first osmotically removed by submersion of the fruit in a 70% sugar solution containing citric and ascorbic acids; additional water is subsequently removed in a 80°C fluidized bed dryer.

F. MISCELLANEOUS PRODUCTS

In Cruess's classic text,[129] processes are described for manufacturing a variety of products which utilize pears including pear butter, jams, jellies, preserves, vinegar, pickles, and jelly candies. Pears are customarily blended with other fruits in jam and jelly manufacture because of their relatively low acid and pectin content. Pear juice concentrate is often used as the fruit base for "mixed-fruit" jellies. Candied pears and glaced fruit should be mentioned, but their production is rather minor. Pear slices can be used to advantage in fruit compotes. Aminov et al.[151] showed that blanching and vacuum heating were necessary to eliminate gas formation in pear slices.

Recent years have witnessed a decline in the consumption of canned pears and an increased demand for fresh market pears. Popular varieties are Bosc, Comice, and Anjou as well as the new red-skinned varieties have been replacing many of the Bartlett plantings which has been the preferred variety for canning. The Asian pear is also going into increased cultivation. While the marketing objective is for high quality fresh fruit, there needs to be processing methodologies for utilizing the sort-outs and surplus. The compositional data base for many of these varieties is incomplete. Such information is needed so that processors can capitalize on the fruit's compositional characteristics and process it into high-quality traditional and nontraditional products. Such work needs to be carried out in concert with processing evaluation trials.

REFERENCES

1. **Lombard, P. B. and Westwood, M. N.,** Pear rootstocks, in *Rootstocks,* Rom and Carlson, Eds., John Wiley & Sons, 1987, chap. 5.
2. **Kikuchi, A.,** Speciation and taxonomy of Chinese pears (in Japanese), *Col. Records of Hortic. Res.,* 3, 1946.
3. **Hedrick, U. P.,** *The Pears of New York,* New York Agr. Exp. Sta., 1921, chap. I.
4. **Challice, J. S. and Westwood, M. N.,** Numerical taxonomic studies of genus *Pyrus* using both chemical and botanical characters, *Bot. J. Linn. Soc.,* 67, 121, 1973.
5. **Lombard, P., Hull, Jr., J., and Westwood, M. N.,** Pear cultivars of North America, *Frt. Var. Jour.,* 34, 4, 74, 1980.
6. **Westigard, P. H., Lombard, P. B., and Berry, D. W.,** Integrated pest management of insects and mites attacking pears in southern Oregon, *Ore. St. University Agr. Exp. Sta. Bul.,* 634, 1979, 40.
7. **Lombard, P. B.,** Pear pollination and fruit set, in *The Pear,* Van der Zwet, T. and Childers, N. F., Eds., 1982, 91.
8. **Westwood, M. N.,** Temperate-zone pomology, in *Fruit Growth and Thinning,* New York, 1978, 199.
9. **Lombard, P. B. and Westwood, M. N.,** Performance of six pear cultivars on clonal Old Home, double rooted, and seedling rootstocks, *J. Am. Soc. Hortic. Sci.,* 101, 214, 1976.
10. **Lombard, P. B. and Williams, R. R.,** The hard side of cider, *HortScience,* 9, 5, 420, 1974.
11. **Kader, A. A., Ed.,** Quality factors: Definition and evaluation for fresh horticultural crops, in *Postharvest Technology of Horticultural Crops,*
12. **Arthey, V. D.,** *Quality of Horticultural Products,* Halstead Press, John Wiley & Sons, New York, 1975, chap. 20.
13. **Fallahi, E. and Larsen, F. E.,** Rootstock influences on 'Bartlett' and 'd'Anjou' pear fruit quality at harvest and after storage, *HortScience,* 16, 650, 1981.
14. **Luton, M. T. and Holland, D. A.,** The effects of preharvest factors on the quality of stored Conference pears, I. Effects of orchard factors, *J. Hortic. Sci.,* 61, 1, 23, 1986.
15. **Raese, J. T., Brun, C. A., and Seeley, E. J.,** Effect of irrigation regimes and supplemental nitrogen on alfalfa greening, cork spot, and fruit quality of d'Anjou pears, *HortScience,* 17, 666, 1982.
16. **Mellenthin, W. M. and Wang, C. Y.,** Preharvest temperatures in relation to postharvest quality of 'd'Anjou' pears, *J. Am. Soc. Hortic. Sci.* 101, 302, 1976.
17. **Leonard, S., Luh, B. S., Hinreiner, E., and Simone, M.,** Maturity of Bartlett pears for canning, *Food Tech.,* 8, 478, 1954.

18. **Chen, P. M. and Mellenthin, W. M.,** Maturity, chilling requirement, and dessert quality of 'd'Anjou' and 'Bosc' pears, *Acta. Hort.*, 124, 203, 1982.
19. **Boonyakiat, D., Chen, P. M., Spotts, R. A., and Richardson, D. G.,** Effect of harvest maturity on decay and postahrvest life of 'd'Anjou' pear, *Scientia Hortic.*, 31, 131, 1987.
20. **Yamaki, S.,** Biochemical changes of cell wall elements with fruit development and ripening and the harvesting period in Japanese pear fruit (*Pyrus serotina*, Rehder var. *culta* Rehder) (in English), *Jpn. Agric. Res. Q.*, 17, 34, 1983.
21. **Mitchell, F. G.,** Preparation for fresh market, I. Fruits, in *Postharvest Technology of Horticultural Crops*, Kader, A. A., Kasmire, R. F., Mitchell, F. G., Redi, M. S., Sommer, N. F., and Thompson, J. F., Eds., Cooperative Extension Division of Agric. and Natural Resources, University of California, Berkeley, CA, *Special Publ. 3311*, chap. 5.
22. **O'Brien, M., Cargill, B. F., and Fridley, R. B.,** *Principles and Practices for Harvesting and Handling of Fruits and Nuts*, AVI Publishing, Westport, CT, 1983.
23. **Kader, A. A.,** Influence of harvesting methods on quality of deciduous tree fruits, *HortScience*, 18, 4, 409, 1983.
24. **Anon.,** U.S. Standards for Summer and Fall Pears, Fed. Register, 20, 5620, 1955.
25. **Anon.,** U.S. Standards for Winter Pears, Fed. Register, 20, 5752, 1955.
26. **Porritt, S. W.,** Effect of temperature on postharvest physiology and storage life of pears, *Can. J. Plant Sci.*, 44, 568, 1964.
27. **Mitchell, F. G., Guillou, R., and Parsons, R. A.,** Commercial Cooling of Fruits and Vegetables, University of Calif. Agric. Publ. Manual 43, 1972.
28. **Richardson, D. G.,** CA recommendations for pears (including Asian pears), in *Controlled Atmospheres for Storage and Transport of Perishable Agricultural Commodities, Proc. of the 4th Natl. CA Conf.*, Blankenship, S. M., Ed., North Carolina State University, Raleigh, NC, 422, 1985.
29. **Chen, P. M., Yoshida, T., and Borgic, D. M.,** Effect of CO_2 concentration on ethylene production, organic acid retention, and internal disorders of pear fruit in low O_2 storage, in *Controlled Atmospheres for Storage and Transport of Perishable Agricultural Commodities, Proc. of the 4th Natl. CA Conf.*, Blankenship, S. M., Ed., North Carolina State University, Raleigh, NC, 135, 1982.
30. **Claypool, L. L.,** Further studies on controlled atmosphere storage of 'Bartlett' pears, *J. Am. Soc. Hortic. Sci.*, 98, 289, 1973.
31. **Gerhardt, F. and Ezell, B. D.,** Effect of carbon dioxide storage on Bartlett pears, *J. Agr. Res.*, 56, 121, 1938.
32. **Hansen, E.,** Reaction of Anjou pears to carbon dioxide and oxygen content of the storage atmosphere, *Proc. Am. Soc. Hortic. Sci.*, 69, 110, 1957.
33. **Frenkel, C. and Patterson, M. E.,** Metabolic effects of CO_2 in 'Bartlett' pears, in *Controlled Atmospheres for the Storage and Transport of Perishable Agricultural Commodities, Proc. of the 2nd Controlled Atmos. Res. Conf.*, Dewey, D. H., Ed., Michigan State University, East Lansing, MI, 108, 1977.
34. **Wang, C. Y. and Mellenthin, W. M.,** Effect of short-term high CO_2 treatment on storage of 'd'Anjou' pear, *J. Am. Soc. Hortic. Sci.*, 100, 492, 1975.
35. **Chen, P. M., Spotts, R. A. and Mellenthin, W. M.,** Stem-end decay and quality of low oxygen stored 'd'Anjou' pears, *J. Am. Soc. Hortic. Sci.*, 106, 695, 1981.
36. **Wang, C. Y. and Mellenthin, W. M.,** Effect of short-term high CO_2 treatments on storage of d'Anjou pears, *J. Am. Soc. Hortic. Sci.*, 100, 492, 1975.
37. **Mellenthin, W. M., Chen, P. M., and Kelly, S. B.,** Low oxygen effects on dessert quality, scald prevention, and nitrogen metabolism of 'd'Anjou' pear fruit during long-term storage, *J. Am. Soc. Hortic. Sci.*, 105, 522, 1980.
38. **Li, P. H. and Hansen, E.,** Effects of modified atmosphere storage on organic acid and protein metabolism of pears, *Proc. Am. Soc. Hortic. Sci.*, 85, 100, 1964.
39. **Chen, P. M. and Mellenthin, W. M.,** Storage behavior of 'd'Anjou' pears in low oxygen and air, in *Controlled Atmospheres for Storage and Transport of Perishable Agricultural Commodities, Proc. of the 3rd Natl. Controlled Atmos. Res. Conf.*, Richardson, D. G. and Meheriuk, M., Eds., Oregon State University Symp. Series No. 1, 139, 1982.
40. **Chen, P. M., Mellenthin, W. M., and Kelly, S. B.,** Fruit quality of 'Bosc' pears (*Pyrus communis*, L.) stored in air or one percent oxygen as influenced by maturity, *Scientia Hort.*, 21, 1, 45, 1983.
41. **Blankenship, S. M. and Richardson, D. G.,** Changes in phenolic acids and internal ethylene during long term storage of 'd'Anjou' and 'Bosc' pears, *J. Am. Soc. Hortic. Sci.*, 110, 336, 1985.
42. **Blankenship, S. M. and Richardson, D. G.,** ACC and ethylene levels in 'd'Anjou' pears in air and controlled atmosphere storage, *HortScience*, 21, 4, 1020, 1986.
43. **Richardson, D. G. and Lombard, P. B.,** Cork spot of 'd'Anjou' pear: Control by calcium sprays, *Commun. Soil Sci. and Plant Anal.*, 10, 383, 1978.
44. **Richardson, D. G. and Al-Ani, A. M.,** Cork spot of 'd'Anjou' pear fruit relative to critical calcium concentration and other minerals, *Acta Hortic.*, 124, 113, 1982.

45. **Richardson, D. G. and Al-Ani, A. M.,** Calcium and nitrogen effects on 'd'Anjou' pear fruit respiration and ethylene evolution, *Acta Hortic.*, 124, 195, 1982.
46. **Al-Ani, A. M. and Richardson, D. G.,** Seasonal uptake of calcium and nitrogen by pear fruit in relation to the seasonal development of cork spot, *Iraqi J. Agr. Sci.*, 4, 51, 1986.
47. **Al-Ani, A. M. and Richardson, D. G.,** Investigations of radioactive calcium uptake and translocation after branch injection and painting on the surface of fruits of pear trees under orchard conditions, *Arab Gulf J. Sci. Res.*, 4, 693, 1986.
48. **Al-Ani, A. M. and Richardson, D. G.,** Effect of calcium and nitrogen on respiration and ethylene evolution of Anjou pear fruit (Arabic, English summary), *Iraqi J. Agr. Sci.* "Zanco", 3, 4, 81, 1985.
49. **Vaz, R. L. and Richardson, D. G.,** effect of calcium on respiration rate, ethylene production, and occurrence of cork spot in 'd'Anjou' pears, *Acta Mortic.*, 157, 227, 1985.
50. **Vaz, R. L. and Richardson, D. G.,** Calcium effects on the postharvest quality of 'Anjou' pears, *Proc. Wash. St. Hortic. Assoc.*, 74, 151, 1983.
51. **Raese, J. T., Pierson, C., and Richardson, D. G.,** Alfalfa greening of 'Anjou' pear fruit, *HortScience* 14, 232, 1979.
52. **Boonyakiat, D., Spotts, R. A., and Richardson, D. G.,** Effects of chlorogenic acid and arbutin on growth and spore germination of decay fungi, *HortScience*, 21, 309, 1986.
53. **Sugar, D.,** Enhanced resistance to postharvest decay in Bosc pears treated with calcium chloride, *Acta Hortic.*, 256, 201, 1989.
54. **Gerasopoulos, D.,** Differential storage temperature and duration effects on ethylene synthesis and firmness of 'Anjou' pear, Ph.D. thesis, Oregon State University, Corvallis, OR, 1988.
55. **Little, C. R., Taylor, H. J., and Peggie, I. D.,** Multiformulation dips for controlling storage disorders of apples and pears, I. Assessing fungicides, *Scientia Hort.* 13, 213, 1980.
56. **Meheriuk, M. and Lau, O. L.,** Effect of two polymeric coatings on fruit quality in 'd'Anjou' and 'Bartlett' pears, in *Controlled Atmospheres for Storage and Transport of Perishable Agricultural Commodities*, S. M. Blankenship, Ed., *Proc of the 4th Natl CA Conf.*, North Carolina State University, Raleigh, NC, 1982, 270.
57. **Chen, P. M., Mellenthin, W. M., and Borgic, D. M.,** Changes in ripening behavior of 'd'Anjou' pears (*Pyrus communis*, L.) after cold storage, *Scientia Hort.*, 21, 2, 137, 1983.
58. **Claypool, L. L., Leonard, S., Luh, B. S., and Simone, M.,** Influence of ripening temperature, ripeness level, and growing area on quality of canned Bartlett pears, *Food Tech.*, 12, 375, 1958.
59. **Knee, M.,** Development of ethylene biosynthesis in pear fruits at − 1°C., *J. Expt. Bot.*, 38, 1724, 1987.
60. **Hansen, E.,** Postharvest physiology of fruits, *Ann. Rev. Plant Physiol.*, 17, 459, 1966.
61. **Blankenship, S. M. and Richardson, D. G.,** Development of ethylene biosynthesis and ethylene-induced ripening in 'd'Anjou' pears during cold requirement for ripening, *J. Am. Soc. Hortic. Sci.*, 110, 520, 1985.
62. **Hansen, E. and Blanpied, G. D.,** Ethylene induced ripening of pears in relation to maturity and length of treatment, *Proc. Am. Soc. Hortic. Sci.*, 43, 807, 1968.
63. **Wang, C. Y. and Hansen, E.,** Differential response to ethylene in respiration and ripening of immature 'Anjou' pears, *J. Am. Soc. Hortic. Sci.*, 95, 314, 1970.
64. **Jermyn, M. A. and Isherwood, C. A.,** Changes in the cell walls of pear during ripening, *Biochem. J.*, 64, 123, 1956.
65. **Bourne, M. C.,** Texture profile of ripening pears, *J. Food Sci.*, 33, 223, 1968.
66. **Ahmed, A. and Labavitch, J. M.,** Cell wall metabolism in ripening fruits, I. Cell wall changes in ripening 'Bartlett' pears, *Plant Physiol.*, 65, 1009, 1980.
67. **Ahmed, A. E. and Labavitch, J. M.,** Cell wall metabolism in ripening fruit, II. Changes in carbohydrate-degrading enzymes in ripening 'Bartlett' pears, *Plant Physiol.*, 65, 1014, 1980.
68. **Ben-Arie, R., Sonego, L., and Frenkel, C.,** Changes in pectic substances in ripening pears, *J. Am. Soc. Hortic. Sci.*, 104, 500, 1979.
69. **Abu-Goukh, A. A. and Labavitch, J. M.,** The *in vivo* role of 'Bartlett' pear fruit polygalacturonase inhibitors, *Physiol. Plant Pathol.*, 23, 123, 1983.
70. **Frenkel, C., Klein, I., and Dilley, D. R.,** Protein synthesis in relation to ripening of pome fruits, *Plant Physiol.*, 43, 1146, 1968.
71. **Klein, I. and Dilley, D. R.,** Malic enzyme isozymes in pome fruits, *FEBS Lett.*, 29, 305, 1973.
72. **Toumadje, A. and Richardson, D. G.,** Endogenous polyamine concentrations during development, storage, and ripening of pear fruits, *Phytochemistry*, 27, 335, 1988.
73. **Romani, R., Labavitch, J. M., Yamashita, T., Hess, B., and Rae, H.,** Preharvest AVG treatment of 'Bartlett' pear fruits: Effects on ripening, color change, and volatiles, *J. Am. Soc. Hortic. Sci.*, 108, 1046, 1983.
74. **Creveling, R. K. and Jennings, W. G.,** Volatile components of Bartlett pear high boiling esters, *J. Agr. Food Chem.*, 18, 19, 1970.
75. **Russell, L. F., Gray, J. I., and Quamme, H. A.,** Qualitative aspects of pear flavor, *Am. Chem. Soc.*, 17, (Abstr.) 6, 1978.

76. **Quamme, H. A.,** Decadienoate ester concentrations in pear cultivars and seedlings with Bartlett-like aroma, *HortScience*, 19, 822, 1984.
77. **Ranadive, A. S. and Haard, N. F.,** Changes in polyphenolics on ripening of selected pear varieties, *J. Sci. Food Agr.*, 22, 2, 86, 1971.
78. Codex Alimentarius Commission, Codex Standards for Fruit Juices, Concentrated Fruit Juices, and Fruit Nectars, Food and Agriculture Organization of the United Nations, World Health Organization, Rome., Vol. X, p. 83, 1982.
79. **Anon.,** Canned fruit juices, Code of Federal Regulations, Office of the Federal Register, National Archives and Records Admin., Part 146, 335, 1985.
80. **Anon.,** The Almanac of the Canning, Freezing, and Preserving Industries, Edward E. Judge and Sons, Inc., Westminster, MD, 70th Ed., 407, 1985.
81. **Luh, B. S., Leonard, S. J., and Patel, D. S.,** Pink discoloration in canned Bartlett pears, *Food Technol.*, 14, 53, 1960.
82. **Chandler, B. V. and Clegg, K. M.,** Pink discoloration in canned pears, I. Role of tin in pigment formation, *J. Sci. Food. Agric.*, 21, 315, 1970.
83. **Chandler, B. V. and Clegg, K. M.,** Pink discoloration in canned pears, II. Measurement of potential and developed colour in pear samples, *J. Sci. Food Agric.*, 21, 319, 1970.
84. **Chandler, B. V. and Clegg, K. M.,** Pink discoloration in canned pears, III. Inhibition by chemical additives, *J. Sci. Food. Agric.*, 21, 323, 1970.
85. **Czerkaskyj, A.,** Pink discoloration in canned Williams' Bon Chretien pears, *J. Food Sci.*, 35, 608, 1970.
86. **Koeppen, B. H.,** Pigments in Bon Chretien pears, *Food Ind. of South Africa*, 24, 17, 1971.
87. **Luh, B. S., Tate, J. N. and Villarreal, F.,** Polyphenolase activity and browning in Bartlett pears, *Fruchtsaft-Ind.*, 8, 274, 1963.
88. **Sioud, F. B. and Luh, B. S.,** Polyphenolic compounds in pear puree, *Food Technol.*, 20, 534, 1966.
89. **Tate, J. N., Luh, B. S. and York, G. K.,** Polyphenoloxidase in Bartlett pears, *J. Food Sci.*, 29, 829, 1964.
90. **Halim, D. H. and Montgomery, M. W.,** Polyphenol oxidase of d'Anjou pears *(Pyrus communis* L.), *J. Food Sci.*, 43, 603, 1978.
91. **Wissemann, K. W.,** Purification and properties of d'Anjou pear *(Pyrus communis, L.)* polyphenol oxidase, Ph.D. thesis, Oregon State University, Corvallis, OR, 1983.
92. **Montgomery, M. W. and Petropakis, H. J.,** Inactivation of Bartlett pear polyphenol oxidase with heat in the presence of ascorbic acid, *J. Food Sci.*, 45, 1090, 1980.
93. **Petropakis, H. J. and Montgomery, M. W.,** Improvement of colour stability of pear juice concentrate by heat treatment, *J. Food Technol.*, 19, 91, 1984.
94. **Skorikova, Y. G. and Lyashenko, E. P.,** The effect of thermal processing on polyphenolic substances in apple and pear juice, Izvestiya Vysshikh Uchebnykh Zavedenii, Pishchevaya Tekhnologiya, 3, 80, 1972, *Food Sci. Technol. Abstr.*, 5, 07-J1021, 1973.
95. **Montgomery, M. W.,** Cysteine as an inhibitor of browning in pear juice concentrate, *J. Food Sci.*, 48, 951, 1983.
96. **Cornwell, C. J. and Wrolstad, R. E.,** Causes of browning in pear juice concentrate during storage, *J. Food Sci.*, 46, 515, 1981.
97. **Matsuoka, T., Tamura, T., Satake, H., Sawada, Y., and Branch, S.,** Amino nitrogen and free amino acids in Bartlett pear juice, J. *Jpn. Soc. Food Sci. Tech.*, 26, 487, 1979.
98. **Akhavan, I. and Wrolstad, R. E.,** Variation of sugars and acids during ripening of pears and in the production and storage of pear concentrate, *J. Food Sci.* 45, 499, 1980.
99. **Beveridge, T., and Harrison, J. E.,** Nonenzymatic browning in pear juice concentrate at elevated temperatures, *J. Food Sci.*, 49, 1335, 1984.
100. **Beveridge, T., and Harrison, J. E.,** Storage browning of pear juice concentrate, *Can. Inst. Food Sci. Technol.*, 20, 170, 1987.
101. **Eskin, N. A. M., Henderson, H. M., and Townsend, R. J.,** *Biochemistry of Foods*, Academic Press, New York, 97, 1971.
102. **Chang, H.,** Chemical analysis of acid precipitated colloid of ripe Bartlett pear juice, M. S. thesis, Oregon State University, Corvallis, OR, 1979.
103. **Babsky, N. E., and Schobinger, U.,** An arabinan in pear juice concentrate, *J. Food Sci.*, submitted.
104. **Claypool, L. L., Leonard, S., Luh, B. S., and Simone, M.,** Influence of ripening temperature, ripeness level, and growing area on quality of canned Bartlett pears, *Food Technol.*, 12, 375, 1958.
105. **Gee, M., Reeve. R. M., McCready, R. M.,** Reaction of hydroxylamine with pectinic acids, Chemical studies and histochemical estimation of the degree of esterification of pectic substances in fruit, *J. Agric. Food. Chem.*, 7, 34, 1959.
106. **Ranadive, R. S. and Haard, M. F.,** Peroxidase localizaton and lignin formation in developing pear fruit, *J. Food Sci.*, 37, 381, 1972.
107. **Casimir, D. J.,** Production of pear puree and concentrated pear puree, *CSIRO Food Res. Q.*, 33, 64, 1973.

108. **Yamane, A.,** Studies on Japanese pear (Nijisseiki) juice, I. Separation of stone cells from pear puree and factors involved in the separation, *J. Agric. Chem. Soc. Jpn.*, 43, 434, 1969.
109. **Yamane, A.,** Studies on Japanese pear (Nijisseiki) juice, II. Quality and separation of stone cells of pear, *J. Agric. Chem. Soc. Jpn.*, 43, 441, 1969.
110. **O'Brien, M. and Singh, R. P.,** Effect of drying on dried fruit quality, *Proc. 3rd Intl. Drying Symp. Food Sci. Technol. Abstr.* 10-j0034, 17, 2, 1985.
111. **Anon.,** Rictwerte und Schwankungsbreiten bestimmter Kennzahlen für Birnensaft, *Confructa*, 28, 74, 1984.
112. **Wrolstad, R. E. and Shallenberger, R. S.,** Free sugars and sorbitol in fruits-a compilation from the literature, *J. Assoc. Off. Anal. Chem.*, 64, 91, 1981.
113. **Weiss, J. and Saemann, H.,** Ergebnisse von Untersuchungen über die D-Sorbit-Gehalte von Fruchtsaften, *Mitt. Klosterneuburg*, 29, 81, 1979.
114. **Fernandez-Flores, E., Kline, D. A., and Johnson, A. R.,** *J. Assoc. Off. Anal. Chem.*, 53, 17, 1970.
115. **Lea, A. H. and Timberlake, C. F.,** The phenolics of ciders, 1. The procyanidins, *J. Sci. Food. Agric.*, 25, 1537, 1974.
116. **Jennings, W. G.,** Volatile esters of Bartlett pears, *J. Food Sci.*, 26, 564, 1961.
117. **Jennings, W. G.,** New fruit esters and the flavor of Bartlett pears, *Int. Fruchtsaft-Union, Ber. Wiss. Tech Komm.*, 6, 277, 1965.
118. **Jennings, W. G.,** Peaches and pears, in *Symposium on Foods: The Chemistry and Physiology of Flavors*, Schultz, H. W., Day, E. A., and Libbey, L. Eds.,
119. **Jennings, W. G. and Creveling, R. K.,** Volatile esters of Bartlett pears, II., *J. Food Sci.*, 28, 91, 1963.
120. **Jennings, W. G., Leonard, S. and Pangborn, R. M.,** Volatiles contributing to the flavor of Bartlett pears, *Food Technol.*, 14, 587, 1960.
121. **Jennings, W. G. and Sevenants, M. R.,** Volatile esters of Bartlett pear, III., *J. Food Sci.*, 29, 158, 1964.
122. **Jennings, W. G. and Tressl, R.,** Production of volatile compounds in the ripening Bartlett pear, *Chemie Mikrobiologie Technologie der Lebensmittel*, 3, 52, 1974.
123. **Heinz, D. E., Pangborn, R. M. and Jennings, W. G.,** Pear aroma: Relation of instrumental and sensory techniques, *J. Food Sci.*, 29, 756, 1964.
124. **Russell, L. F., Quamme, H. A., and Gray, J. I.,** Qualitative aspects of pear flavor, *J. Food Sci.*, 46, 1152, 1981.
125. **Strandzhev, A.,** Effect of blanching on the content of aroma compounds in pears, *Bulgarski Plodove Zelenchutsi I Konservi*, 7, 21, 1977.
126. **Gebhardt, S. E., Cutrufelli, R., and Matthews, R. H.,** Composition of foods, fruits and fruit juices, raw, processed, prepared, USDA Human Nutrition Information Service, Agriculture Handbook, 8, 211, 1982.
127. **Holloway, W. D. and Greig, R. I.,** Water holding capacity of hemicelluloses from fruits, vegetables, and wheat bran, *J. Food Sci.*, 49, 1632, 1984.
128. **Luh, B. S. and Kean, C. E.,** Canning of fruits, in *Commercial Fruit Processing*, Woodroof, J. G. and Luh, B.S., Eds., AVI Publishing, Westport, CT, 1975.
129. **Cruess, W. V., Ed.,** *Commercial Fruit and Vegetable Products*, 4th ed., McGraw-Hill, New York, 1958, 178.
130. **Graham, R. P., Hart, M. R., Ralls, J. W., Mercer, W. A., and Maagdenberg, H. J.,** Minimizes fruit peel pollution, *Food Eng.*, 43, 52, 1971.
131. **Heil, J. R., Carroad, P. A., Merson, R. L., and Leonard, S.,** Development of high vacuum flame processes for sliced peaches and pears, *J. Food Sci.*, 48, 1106, 1983.
132. **Beauchat, L. R. and Heaton, E. K.,** Sensory qualities of canned peaches and pears as effected by thermal process, sorbate, and benzoate, *J. Food Prot.*, 45, 942, 1982.
133. **Schobinger, U. and Mueller, W.,** Evaluation of apple and pear cultivars for beverage manufacture from the viewpoints of harvesting and processing characteristics, *Flussiges Obst*, 42, 414, 1975.
134. **Spayd, S. E., Nagel, C. W., Hayrynen, L. D., and Drake, S. R.,** Color stability of apple and pear juices blended with fruit juices containing anthocyanins, *J. Food Sci.*, 49, 411, 1984.
135. **Lauren, J.,** Manufacture of apple juice by DDS diffusion, *Flussiges Obst*, 41, 284, 1974.
136. **Schobinger, U., Durerr, P., and Akesson, A.,** Technological and analytical data on the enzymatic liquefaction of apples and pears, *Alimenta*, 20, 37, 1981.
137. **Kirk, D. E., Montgomery, M. W., and Kortekaas, M. G.,** Clarification of pear juice by hollow fiber ultrafiltration, *J. Food Sci.*, 48, 1663, 1983.
138. **Kaupert, N. L., Lescano, H. G., and Kotliar, N.,** Conservation of apple and pear juice concentrates, Synergic effect of heat and radiation, in *Combination Processes in Food Irradiation*, Comision Nacional de Energia Atomica, Buenos Aires, Argentina, 205, 1981.
139. **Beech, F. W. and Carr, J. G.,** Cider and perry, in *Alcoholic Beverages*, Rose, A. H., Ed., Academic Press, NY, 1977, chap. 3.
140. **Amerine, M. A., Berg, H. W., and Cruess, W. V.,** *The Technology of Wine Making*, 3rd Ed., AVI Publishing, Westport, CT, 540, 1972.

141. **Tanner, H.,** Application of mash heating in the manufacture of red wine and spirits, *Mitteilungen aus dem Gebiete der Lebensmitteluntersuchung und Hygiene,* 61, 324, 1970.
142. **Tanner, H.,** Production of fruit brandies with low methanol content, *Schweizerische Zeitschrift fuer Obst- und Weinbau,* 106, 625, 1970.
143. **Luh, B. S.,** Nectars, pulpy juices, and fruit juice blends, in *Fruit and Vegetable Juice Processing Technology,* Nelson, P. E. and Tressler, D. K., Eds., AVI Publishing, Westport, CT, 3rd Ed.,436.
144. **Leonard, S., Pangborn, R. M., and Tombropoulos, D.,** Influence of processing variables on flavor stability of pear puree, *J. Food Sci.,* 41, 840, 1976.
145. **Luh, B. S. and Sioud, F. B.,** Aseptic canning of foods, IV. Stability of pear puree with essence recovery, *Food Technol.,* 20, 1590, 1966.
146. **Somogyi, L. P. and Luh, B. S.,** Dehydration of fruits, in *Commercial Fruit Processing,* Woodroof, J. G. and Luh, B. S., Eds., AVI Publishing, Westport, CT, 374, 1975, chap. 8.
147. **McBean, D. M.,** Improved dried pears, *Food Preservation Q.,* 29, 49, 1969.
148. **Anon.,** Quality dried fruit without sulphur—Ascorbic acid/malic acid dip maintains color, flavor, *Food Process.,* 33, 25, 1972.
149. **Anon.,** Delicate drying process minimizes fruit and vegetable aroma, color changes, *Food Prod. Dev.,* 10, 37, 1976.
150. **Grishin, M. A.,** Drying fruit in a turbulent current after osmotic dehydration, Izvestiya vysshikh uchebnykh zavedenii, *Pishchevaya Tekhnologiya* 4, 146, 1970, Food Sci. and Tech. Abstracts 2, 06-J0681, 1971.
151. **Aminov, M. S., Dibirov, A. M. and Azizov, G. G.,** The effect of preliminary fruit treatment method on the quality of compotes, Konservnaya I Ovoschchesushil 'Naya Promyshlennost', 8, 20, 1976, *Food Sci. and Tech. Abstracts,* 9, 08-J1209, 1977.

Chapter 6

THE PEACH, THE NECTARINE, AND THE PLUM

G. O. Couvillon and G. Krewer

TABLE OF CONTENTS

I. INTRODUCTION

A. PEACHES AND NECTARINES

The early literature credits Persia as the native area of the peach, but Decandolle[1] found reference to the peach in Chinese literature of 2000 B.C., Li[2] reported mention of the peach in Chinese literature of 3000 B.C., while Wang[3] wrote that the peach has a written Chinese history of 4000 years and today one finds several wild *Prunus* species growing there. Also, there is no mention of the peach in the early Hebrew literature or in Sanskrit which would indicate that it was unknown from Persia to Western India about 1500 B.C.[1] The confusion in the literature as to the origin of the peach probably resulted from the first mention of the peach in western literature by Theophrastus (332 B.C.) [1,4] who described this fruit, as it was growing at that time in Greece where it was brought from Persia. The fruit probably reached the West as a result of Alexander the Great who brought this fruit to Greece following one of his first expeditions to the east. Li[2] suggested that the peach spread westward into Persia during the Zhou Dynasty (1100 to 221 B.C.) and was rapidly disseminated, probably as a result of the numerous trade expeditions between China and the west during the Western Han Dynasty (206 B.C. to 24 A.D.). From Greece, where the peach was not a common fruit,[1] it was disseminated to Europe and North Africa where its improvement was initiated. According to Magness,[5] Virgil (71 to 91 B.C.) was the first to give an account of the peach in Rome. However, Knight[6] believed that the peach was not known in Rome until the reign of Emperor Claudius (1st Century) and was first mentioned by the noted agricultural scientist Columelia. Pliny[7] (23 to 79 A.D.) wrote that the peach reached Italy from Persia through Egypt and Rhodes and gave the first accurate description of this fruit in Roman literature. He also listed 6 types (cultivars) of peaches growing in Rome, indicating that some germplasm selection had occurred in the 80 years between the time of Virgil's report and that of Pliny. The peach was disseminated to other parts of Europe by the Romans during their various conquest. From Europe the peach reached North America probably due to its introduction into Florida and Mexico by the Spanish.[5] By 1571, 3 types of peaches were being grown in Mexico. The English introduced this fruit into the northeast U.S. and the French into Louisiana.[8] Indians disseminated the peach inland and throughout the U.S. which resulted in the development of a distinctive type referred to as the "Indian Race".[5]

Prunus germplasm is varied and consists of numerous species and types within a species. In China, *P. daviaina* (Carr) Franch is the most common "wild" species and is found in abundance in Hopei, Honan, Shansi, Shensi, Kansu, and Szechuan Provinces. In more limited areas of China one finds *P. kansuensis*, *P. mira* (Koehne) Kov. et Kost., and *P. ferganensis*. These are thought to be ancestors of today's peach, *P. persica* (L.) Bastch. There are trees of *P. mira* that are over 100 years of age in China which are growing vigorously and producing fruit.[3] Although the endocarp of this species is almost always smooth, there are individuals that produce grooved endocarps, fruit approximately 90 gms in weight with white, yellow, or whitish-green flesh. The flesh may be melting or non-melting and of good flavor.[3] These characteristics have led some to suggest that *P. mira* is the ancestor of *P. persica* Batsch (Wang).[3] Today there is tremendous variation in the cultivated cultivars of *P. persica* in China which has led Chinese pomologists to classify peach cultivars into 6 groups. The *North China Group* consists of hardy trees of high drought tolerance that produce fruit with deep sutures, long beaks, and non-melting flesh of good quality. In eastern, southern, and western China is found the *South China Group* often referred to as the "Water Honey Peach". Trees of this group are adapted to subtropical climates (low chilling requirement) and produce fruit with a round apex and melting flesh of good quality. The *Yellow Peach Group* has fruit with prominent beaks, free stones, and yellow flesh (yellow fleshed peaches are thought to have originated in Europe, but have been described in Chinese literature for centuries). The *Flat Peach or Pan-Tao* group consists of approximately 20 cultivars that are adapted to subtropical climates. Fruit of this group are flat in shape, with melting flesh, and very high quality, but produce very low yields. The *Tung-Tao or Winter Peach Group* produces very late ripening fruit with certain cultivars that mature in October, November, or December. The fruit are small, of satisfactory quality with a long storage life. Nectarines are grouped into a 7th and separate group, *The Nectarine Group*[3]

On the other hand, Li[2] classifies the Chinese peaches into *Northern*, *Southern*, and *European or Persian* ecotypes. He divides the Northern Group into the *Miantao* and *Mitao* subgroups while the Southern Group is subdivided into the *Yingtao*, *Shumitao*, and *Pentao* subgroups. The European or Persian group is composed of those cultivars developed from genotypes introduced from China to the Western World.

The nectarine [*P. persica* var. *Nectarina* (Ait) Maxim] is a single gene mutant (the nectarine lacks the gene for fruit pubescence) of the peach so it can be assumed with some degree of certainty that nectarines have been around as long as the peach. It has been suggested by Knight [6] that the fruit Pliny described as duracinus was undoubtedly the nectarine. According to Hedrick, Peter Collinson in 1784 was the first to record the case of a peach tree bearing a nectarine. Thus, one can assume that the nectarine was introduced from China and distributed throughout the world in the same manner as the peach or that nectarines appeared as mutants on peach trees wherever they were introduced and grown.

There are two distinct horticultural forms of the peaches (nectarine), distinguished by the relationship of the fruit mesocarp to the endocarp. The ''clingstones'' or pavies (*persica vulgaris*, Risso) in which the mesocarp adheres to the endocarp and the ''freestones'' (*persica domestica*, Risso) which have a separated mesocarp.[8] The freestone cultivars have a melting flesh texture (thin cell walls that rupture and allow juice to flow into the intercellular spaces) which gives a pleasant flavor and good eating qualities while most clingstones have non-melting, firm flesh.

Cultivar development of the peach and nectarine has been extensive in recent years and today there exists over 3000 peach and several hundred nectarine cultivars. Increased consumption of nectarines in recent years has resulted in increased interest among breeders to produce new cultivars of this fruit. Cultivar development has occurred through either the selection of chance seedlings, gene mutations, or hybridization. Since the production of cultivars through hybridization is a relatively recent technique, early cultivars were the result of chance seedling selection or genetic mutations. Pliny[7] mentioned 6 peach cultivars that were grown in Rome which without doubt originated from either chance seedling selection or gene mutation. During the period from 1850 to 1900 a large number of cultivars were selected from chance seedlings by peach growers. Two of the most prominent were 'Elberta' and 'Georgia Belle' which were selected from an orchard of 'Chinese Cling' seedlings planted in 1884 by Mr. Samuel Rumph of Marshallville, Georgia. Peach breeding through controlled crosses began in 1895 at the New York Agricultural Experiment Station at Geneva, NY.[4] Since that time many peach, nectarine, and plum breeding programs have been developed by university experiment stations, the USDA, and private breeders. Modern breeding programs have as their objectives; (1) extension of the maturity season, (2) development of early maturing cultivars, (3) improved processing characteristics, (4) improvement of flavor and quality, (5) modifying tree size and shape for mechanization and high density plantings, (6) disease and insect resistance, and (7) improved shipping characteristics.[9]

B. PLUMS

Plums have the greatest diversity of all stone fruits and rank next to the peach in commercial production. In geographical distribution, both wild and commercial plums circle the globe in the north temperate zone and commercial cultivars belong to 15 major species as well as interspecific crosses. Although plums were cultivated in the old world for over 2000 years, they did not attain horticultural importance in the U.S. until after the American Revolution and did not reach commercial importance until the settlement of the Pacific states.[4,5] Of the most important commercial species, *P. domestica* and *P. domestica* var. *insititia*, which are members of the Eur-Asian group (often referred to as the European Plum) were the first plums imported and grown in the U.S., and introduced from Europe as seeds, trees, or scions. This group originated in Western Asia.[4,8] Certain cultivars of *domestica*, those with a high solids (sugar) content, can be dried and made into prunes. The species *P. salicina, Lindl.* and *P. Simonii, Carr.*, are members of the Oriental Plum Group which were introduced into the U.S. from Japan in the late 19th century by Luther Burbank and others. Many of the fresh market plum cultivars belong to this group with *P. salicina* the major commercial species. Although the Oriental plums were introduced from Japan, they are thought to be native to China. Many cultivars have been developed from interspecific crosses between *P. salicina* and several native American species.[8] Certain species of plums are important as rootstocks such as Myrobalan (*P. cerasifera* Ehrh.) or in breeding regionally adapted cultivars as *P. angustifolia* Marsh. (Chickasaw plum), *P. munsonian* Wright and Hedr., (wild Goose plum), and *P. simonii* (Apricot plum).[11]

II. AGRONOMIC CHARACTERISTICS

A. PEACHES AND NECTARINES

Peaches and nectarines (*Prunus persica* L. Batsch) are adapted to tropical, subtropical, and temperate zone areas but individual cultivars usually have a limited range of best performance. Chilling requirements of commercial cultivars vary from 50 to 1050 h below 7.5°C and disease resistance is variable. Fruit flesh color may be white, yellow, or red and the skin either pubescent or glabrous. As previously mentioned, seeds may be attached to the flesh (clingstone) or loosely attached (freestone) and the flesh may be soft-melting, firm-melting, or non-melting (Table 1).[9,11]

As a general rule early season fresh market peaches are clingstones and late maturing cultivars are freestone types, both with melting flesh. Simple dominant genes are responsible for the freestone types, both with melting flesh. Simple dominant genes are responsible for the freestone condition, white flesh, and pubescent fruit epidermis.[12] Late season peaches used for processing may be freestone or clingstone, but nonmelting clingstone types are more widely used for canning in California (Table 2).[13] Certain of melting flesh freestone cultivars, such as Fay Elberta, are preferred for freezing[12]

Peaches are vigorous but short lived trees. In some areas of the world such as the southeastern U.S. the trees may live only 8 to 10 years due a condition known as Peach Tree Short Life, thought to be due to nematodes, diseases, soil problems, and winter injury. However, in desert regions with alkaline soils orchards 30 years old are common. The peach is usually a small tree, but some current breeding programs are producing dwarf, shrub-like forms that may become commercial cultivars in the near future. Due to the many active peach breeding programs peach cultivars are often replaced about every 15 to 20 years.

Breeding programs for peaches and nectarines in humid areas actively screen for disease resistance. In the northeast U.S. and Canada peach canker (*Valsa* sp.) disease resistance is an important consideration. In the eastern U.S. resistance to fruit cracking and brown rot (*Monilinia fructicola* [Wint.] Honey) is very important in selection of individuals to be released as new cultivars. Bacterial spot (*Xanthomonas Campestris* pv. *pruni* [Smith] Dye) resistance varies greatly among cultivars. Susceptible cultivars often have heavily damaged leaves and fruit in humid years in the southeastern U.S.[9]

1. Production of Peaches and Nectarines

World production of fresh peaches and nectarines is about 5,400,000 metric tons (t)

TABLE 1
Some Peach Cultivars Grown in the Eastern U.S.[a]

Melting Type Flesh

Yellow-Freestone

Afterglow
All Red Elberta
Autumn
Biscoe
Blake
Candoka
Colora
Cresthaven
Daroga Red
Dixigold
Dixiland
Earlihale
Early Elberta
Early Golden Jubilee
Early Halehaven
Early Triogem
Elberta
Emery
Envoy
Fairhaven
Fay Elberta
Fayette
Fireball
Fireglow
Gage Elberta
Garnet
Glohaven
Goldeneast
Golden Globe
Golden Jubilee
Goldenred
Goodcheer
Halberta Giant
Halegold
Hale Harrison Brilliant
Halehaven
Harbrite
Harken
Harmony
Herb Hale
Jefferson
Jerseyqueen
J. H. Hale
Jubilant
July Elberta
Keystone
Kirkman Gem
LaGold
LaPremiere
LaRed
Lizzie
Loring
Madison
Marhigh
Marqueen
Midway
Monroe
Olinda
Ozark
Poppy
Primrose
Rnager
Redcrest
Redelberta
Redglobe
Red Gold
Redskin
Redtop
Richhaven
Rio Oso Gem
Scott Elberta
Shippers Late Red
So Good
Somerset
Southern Glow
South Haven
Southland
Stark Lategold
Sullivan Elberta
Summercrest
Summergold
Summerqueen
Summerset
Suncrest
Sunhigh
Sunshine
Suwanee
Texaberta
Triogem
Troy
Tyler
Valiant
Vanguard
Vedette
Velvet
Vesper
Veteran
Viceroy
Washington
Wilma

Yellow-Semi-Free to Free

Comanche
Coronet
Dixiegem
Fair Beauty
Globehaven
Golden Beauty
Goldfinch
Goldgem
Kalhaven
LaGem
Newday
Prairie Daybreak
Prairie Sunrise
Redhaven
Regina
Sentinel
Shinns Delicious
Sunbeam
Sunhaven
Veefreeze

Yellow-Semi-Cling

Buttercup
Candor
Colins
Early East
Fisher
Garnet Beauty
Harbelle
Jerseyland
Marigold
Missouri
Norman
Pacemaker
Pekin
Prairie Dawn
Prairie Rose
Simpson Red
Sunbeam
Tulip

Yellow-Cling

Cardinal
Dawne
Dixiered
Earliglo
Earlired
Early Coronet
Early Vedette
Golden Elberta Cling
Goldray
Hiland
Junegold
Magnolia
Marcus
Maygold
Mikado
Redcap
Royalvee

TABLE 1 (continued)
Some Peach Cultivars Grown in the Eastern U.S.[a]

Melting Type Flesh		
Shephard's Beauty	Summerrose	**Nonmelting Flesh**
Springcrest	White Hale	
Springold	Wildrose	**Yellow-Cling**
Sunrise		
Whynot	**White-Semi-Cling**	Ambergem
		Babygold 5
White-Semi-Free to Free	Cumberland	Babygold 6
	Erly-Red-Fre	Babygold 7
Babcock	Pioneer	Babygold 8
Belle	Radiance	Babygold 9
Delicious	Rosebud	Cherryred
Early Hiley		Melogold
Hiley	**White-Cling**	Merrill Gem
Laterose		Mountaingold
Nectar	Duke of Georgia	Piedmontgold
Newcheer	Early Rose	Suncling
Pearson Hiley	Maybelle	
Raritan Rose	Mayflower	**White-Cling**
Redrose	Redbird	
R.L. Stoner	Springtime	July Heath
	World's Earliest	White English

[a] Classification of cultivars according to flesh color, flesh type, and Greeness of pit.

Adapted from Savage, E. F. and Prince, V. E., *University of GA Res. Bull.*, 114, 1972. With permission.

year. Approximate production, in metric tons, by region is: North America — 1,600,000, South America — 370,000, Europe — 3,000,000, Africa — 150,000, Asia — 230,000, and Oceania — 100,000. Canada produces about 37,000 t of fresh peaches, Mexico 163,000 t, and the U.S. 1,427,000 t. The value of the crop in the U.S. averages about $350 million/year[14] produced on 88,962 ha.[21] The largest producing states are California, South Carolina, Georgia, New Jersey, and Pennsylvania.[14] Utilization of peaches for processing in 1984 was canned: 445 , dried: 20, frozen: 32, pureed, etc.: 11 million kg. There are about 13,354 ha of nectarines in the U.S. on which approximately 20 million kg are produced annually, about 95% in California. Approximately 1% of the crop is processed and the rest is sold fresh.[15,16] Peach and nectarine yields vary depending upon many factors but range from 40 to 200 kg of fruit per tree at standard spacings (6 × 6 m.[17] Yields of 20 to 30,000 kg/ ha are possible with mature high and low density orchards but high density orchards are more precocious.[18]

B. PLUMS AND PRUNES

Of the 15 species of plums used for fruit production in the world, the most important fresh market species is the Japanese plum (*Prunus salicina* Lindl) noted for good size, attractiveness, firmness, and good keeping qualities (Table 3). Most cultivars have a chilling requirement of 600 to 800 h below 7.7°C and have flower buds about as hardy to winter cold as peaches. Very few of the important commercial cultivars of the Japanese type are derived strictly from *P. salicina*. Those recommended for culture in the Southeast U.S. are usually complex hybrids between Japanese and native plums (Table 4).

There is considerable variation among Japanese plums in character of tree growth, some having a low, spreading habit, while others grow upright. Leaves are medium-sized and

TABLE 2
Popular Cultivars of Peaches and Nectarines in California

Nonmelting flesh clingstone peaches (over 240 ha)	Melting flesh clingstone and freestone peaches (over 160 ha)	Nectarines (over 160 ha)
Andross	Angelus	Armking
Bowen	Cal Red	Aurelio Grand
Carolyn	Carnival Elberta	Autumn Grand
Carson	Elegant Lady	Early Sun Grand
Corona	Fairtime	Fairland
Halford	Fay Elberta	Fantasia
Klamt	Fayette	Firebrite
Loadel	Flamecrest	Flamekist
Monaco	Flavor Crest	Flavortop
Paloro Peak	Gemfree	Independence
Starn	June Lady	Late Le Grand
Sullivan #4	May Crest	May Glo
Wiser	Springcrest	May Gran
	Springgold	Maybelle
	Suncrest	Red Diamond
		Red Grand
		Red Jim
		Red June
		Royal Giant
		Spring Red
		Summer Grand
		Sun Grand

Adapted from Odell, L. L., Williams, L. S., and Severson, S., *CA Crop and Livestock Reporting Serv.*, Sacramento, CA, 1985. With permission.

sharp-pointed and free of pubescence on the lower surface. The fruit are produced on both spurs and one year old shoots.[17]

The European plum (*Prunus domestica* L.) is noted for good flavor and quality and is used for fresh market and processing (Table 5). Most cultivars have a chilling requirement of 800 to 1200 hours below 7.7°C and flower buds about as hardy to winter cold as apples. European plums are characterized by large, strong vigorous trees with thick leaves. The leaves are glossy, have a dark green upper surface and a pale green lower surface with considerable pubescence. The fruit are generally produced on spurs. [17] Prunes are firm-flesh plums with a high sugar content that can be dried whole without fermenting at the pit. All prune cultivars are European plums (*Prunus domestica*). Non-prune type plums may be dried when halved and the product is known as dried plums. Damson plums (*Prunus insitita* L.) produce small purple fruit which are used for processing into preserves and jellies. Native species are valued locally for their fruit such as *P nigra*. Ait. (Canadian plum) and *P. american* Marsh. (American plum).[10]

Plums are extremely variable in color and fruit size. The pits may be freestone, clingstone, or "air free". *Prunus domestica* fruits have both yellow and green flesh and also red or blue skin colors with an unlimited range in variations. *Prunus domestica* is sometimes divided into sub-groups: (1) Prune Group; these can be dried without removal of the pit (2) Green Gage Group; green-yellow, sweet, juicy plums, important canning cultivars, (3) Yellow Egg Group, small group of large yellow or purple plums desirable for canning only, (4) Imperatrice Group; generally blue fruits used mostly for fresh market, and (5) Lombard

TABLE 3
Leading Commercial Plum Cultivars in California

Cultivar	Matures week/month	Requires pollenizer	Tree shape	Fruit shape	Skin color when ripe	Flesh color
Beauty	4 May	No[a]	Upright	Heart	Greenish-yellow to bright crimson	Amber streaked with scarlet
Burmosa	1 June	Yes[b]	Spreading	Semi-heart to round	Yellow with red blush	Light amber
Casselman	2 Aug.	No.[a]	Upright	Conical	Purplish-crimson	Yellow
Durado	3 May	Yes	Spreading	Flat-oval	Purple	Greenish-amber
El Dorado	4 June	Yes	Very Upright	Flat-oval	Dark red with purple tint	Yellow
Friar	4 July	Yes	Upright	Flat-oval	Black	Amber
Frontier	4 June	Yes	Moderate upright	Heart	Purple-black	Red
Grand Rosa	2 July	No	Upright	Conical	Red to purplish-red	Yellow
July Santa Rosa	4 June	No	Upright	Conical	Bright purplish-red	Yellow
Kelsey	3 July	Yes	Moderate upright-moderate vigor	Heart	Green to yellow	Greenish-yellow
Laroda	2 July	Yes	Upright	Conical	Dark red to purple	Yellow
Late Santa Rosa	2 July	No[a]	Upright	Conical	Resembles Santa Rosa with dark suture line	Yellow with red near skin
Mariposa	4 June	Yes[b]	Upright	Heart	Mottled greenish-yellow with red spots	Red
Nubiana	2 July	Yes	Moderate upright	Flat-oval	Purplish-black	Yellow
President[c]	4 July	Yes	Upright-spreading	Oval	Dark purple	Yellow
Queen Ann	3 July	Yes	Moderate upright-moderate vigor	Heart	Purplish-black	Light yellow
Queen Rosa	3 June	Yes[b]	Moderate upright	Round	Red to purplish-red	Light yellow, reddish near skin
Red Beaut	3 May	Yes[b]	Upright-spreading	Conical	Bright red	Yellow
Redroy	1 July	No[c]	Upright-moderate vigor	Conical	Bright red	Yellow
Roysum	2 Sept.	No	Upright	Conical	Light purple, white ground color	Yellow with red near skin
Santa Rosa	2 June	No[c]	Upright	Conical	Purplish-crimson	Amber with red near skin
Simka	2 July	No	Moderate spreading	Heart	Ebony	Yellow

Standard[c]	1 Aug.	Yes	Upright—moderate vigor	Oval	Purple, heavy blue bloom	Yellow
Tragedy[c]	3 June	Yes	Spreading	Oval	Dark blue to purple	Greenish-yellow
Wickson	4 June	Yes	Moderate upright—moderate vigor	Heart	Greenish-yellow with red tinge, white bloom	Bright yellow

[a] Cross pollination may increase set in some areas.
[b] Early blooming cultivar.
[c] European, Plum cultivar.

Adapted from J. H. LaRue and Gerdts, M., *Commercial Plum Growing in California, University of CA Coop. Exp. Ser. Leaf.* 2458, 1976. With permission.

TABLE 4
Plum Cultivars Recommended for Georgia

Cultivar	Commercial growers	Chill Hours Approx.	Pollen shed	Skin color	Flesh color	Fruit size (cm)	Fruit shape	Fruit firmness	Peak harvest at byron	Comments
6 weeks	Yes	650	Moderate	Red	Yellow/Red	3.8	Round/Ovate	6.4	May 25	
Bruce	Yes	550	Weak	Green/Red Orange	Yellow/Red	4.3	Round/Ovate	6.5	May31	Poor flavor
Robusto	Yes	650	Weak	Bright Red	Red	4.1	Round	7.4	June 5	Vigorous tree
Wade	Maybe	750	Moderate	Deep Red	Yellow Red	4.3	Round/Oblate	7.3	June 7	S.C release
Segundo	Yes	700	Weak	Yellow/Red	Yellow Red	4.8	Round	7.7	June11	
Morris	Maybe	800	Moderate	Purple	Blood Red	4.6	Round/Ovate	7.5	June16	Usually Shortlived
AU Roadside	Yes	700	Moderate	Magenta	Red	4.8	Round/Ovate	6.8	June 19	
AU Producer	Yes	750	Moderate	Dark Red	Red	4.3	Round	7.5	June 20	
Byron Gold	Yes	700	Strong	Yellow	Yellow	5.1	Round	8.5	June 30	New release
Explorer	Yes	650	Moderate	Purple	Amber	5.1	Rond/Oblate	7.3	July 3	Erratic producer in some areas
Crimson	Yes	750	Moderate	Bright Red	Red	4.1	Round	8.3	July 6	

Note: Fruit size in centimeters. Fruit firmness, 6.5 = Borderline for local sales; 7 = Borderline for distant shipping; 8 = very firm.

Information from Dr. Jim Thompson (Retired), Byron USDA Station, Byron, GA., 1985.

TABLE 5
European Plums and Prunes of Commercial Importance

Cultivar	Ripening season	Needs pollinizer	Fruit shape	Fruit size	Free-stone	Skin color[a]	Flesh color[a]	Use[b]
Tragedy	Early	Yes	Oval	L	Yes	B—P	G—Y	F
California Blue	Early	No	Oval	L	Yes	B	Y	FC
Iroquois	Early	No	Oval	M	Yes	B	G—Y	FC
Early Italian	Early	No	Oval	M	Yes	B	G	FC
Lombard	Early	No	Oval	L	Yes	R	Y	F
Queenston	Early	—	Oval	M	Yes	B	Y	F
Parsons	Early	Yes	Oval	M	Yes	R—B	A—Y	FD
Stanley	Mid	No	Obovate	M	Yes	B	Y	FC
Bluefre	Mid	No[c]	Oval	L	Yes	B	Y	F
Damson	Mid	No	Oval	S	Yes	B	Y	FC
Bluebell	Mid	Yes	Oval	L	Yes	B	Y	FC
Grand Duke	Mid	Yes	Oval	L	No	B	Y	F
Agen (French)	Mid	No	Obovate	M	Yes	B	Y	FCD
German	Mid	Yes	Oval	M	Yes	B	Y	FCD
Italian (Fellenberg)	Mid	No	Oval	M	Yes	B	A—Y	FCD
Imperial	Mid	Y	Oval	M	Yes	B	Y	FD
Brooks	Mid	No	Oval	VL	Yes	B	Y	FD
Sugar	Mid	No	Oval	M	Yes	B	Y	FD
Sargeant	Mid	—	Oval	M	Yes	B	Y	FD
Reine Claude	Mid	No	Oval	M	Yes	G—Y	G—Y	FC
President	Late	Yes	Oval	L	Yes	P	Y	F
Pozegaca	Late	—	Oval	S	—	B	—	Br
Standard	Late	Yes	Oval	L	Yes	B—P	G—Y	F
Vision	Late	Yes	Oval	L	Yes	B	Y	F
Victoria	Late	No	Oval	M	Yes	R	Y	F
Moyer Pefecto	Late	No	Oval	L	Yes	B	Y	FD

[a] B = Blue, P = Purple, G = Green, Y = Yellow, R = Red, A = Amber.
[b] F = Fresh, C = Canned, D = Dried, Br = Brandy.
[c] Cross-pollination may improve fruit set in some areas.

Adapted from M. N. Westwood, *Temperate Zone Promology*, Freeman and Co., New York, 1978. With permission.

Group; generally reddish fruits of small size used mostly for fresh market.[19] The major prune cultivars (*P. domestica*) have red to purplish skin with yellow flesh. Exterior colors of other plum species and hybrids may be green, black, purple, red, or yellow. Flesh color may be yellow, amber, or red. Red or black plums with amber flesh are in greatest demand for the fresh market. Fruit shape varies from ovate to oblate and fruit firmness ranges from soft to firm. Sugar, tannin, acid levels, and fruit size are also extremely variable. Self-unfruitfulness is common in plums and pollen sterility occurs occasionally.[20] There is also a high degree of pollen and stigma incompatibility among some cultivars. European plums (*P. domestica*) are mostly self-unfertile but little incompatibility is found between cultivars. 'French' prune is self-fertile.[10]

1. Disease Resistance

Disease resistance is important in the regional adaptability of plums. Breeding programs in the southeastern U.S. have developed cultivars fairly resistant to plum leaf scald disease, bacterial leaf spot, and bacterial canker.[21] This has been achieved by crossing native southeastern plums with higher quality Japanese-type plums.[22] The result has been the release of a series of plums which are relatively long-lived (10 to 15 years). Brown rot resistance is difficult to attain, since parental variation in resistance is slight. Susceptibility to bacterial leaf spot has been investigated and *Prunus domestica* cultivars were found to be the most susceptible to leaf and fruit bacterial infection, but no stem cankers developed. In *P. salicina*, stem cankers did not develop in 'Burbank', 'Elephant Heart', 'Satsuma', and 'Kelsey' while most other cultivars were badly damaged. Resistance to stem cankers in *P. salicina* is associated with the spreading growth habit of this species.[10]

2. Production of Plums and Prunes

Fresh plums and prunes are popular in much of the world. North America produces about 700,000; South American, 90,000; Europe, 2,000,000; Oceania, 24,000; Asia, 35,000; and Africa, 17,000 t. North American production is divided between Canada, which produces about 6,000 t of fresh plums and prunes and the western states of the U.S. of which California produces from 150,000 to 220,000 t of non-prune type plums. Only a few thousand tons of these are frozen and none canned. California also produces 125,000 to 175,000 t of prunes (85% of the total U.S. production) for the fresh market, drying, and some for canning.[23] Other major plum and prune producing states are Washington, Michigan, Oregon, and Idaho. Production in states other than California are mostly European plums. There were 14,684 ha of fresh market type plums in California in 1984. Prunes were grown on 32,697 ha in California in 1984 and 31,088 of these ha were of the 'French' prune cultivar with an average farm value of about $100 million.[14,16] The large, sweet cultivars of white plums such as 'Green Gage' and 'Yellow Egg' as well as purple *P. domestica* cultivars such as 'Italian' are used for canning[13] (Table 5).

Yield of plums and prunes is influenced primarily by cultivar, weather, ripening date, tree training, fruit thinning, nutrition, and pest management. Typically, plums trees will produce 50 to 150 kg of fruit per tree at standard spacing (6 × 6 m) and average prune yields in California have varied from 4.45 to 6.7 t/ ha.[24]

III. QUALITY AND STORAGE OF PEACHES, NECTARINES, AND PLUMS

A. QUALITY CHARACTERISTICS

1. Peaches and Nectarines

Fruit quality, storage characteristics, and shelf life of fresh market peaches and nectarines depend upon cultivar and cultural practices. Cultivars vary in yield potential and fruit quality. As mentioned, peaches and nectarines are classified as "yellow' or "white" fleshed and although "yellow" fleshed fruit dominate the market, "white" fleshed fruit have gained in popularity in Europe and demand a premium price. Generally, "white" fleshed fruit are lower in acid and are more easily bruised, thus do not maintain quality during handling, storage, and shipment. Early maturing cultivars are of low quality (low solids), small in size, clingstone types with "white" or "yellow" melting flesh, and produce low yields. These early maturing fruit are prone to a condition known as "split pit", an opening or cracking of the endocarp (stone or pit) during fruit maturity, which lowers quality and renders the fruit unacceptable for market.[25] This condition is less pronounced, and in many cases absent, in mid and late season maturing cultivars. Also, fresh market cultivars which mature their fruit in mid and late season are freestone types with melting flesh, produce fruit of greater size and higher percent solids. Although mid to late season clingstone cultivars are generally not used for fresh market fruit, there are exceptions. In the Bajo Aragon region of Spain, clingstones known as Calande cultivars, are preferred to freestones as fresh market fruit and demand a premium price. These cultivars produce fruit on spurs on 2 year old wood and the fruit mature from September to November. Approximately 25 d following bloom the individual fruit are enclosed in small paper bags which remain to maturity producing pesticide free, clean, firm non melting flesh, and uniformally colored fruit of high quality. Clingstones are also preferred as fresh market fruit in Brazil and some other parts of the world. Also, fresh market freestone peaches are produced with bag covers in Japan.

In all cases, insect and disease control during the production cycle is mandatory since fruit with the slightest damage or blemish are considered of unacceptable quality for market. Nectarines have greater susceptibility to Brown Rot than the peach, probably since the peach "fuzz" prevents some spore contact with the fruit surface. All stone fruit are very delicate and are easily bruised, thus extreme care must be taken in harvesting, packing, storage, and shipping if quality is to be maintained. Although some clingstone fruit for processing are mechanically harvested, as a general rule peaches and nectarines for fresh market or processing are harvested manually and the fruit placed in containers of various sizes (from 1 to 40 bushels; 1 bushel = 22.68 kg) for transport to the packing area. Fruit must be picked at a stage of maturity that will allow the fruit to ripen to a high culinary quality yet be firm enough (7.3 to 7.7 kg pressure) to withstand bruising and avoid premature softening during storage and marketing. Fruit harvested too early (greater than 7.3 to 7.7 kg pressure) will not ripen properly, thus are of low quality. Ground color chips (color charts) are available which aid the grower in determining proper harvest dates, but most growers rely on experience.[26] Since fruit in Europe and Japan are shipped shorter distances than those produced in the U.S., they are harvested in a more mature stage and small "in field" containers are used to minimize bruising. Care must be exercised during harvest to minimize bruising and to insure that only blemish-free fruit of adequate size and the proper maturity stage are harvested. The fruit are harvested into field picking bags or buckets and carefully transferred to field bins or boxes for transport to the packing facility. Once at the packing shed, the fruit are hydrocooled in a chlorine solution then washed by dumping the field containers into a water tank into which is dissolved a fungicide for postharvest fungi rot protection. The fruit are graded (removal of defective fruit), in most cases "defuzzed" by brushing, sized, waxed (a fungicide is usually added to the wax), and packed into fiberboard or wooden containers then placed into cold storage. The containers vary from 1/2 to 3/4 bushel fiberboard boxes or wire bound crates to small single layer wooden lugs that contain from 16 to 36 fruit depending upon fruit size. The single layer shipping containers are used primarily in Europe and other areas where soft mature fruit are shipped. Fruit that are shipped long distances are usually packed in boxes in which trays, that contain individual fruit cells (tray packs), are added. These eliminate fruit to fruit contact, thus minimize bruising. Care must be taken in handling fruit during the hydrocooling, grading, and sizing stages of the postharvest cycle since surface discoloration can occur. This condition is thought to be brought about due to several factors. Iron (10 ppm or greater) in water used in the postharvest process, leachate from latex-rubber drying rollers, excessive brushing and/or vibration, ammonia, and certain fungicide sprays have been shown to be responsible. This condition is referred to as inking, ink spot, black streak, purple streak, or purple spot. Initially burgundy colored areas appear within the dark red pigmented areas of the fruit epidermis. These areas can turn purplish-black in the latter stages of the disorder. A similar disorder may develop within the yellow-orange colored area of the epidermis and is referred to as grease spot.[27] Following packing, the fruit are then placed in cold storage until shipped. It is important that the temperature of the internal fruit tissue is dropped to 0°C within 24 h of harvest.[17]

Should a grower desire (it is not mandatory that they do so) to pack peaches or nectarines by USDA standards, certain maturity and quality standards must be followed. For peaches and nectarines these standards fall into 4 classification categories (grades): (1) U.S. Fancy; (2) U.S. Extra No. 1; (3) U.S. No 1; and (4) U.S. No. 2. Separation into the Fancy and Extra No. 1 categories is determined by percent of the fruit epidermis showing red or pink color (1/3 and 1/4, respectively), and the percentage of soft and/or overripe fruit (no more than 2%). The two lower grades have no color standards, but retain the soft and overripe fruit limitation. Any misshapen fruit are placed in the No. 2 grade.[27] Although 4 grades exist, the primary market is for only the two highest grades and one seldom finds the lower grades on the retail grocery counter. Printed on the container is the grade designation and the fruit size of the contents. Fruit from early maturing cultivars as small as 1 7/8 in are shipped, but there is generally a 2 inch minimum required for fruit of later maturing cultivars. Current consumer demands are for large (2 1/4 to 2 1/2 in. and larger) fruit, thus small fruit bring low prices and premium prices are paid for the larger fruit categories.

2. Plums and Prunes

Plums are harvested manually while prunes are mechanically harvested. Plums may color 10 d or more before the fruit are mature. Since Oriental plums may vary in skin color from light green, to yellow, to red blushed, or solid red, a color-flesh firmness standard, for each cultivar, is used to determine harvest time.[17] The standard uses skin and flesh color as maturity indexes. Since European plums have a firm flesh, soluble solids (17%) has been found to be the most reliable maturity index for this species. The fruit are placed in baskets, buckets, boxes, or bins and are brought to the packing facility in these containers. A great amount of care should be exercised when harvesting and handling plums for fresh market since the fruit are delicate and highly perishable. It is important to preserve as much of the bloom (natural waxy fruit coating) as possible. Fruit are mechanically sized, graded (according to USDA standards), and usually packed into fiberboard boxes that contain from 112 to 340 fruit per box, depending upon fruit size.[17]

Plums and prunes, as is the case for peaches, carry the U.S. Fancy, U.S. No. 1, and U.S. No. 2 grades, but lack the U.S. Extra No. 1 category. A U.S. Combination grade (combination of U.S. No. 1 and 2 fruit), exists and is considered a superior pack than the U.S. No. 2 grade.[28] The criterion for separating fruit into the various grades are too numerous to discuss here.

B. STORAGE

1. Peaches and Nectarines

Peaches and nectarines are extremely perishable and do not lend themselves to prolonged periods of cold storage. For maximum storage life, fruit should be held at 0°C and in many cases freezing of the mesocarp tissue will occur at −1°C. Should the temperature during storage range from 2 to 10°C or higher, some deterioration of flavor and texture may occur. At 0°C, the storage life of fruit of most cultivars is 2 to 3 weeks and when held for longer periods, chilling injury usually occurs. Fruit with chilling injury appear normal when removed from storage, but do not ripen satisfactorily when moved to room temperature and develop a physiological or internal breakdown which is characterized by the development of dry and mealy or "wooly" flesh (mesocarp). In advanced stages the flesh becomes badly discolored and a watery translucent breakdown occurs around the endocarp.[29] It has been suggested that internal breakdown is due to an accumulation of toxic materials in the fruit during cold storage. Following harvest, there is synthesis and degradation of toxic materials in the peach fruit and when fruit are maintained at ambient temperatures, the two reactions are at equilibrium. When the mesocarp tissue is dropped to 0°C during storage, the accumulation of toxic substances is more rapid, thus tissue damage will occur.[30] Exposure of peach fruit for 48 h to 23 to 25°C following 2 and 4 weeks storage reduced internal breakdown and resulted in a 6 week storage life.[31,32] Although not a common practice, controlled atmosphere storage (1% 0_2 and 5% CO_2) has been reported[29] to increase the storage life of peaches when compared to normal low temperature storage. Controlled atmosphere stored fruit held at 0°C and intermittently warmed to 18.3°C for 2 d during storage retained higher quality and had lower respiration rates than those held in 0°C air. However, skin browning frequently occurred during subsequent ripening at room temperature. Also, the intensity of sensory attributes

TABLE 6
Nutritional Value of Peaches, Nectarines, and Plums

	Composition per 100 g edible portion							
Fruit	Water (%)	Energy (cal)	Protein (%)	Fat (%)	Carbohydrate (%)	Ca (mg)	Phos. (mg)	Iron (mg)
Peach & nectarine	89.1	38	0.6	0.1	9.7	9	19	0.5
Plum	81.1	66	0.5	0.2	17.8	18	17	0.5

Adapted from Salunkhe, D. K., Ed., *Storage, Processing, and Nutritional Quality of Fruits and Vegetables*, CRC Press, Boca Raton, FL, 1976. With permission.

such as low boiling point flavor volatiles were lower in CA stored fruit that had been intermittently warmed and the intensity of these attributes decreased with length of CA storage.[33] Nectarines have been reported to develop "off flavor" when stored under controlled atmosphere. Peach fruit allowed to soften (ripen) at room temperature will reharden when returned to 0°C storage.[34] A relative humidity of at least 90% should be maintained in the cold storage unit to prevent shrinkage (weight loss) of peaches.[27] Fruit are generally shipped in refrigerated highway trailers in which the temperature is held at 2.8°C.

2. Plums and Prunes

Plums are usually stored at 0°C, but storage at higher temperatures (2.8 to 10°C) will result in higher flavored fruit with a shorter shelf life. Certain cultivars will maintain quality up to 9 weeks at 0°C if mature fruit of high sugar content are placed in storage. Greener fruit will show an internal breakdown when stored for long periods at 0°C. Plums are shipped in the same manner as peaches.

C. NUTRITIONAL VALUE

Peaches, nectarines, and plums are high in fiber and of the 12 most nutritious fruits rank within the middle percentile. Plums have higher energy, fat, carbohydrate, and calcium levels than peaches or nectarines, but are lower in water, protein, and phosphorous. Although plums are advertised by the industry to have a high iron content, they do not differ from peaches in this respect (Table 6).[35] When compared to 42 of the major fruits and vegetables, peaches ranked 12th in Vitamin A (1300 IU) and plum 15th in potassium content, but neither ranked within the top 20, of the fruits and vegetables studied, in content of 13 other nutritional parameters.[35]

IV. FACTORS INFLUENCING THE QUALITY OF PEACHES, NECTARINES, AND PLUMS FOR PROCESSING

A. PEACHES AND NECTARINES

Fruit yield and quality are influenced by many external factors but fruit type, weather, ripening date, tree training, fruit thinning, soil type, nutrition, diseases, insects, nematodes, and handling have the greatest impact. The quality of peach fruit is controlled by the following: (1) citric and malic are the major acids; (2) sucrose is the major sugar; (3) carotenoids are responsible for the yellow-orange color; (4) the volatile aroma compounds have been determined to be largely a series of lactones; (5) poly-phenols that contribute to the astringency and flavor; and (6) texture, which is a function of cultivar and maturity and controlled by alterations in the pectic substances.[36]

1. Clingstone Types

Several hundred important peach cultivars are grown in the world. Both clingstone and

TABLE 7
Cultivars of Peaches for Processing in South Carolina

Weeks before or after Elberta	Standard varieties for processing	Other varieties	Promising new varieties	Firm clings for canning
Before				
5	Coronet	Dixiegem		
4	Redhaven			Merrill Gem
3	Ranger	Triogem	Daroga red	Babygold 5
		Golden Jubilee	Redglobe	
2	Southland	Loring	Earlihale	Ambergem
	Sunhigh	Blake	Summerqueen	Suncling
			Madison	
1	Sullivans Early Elberta			Babygold 7
0	Elberta		Jefferson	Babygold 8
	Redskin		Dixiland	
After				
1	Rio Oso Gem	Shippers Late Red		Babygold 9

[a] Elberta ripens about July 25 at Clemson.

Adapted from VanBlaircom, L. O., Sefick, R. A., Baumgardner, and Ferree, R. J., *SC Agri. Exp. Stn. Bull.*, 539, 1968. With permission.

TABLE 8
Proportion of Tree Population and Ripening Dates of 16 Varieties of Peaches Suitable for Canning in the Southeastern U.S.

Variety	Tree population (%)	Suitability for canning
Redhaven	0.4	Good
Ranger	0.8	Fair
Keystone	8.3	Good
Southland	6.2	Good
Sunhigh	1.1	Good
Redglobe	2.8	Very good
Loring	4.7	Fair
Sullivan Elberta	6.5	Fair
Blake	2.5	Poor
Dixiland	2.8	Fair
Redskin	3.6	Fair
Elberta	8.6	Fair
Jefferson	—	Fair
Fay Elberta	—	Very good
Rio Oso Gem	1.0	Fair
Late Gold	—	Very good

Adapted from Boggness, T. S., Jr., and Heaton, E. K., *University of GA Res. Rep.*, 1973, 159. With permission.

freestone peaches are used for processing (Table 7 and 8). Non-melting flesh clingstones are used for canning and puree, melting and non-melting flesh clingstones are used for pickling, and melting flesh freestones are used for canning, freezing, drying, and the manufacture of peach nectar and pulp (puree). The clingstone cultivars are firmer in texture and hold their shape during processing. The popular clingstone cultivars with non-melting flesh are attractive to processors because they have height and color in both a sliced or half pack. Freestone cultivars have a "ragged" look after processing and constitute less than 20% of the U.S. peach pack.[36] Desirable attributes for high quality clingstone peach cultivars include fruit of good shape and symmetrical size. Fruit size should be 5.08 to 6.03 cm or larger depending on the area of the U.S.[13] In California, a No. 1 cling peach is not less than 6.03 cm in diameter, ring measured, and of good shape, color, texture, and quality, and in good condition for canning or freezing.[37]

A peach cannot be classed as a No. 1 cling peach if it has worms or worm damage, a gum pocket, rot, brown rot, black flesh, green core, is overripe, if it is a windfall, a visible split-pit, or has more than a reasonable amount of mildew, scab, red streaks in the flesh, bruises, tears, or punctures, parasite or insect damage, hail damage, San Jose scale, sunburn, shothole fungus, or other imperfections. Imperfections must be limited to the extent that the bin of clingstone peaches can still make the choice grade.[38] Clingstone peaches should have a small pit with little or no red flesh color. Ideally, the flesh should be clear and yellow-colored. It should be able to withstand hauling, refrigeration, handling, and have a pleasing taste and aroma which is retained in the processed product. Good production is another important attribute.[14]

Cultivars of peaches show differences in color, texture, and flavor while fresh, and show even greater differences in appearance and quality after freezing and thawing.[13] Several researchers have rated clingstone cultivars to their suitability for production and processing.[39,40] Peach pickles are best made from non-melting, medium-sized, small seeded clingstone peaches. However, some cultivars of melting flesh clingstones are suitable for peach pickles.[41]

2. Freestone Types

Freestone peaches for processing are used for canning, freezing, preserves, drying, and the manufacture of peach nectar. Freestone peach cultivars should have the following characteristics for a desirable canned product. Uniform flesh color, preferably orange-yellow and free from excessive red color at the pit, moderate resistance to bruising, resistance to browning or darkening during or after processing, and well-balanced aroma and flavor. The cultivar should be small seeded, easily pitted, and free of split pits. It should peel smoothly, retain its shape, and be free from raggedness. The texture should be firm and free from excess fiber. Size should be uniform, preferably 5.72 to 6.99 cm in diameter.[40,42,43,44] To meet the U.S. No. 1 grade, freestone peaches for canning, freezing, and pulping must be of a similar cultivar which is mature, fairly well-formed, firm, firm ripe, or ripe, but not hard or soft; free from decay, worms, worm holes, and split pits, and free from damage caused by scab, bacterial spot, other diseases, insects, bruises, or other means. Ground color shall not be greener than greenish-yellow. There is a 10% tolerance for the grade.[37]

The best frozen peaches are made from mid to late season peaches.[45] Three cultivar types of frozen peaches are recognized: white freestone, red freestone, and yellow freestone.[46] Unlike canned peaches, the industry prefers retail packages of frozen slices to contain yellow cultivars with red around the pit cavity.[13] 'Fay Elbertas' used for freezing in California should have a fruit size of 6.03 to 6.35 cm with a 5% constant tolerance, zero tolerance for worms, rot, and mold, and a 5% variable tolerance for general defects.[44]

For drying, Fay and Regular Elberta should be 5.72 to 6.03 cm in diameter with a 5% constant tolerance on size and zero tolerance for worms, rot or mold.[44] In the case of the production of peach pulp and nectar later season cultivars are most desirable due to greater soluble solids and flavor. Fully ripe, soft peaches with no red near the pit are best for pulping. They should be free from spoilage and insect infestation to obtain the highest quality product.[47] Cultivars have been rated as to their desirableness for pulping based on freedom from browning, flavor, and sugar acid ratio.[48]

Small quantities of nectarines are processed. For processing nectarines may be clingstone or freestone, but freestone types are preferred for frozen unpeeled halves. Since peeled, frozen nectarines are prone to be mushy, proper maturity is important. Overripe nectarines take on a characteristic disagreeable flavor which is very noticable.[49]

Peach rootstock cultivars or selections vary tremendously in their ability to survive and produce, especially under southeastern U.S. conditions. In a study at six locations in the Southeast 'Lovell' and 'Halford' appeared to be the best rootstocks studied.[50] Although Lovell' seedlings are the most widely used rootstock for peach in the Southeast, 'Nemagard' is the most widely used in California. Interest in own rooted peach cultivars has increased. Peach trees on their own roots had greater tolerance to drought, greater yield efficiency, produced larger fruit, smaller trees, and accumulated greater levels of leaf calcium under conditions of water stress than did trees on 'Lovell' or 'Nemagard' rootstocks.[51] Peach rootstock selection is important in the control of certain root-knot nematode species and calcium induced iron chlorosis.[50,52]

Tremendous fluctuations in the production of peaches can be attributed to freeze damage. For example, production from the southern states varied from 129 million kg in 1983 to 345 million kg in 1984.[53] Low temperatures of $-18°C$ during the late winter season will result in 90% bud kill on 'Elberta'. The flower buds rapidly lose their hardiness as they swell and temperatures of $-4.4°C$ can kill 90% or more of the flowers at full bloom.[54] The peach is more vulnerable to cold temperatures than most deciduous fruits due to its early blooming habit. For this reason good site selection is critical for the production of peaches. Cold damaged peach fruit may contain necrotize tissue which reduces its quality for fresh market or processing.

Although nearly all peach orchards in California and other "dry" states are irrigated, irrigation is not as common in the southeast and eastern U.S. Yield in the non-irrigated orchards in the Eastern U.S. is profoundly affected by drought in some years. Peach fruit are over 85% water and much of their growth occurs during final swell of the fruit, the last 25 d before maturity. Yields during dry years may be doubled by irrigation even in a humid state such as Georgia.[55] Over a number of years in Maryland the average increase in size from irrigation was 29% for 'Sunhigh' and 26% for 'Elberta'.[55] Irrigation has been shown to decrease soluble solids, but increases acidity of raw peaches while improving the taste and color of canned peaches.[57] Water shortage and/or high temperatures just prior to harvest can result in a condition known as "pitburn", a browning of the flesh near the pit in clingstone peaches.[58]

Lack of chilling hours reduces yields in the South about one year in ten. If a cultivar receives within 100 h of its chilling requirement (except for the very low chilling cultivars) a half crop may be produced, but total crop failure usually results if less is received. When insufficient chilling occurs lack of foliation causes abortion of developing fruits.[59] Some cultivars which should seemingly be adapted to certain areas are unproductive for unknown reasons or form "button" fruit which remain dwarfed and ripen after the normal season.[11]

Date of maturity has a profound effect on fruit yield, size, and quality. Although peach cultivars begin ripening in the lower South and California in early May maximum fruit size is obtained only in cultivars that ripen in August and September (Table 9). Mid and late season cultivars can easily produce twice the yield of early season cultivars and have higher fruit soluble solids and more acceptable texture. Early season cultivars are also more likely to have split or shattered pits which reduce fruit internal quality.

TABLE 9
Principal Shipping Peach Cultivars Grown in California

Cultivar	Earliest Shipping date (week-month)	Fruit diameter (cm)
Springtime	2, May	5.69
Armgold	2—3, May	5.66
Springold	2—3, May	5.74
Royal Gold	2—3, May	5.79
Sprincrest	4, May	6.05
Early Coronet	1, June	6.40
Royal May	1, June	6.30
Cardinal	1, June	6.27
Gemfree	1—2, June	6.58
Coronet	2, June	6.63
Red haven	2—3, June	6.63
Regina	3, June	6.86
Babcock	3, June	6.81
Redtop	3—4, June	6.91
Redglobe	4, June 1, July	7.37
July Elberta	4, June	7.34
Scarlet Lady	1, July	7.14
Suncrest	1—2, July	7.34
Fortyniner	2—3, July	7.42
Elberta	3, July	7.52
Fay Elberta	3—4, July	7.39
Angelus	4, July	7.62
Pacifica	4, July	7.39
Fayette	4, July	7.57
Rio Oso Gem	4, July	7.85
O'Henry	1—2, August	7.77
Madera Gem	3—4, August	7.80
Summerset	3—4, August	7.39
Parade	3—4, August	7.92
Fiesta	4, August	7.89
Fairtime	4, August 1, September	7.77
Carnival	1—2, September	7.65
Late John Gee	2, September	8.00
Halloween	3, September	8.08

Adapted from Gerdts, M. and LaRue, J. H., *University of CA, Coop. Ext. Serv. Leaf., 2851*, 1976. With permission.

Tree training has a tremendous influence on yield and fruit size. For example, fresh market peach production from the heavily pruned peach orchards in Alabama and Georgia is less than 1/2 that of California peach orchards which contain large trees. Although total acreage between the two areas is equal, California peach production is double.[16,60] In the Southeast, Krewer et al.[61] showed that a 50% yield increase can be obtained on mid season peaches by increasing tree height from 1.83 to 3.05m. Yields can also be increased using high density orchards which have increased yield capacity while maintaining good fruit size.[62] Fruit size, acidity, flesh color, and taste are increased with severe pruning.[36,57]

Fruit thinning always reduces total yield, but is a necessity if large-sized peaches or nectarines are to be obtained. Often peach or nectarine trees will produce over 5,000 blooms but can properly size only 400 to 900 fruit, depending upon tree size. The earlier one thins, the larger the final fruit size. Fruit thinned early also ripen more quickly.[53] Fruit thinning must be adjusted according to the environmental conditions of the orchard and desired fruit size. With irrigation, distance between fruits on the shoots can be reduced and yields increased.[57]

Deep, well-drained soils with a good fertility level and moisture holding capacity are best for peach production. Tree survival was 95% in deep well drained soils verses 65% in imperfectly drained soils. Yield per tree was 80% greater in those growing in favorable soils.[64]

Nitrogen management can have a pronounced effect on fruit quality. Fruit, yellow ground color, and blush are decreased with high nitrogen. On the other hand, fruit size and yields are decreased when nitrogen levels are too low. Generally, this occurs below 2.5% nitrogen in leaf tissue during mid-summer.[55] Reports of phosphorus deficiency are rare in peaches since this nutrient is not proned to leaching and peaches forage for phosphorus.[65] Potassium is an important nutrient in peach quality. Desirable skin and flesh color has been associated with adequate supplies of potassium. In addition, foliar potassium levels below 1.0% may reduce fruit size.[67] Calcium in the form of agricultural lime is very important in maintaining the longevity of peach orchards. Magnesium deficiency is generally not an important problem with peaches but when a deficiency occurs flower bud formation can be reduced.[68]

Zinc is an element commonly deficient in the lighter soils in the southeastern U.S. and California. Leaf loss, premature leaf fall and reduced fruiting often occur.[69] Boron deficiency or toxicity sometimes occurs on peaches. Twig dieback and flower bud drop are often associated with boron problems.[70] Iron deficiency characterized by interveinal chlorosis is an important problem with peaches growing on alkaline soils. Copper is often low in peach foliage in the southeastern U.S. and fruit cracking and dieback of branches may occur at very low copper levels.

Weeds reduce peach growth by removing water, nutrients, and releasing chemicals toxic to peaches. Young trees growing in sod will make only a fraction of the growth of trees growing under weed free conditions. Yield in most years is greatly reduced if weeds are allowed to grow on the orchard floor.[17]

Birds can damage peaches by "pecking" holes in the fruit. Avian species which have been reported to damage ripening peaches includes the Orioles, Blue Jays, Common Grackles, Starlings, Robins, and Red-winged Blackbirds.[71]

In the eastern U.S., peach diseases can cause hugh losses in wet years. Brown rot *Monilinia fructicola* [Wint.] Honey or *laxa* [Aderh. Huhl.] Honey) attacks blossoms, shoots, and fruit. Initially, small, circular, light brown spots develop on the surface of the fruit and expand rapidly under favorable conditions, destroying the entire fruit in a few hours Partly damaged fruit are suitable only for wine making. Bacterial spot (*Xanthomanas campestris* pv. *pruni* [Smith] Dye) can cause severe defoliation and fruit spotting. Infected fruit are culled for fresh market and deep lesions will reduce the quality of peaches for processing. Peach leaf curl (*Taphrina deformans* [Berk.] Tul.) is a problem over much of the U.S. It is a springtime disease of peach and nectarines that can result in severe defoliation, weakened trees, and reduced fruit quality, set, and yield.[72] A host of virus and bacteria diseases attack peach trees and lower fruit quality in many parts of the world. Phony disease of peach and nectarines is caused by a rickettsia-like bacteria spread by leafhoppers. It is mainly a problem in the lower southeastern U.S. Once a tree is infected, shoot growth and fruit size decrease greatly within a few years.[73] The infected fruit are of little or no value except for home preservation of peach slices.

In the eastern U.S. the major peach fruit pests are plum curculio, plant bugs, stinkbugs, oriental fruit moth, and peach twig borer. The plum curculio (*Conotrachelus nenuphar* Herbst) may damage fruit in both the adult and larval stages. Adult damage consists of tiny, circular feeding punctures or small crescent-shaped cuts made by females following egg laying. The larvae tunnel and feed in developing fruit, usually boring to the pit.

Numerous species of plant bugs and stinkbugs feed by sucking sap from plants. When feeding, toxic substances are injected into the peach fruit to "break down" tissue. Cells are destroyed and fruit development inhibited at the feeding site, while surrounding tissues continue to grow and expand reducing the value and quality of the fruit. Oriental fruit moth (*Grapholitha molesta* Busck) is found in much of the U.S. The first generation larvae attack shoot terminals while later generations often attack the fruit by boring near the stem end, leaving little or no external signs of entry.

The peach twig borer (*Anarsia lineatellia* Zeller) attacks the pith of new shoots early in spring. Later generations feed in fruit often penetrating to the pit. Late maturing peach cultivars normally sustain the most fruit damage.[74,75] Other insects which attack the tree may reduce fruit quality indirectly. These include peach tree borer (*Synathedon exitiosa* Say), lesser peach tree borer (*Synanthedon pictipes* Grote and Robinson)' white peach scale (*Pseudaulacaspic pentagona* Targioni-Tozzetti), and various species of aphids, and mites. San Jose scale (*Quadraspidiotus perniciosus* Comst) is primarily a pest of the tree but may also discolor the fruit. In California the most important insects pests are San Jose scale, peach twig borer, oriental fruit moth, and mites.

In much of the U.S. nematodes are serious root pests in peaches and nectarines. Three groups of nematodes are known to attack peaches and nectarines; *Meloidogyne* sp. (root-knot nematodes), *pratylenchus* sp. (lesion nematodes), and *Criconemella* sp. (ring nematodes). Trees attacked by root-knot nematodes are often stunted and produce small fruit while root feeding by lesion nematodes may cause peach trees to be non-vigorous and unproductive. Ring nematodes are often a predisposing factor in tree loss from cold injury and other diseases.[52]

3. Harvest Injuries

About 25% of the cling peach crop used for processing in the U.S. is partially mechanically harvested with shake-catch harvesters. Many cling varieties require one or two hand harvests before the bulk of the crop is removed by machine.[75]

Injury can accumulate throughout all stages of harvesting and trasport. Open wounds, such as cuts and punctures, can be eliminated by proper sorting. Bruises can accumulate and increase in size and number with the aging. Fruit picked during the coolest part of the day are less subject to bruising.[77]

Browning of the peach flesh can be a major quality problem for processors of peaches. When cellular membranes of the peach are injured by cutting, peeling, or bruising, polyphenol oxidase catalyzes the conversion of catechol O-quinones. These quinones can break down to chalcones and oxidize and polymerize to brown pigments. Rapid handling of the peaches during processing is important to minimize browning.[36]

B. PLUMS AND PRUNES

Only certain plum species and cultivars are suitable for processing. Within *P. domestica* the primary canning types are those in the Prune, Green Gage, and Yellow Egg groups. The Damson plum (*P. insititia*) is sometimes used for processing into jellies, jams, and marmalades.[19] In the northwestern U.S. the 'Italian' prune (*P. domestica*) is grown primarily for canning and fresh market. Since early maturing plums of high quality are needed to extend the season of harvest and processing, plant breeders are selecting for early maturing fruit of high quality. Plums for canning should be mature and free from serious defects. They are sorted into sizes 2.62, 3.18, 3.81, and 4.45 cm in diameter. Canned dried prunes should be of large sized (110 to 132 count per kg).

Of the prune type plums for processing the 'French' prune is the dominate (96%) cultivar in California due to its high quality and ease of mechanical harvest. Most of the remaining acreage is planted to 'Imperial', 'Sugar', and 'Robe de Sargeant' (Robes). The advent of mechanical harvesting favored 'French' prunes which show less fruit injury due to impact onto catching frames and the rigor of pickup machines during the harvesting process than non-'French' type cultivars. This "once over" mechanical harvest process has a greater detrimental effect on the quality of 'Imperials', 'Robes', and other non-'French' cultivars. "Once over" mechanical harvest results in smaller fruit size, more underripe and overripe fruits, thus lower overall quality than multiple hand-harvested fruit which has resulted in more rigorous grading of mechanically harvested fruit of the non-'rench' cultivars, thus a lower return to the grower.[24] Small fruit are tougher due to a higher percentage of skin and withstand the rigors of mechanical harvesting better than large fruit, but large fruit command a premium price. Defects most commonly seen in prunes are end cracks, side cracks, scab, sunburn, mold, dirt, and insect injury. High quality fresh prunes should have a reddish-purple skin color with amber flesh. The complete disappearance of green color from the flesh is the best index of maturity. Other maturity indices are used such as shifts in the regions of maximum light transmission which have been correlated with maturity changes of prune-type plums[78] and flesh firmness. Flesh firmness is measured by a apparatus which applies and records the pressure required to penetrate the fruit flesh. Plum fruit that can withstand 1.8 kg or less pressure are considered mature. Soluble solids should be 24% or greater, although in years of a heavy crop lower soluble solids levels occur.[24]

Fruit size and yield of plums and prunes are controlled by several factors, but weather, irrigation, ripening date, fruit thinning, pruning, nutrition, and pest management are most important. Yield can be severly reduced by spring freeze damage. 'Early Italian' has been reported to receive 10% flower bud kill at −11.1°C and 90% flower bud kill at −17.2°C during first swell[79] and 10% bloom kill occurs at 2.2°C and 90% at −5°C during full bloom.[80]

Deep medium, textured soils are best for plum and prune fruit growth. Japanese type plums will grow on poorer soils than the European plums.[81] Adequate soil moisture is necessary for optimim fruit growth since final fruit size is reduced once fruit growth is stunted due to a lack of water.[24] On the other hand, excessive moisture near harvest can result in severe fruit cracking in some plum cultivars. Fruit end cracks of prune type plums are usually caused by tree moisture stress followed by irrigation and in California is subject to occur from late June to mid July.[82] Side cracks are caused by diurnal fluctuations in water. The fruit portion exposed to the sun loses its ability to expand (stretch) during fruit expansion, resulting in the development of cracks on the exposed fruit surface. The side cracking period generally occurs during the first two weeks in July and cool weather during this time will enhance the condition.

Prunes grown in a cool climate drop from the tree as they mature requiring multiple harvests. This is one of the reasons for the shift in the prune industry from the coastal valley of California to the warmer central valley.[24] However, high temperatures during the winter can cause bud abscision on some plums and prunes.[82] During the growing season, prunes often develop pit burn if exposed to temperatures above 40.6°C. This condition is characterized by a browning and disintegration of the mesocarp tissue near the pit.[84] Temperatures of 40.6°C or above seem to be about seven times as injurious as temperatures between 38.9°C and 40.6°C.[85] Since plums are primarily self-unfruitful weather can sometimes disrupt pollination especially if bee activity is low or rain occurs during bloom resulting in reduced fruit set. Variation in bloom date between cultivars can result in reduced yields and is due to temperature during the winter (quantity of low temperature) and early spring (level). As is the case with peaches, date of maturity also affects plum yield. Cultivars that mature fruit in early season produce low yields due to small fruit size because of the reduced fruit growth period (Table 5).

Fruit thinning is important in the production of the plums due to its influence on fruit size and soluble solids.[86] As a general rule, the fruit should be thinned to about 8 cm for maximum size and soluble solids. It is an important tool in increasing drying ratios in prunes.[87] Prunes are usually mechanically thinned and removal of at least 1/3 of the fruit on a tree will result in a decrease of 20 fruit necessary to produce a pound. Mechanical thinning should be done during early pit hardening for best results.[24] With 'Stanley' plum, crop load within the range of 200 to 900g of mature fruit per cm^2 trunk cross-sectional area was negatively correlated with fruit size, soluble solids, and color.[88]

Tree training is required to produce high quality plums. Annual dormant pruning involves the removal of fruiting wood throughout the tree to reduce crop load and encourage renewal growth for fruiting in later years. Interfacing branches, broken limbs, suckers, and water sprouts are also removed. Both Japanese and European plums produce most of their crop on spurs which develop along larger branches. Spurs of Japanese plums are usually 5 to 10 cm long, while those on European plums are shorter.[86]

Adequate nutrition is required to produce plums and prunes of high-quality. Nitrogen, zinc, and iron are often limiting for plums in California soils and nitrogen and/or zinc deficiency have been associated with small fruit size. About 100 kg of nitrogen/ ha are needed per year in California.[86] However, excessive nitrogen promotes shoot growth which increases internal shading and pruning costs.[24,89] Potassium deficiency, also commonly associated with small fruit size in prunes, causes leaf scorch as well[90] and boron deficiency may cause brown sunken areas in the flesh[85]

Diseases can destroy fruit and reduce fruit quality. Brown rot (*Monilinia laxa* [Aderh. & lRuhl.] Honey or *fructicola* [Wint.] Honey) infects blooms or fruit causing severe damage. Russet scab results in reduced bloom (wax) on fruit, producing dried fruit with a heavily damaged netted appearance. The cause is unknown but application of fungicides during bloom reduces the problem.[91] Bacterial canker (*Pseudomonas syringae* Van Hall) may attack young plum leaves and flowers, producing leaf shotholes plus blossom and leaf blast. Bacterial spot (*Xanthomonas campestris* pv. *pruni* [Smith] Dye) is widespread in the eastern U.S. Severely infected leaves abscise resulting in early heavy defoliation which reduces fruit size and weakens the tree. Fruit infected early in the growing season develop cracks or checks on the epidermis.[72] Cystospora canker (*Cystospora leucostoma* Sacci) is a serious problem on European plums.[86]

Insect pests can totally destroy a plum or prune crop if not controlled. The most destructive fruit pest in the eastern U.S. is the plum curculio (*Conotrachelus nemuphar* Herbst). It may damage fruit in the adult stage by feeding and egg laying. The larvae tunnel and feed in developing fruit, usually boring to the pit.[75,92,93]

Common tree pests in the eastern U.S. are the peachtree borer (*Synanthedon exitiosa* Say) and San Jose scale (*Quadraspidiotus perniciosus* Comst.). The peachtree borer often attacks plums grafted on peaches. San Jose scale may kill trees and also damage and discolor fruit.[75]

In California, the primary tree pests are mealy plum aphid (*Hyalopterus arundinis* Geoffrey), San Jose scale, mites (*Tetranychus sp.*), peach twig borer (*Anarsia lineatella* Zeller), and various wood borers while the primary fruit damaging pests are the peach twig borer, codling moth (*Carpocapsa pomonella* Linn), leaf rolling caterpillars, and the green fruit worms (*Xylina sp.*). The peach tree twig borer can burrow into growing shoots resulting in shoot death or feed on the fruit, thus reducing fruit grade. Codling moth larvae enter both immature and mature fruit and tunnel deeply into fruit tissue. Leaf rolling caterpillars feed on leaves and buds, and tunnel throughout the fruit causing severe damage as well as excrement contamination. Green fruit worms can damage young plums and prune by eating holes in the fruit.[86,94]

V. PROCESSING

A. PEACHES AND NECTARINES

Although only 1% of the total nectarine production is processed, a significant portion of the peach crop is processed by canning, drying, pureeing, freezing, production of refrigerated fresh slices, or preparation into preserves. Other processed products such as juices (nectar), cloths, bars, molds, pickles, and etc. are made from peaches,[35] but these are in low demand. An overwhelming majority of the clingstone peach crop is processed by canning and it is one of the most popular of all processed fruit.

Refrigerated fresh slices were a popular product several years ago, but have declined in popularity due to difficulty in maintaining flavor and the short shelf life of the product. Fruit to be processed in this manner must be of high quality, table ripe, medium to large in size, resistant to browning, deep yellow with a minimum of red flesh color, and of good texture, flavor, aroma, and color. It is important that fruit to be processed in this manner have a high level of aromatic volatiles so that flavor is maintained. The fruit are lye peeled (immersing in a 5% lye solution at 98.5°C for 30 s, placed in a rotary washer to remove the peel, rinsed in an acid wash (0.5% citric acid, 0.15% ascorbic acid, and 0.5% sodium benzoate), de-seeded, sliced, and packed in a container filled with a suitable medium (syrup of 20 to 30 ° Brix) and refrigerated immediately. The containers used are generally glass jars which when filled with fruit must be cooled to 0°C immediately. Fruit processed in this manner will maintain quality for as long as 12 weeks.[49]

Late maturing freestone peaches are generally used for the production of frozen peach slices. The fruit are prepared in much the same way as are refrigerated peach slices except a syrup is not added and the slices are frozen. It is very important to feeeze fruit with good aromatic characteristics since these will not increase during processing and storage and cannot be added to the pack.

Fruit to be canned are washed and sized then mechanically pitted (removal of the endocarp). During the pitting process the fruit are oriented so that they are cut along the suture and halved. The cut fruit sections are lye peeled with a spray of hot sodium hydroxide. The temperature and concentration of the sodium hydroxide solution is dependent upon cultivar and stage of maturity. The cut and treated halves can be sliced if so desired and are washed with a high pressure spray to remove the peel. The peeled fruit are graded for defects, color, and maturity and separated into A, B, or C grades. The graded halves or slices are then placed into enamel coated cans into which a hot 25 to 30% Brix syrup is added. In the case of low calorie packs, a fruit juice mixture is added in place of the syrup. Syrups utilizing fructose instead of sucrose are sometimes used in low calorie packs since fructose is sweeter than sucrose thus smaller quantities are need to maintain the same level of sweetness. The temperature of the pack is raised (79.5°C) by exposing the cans to heat in a steam tunnel to eliminate gasses from the intercellular spaces of the fruit. The non-enameled cover is then hermetically sealed onto the can and the can and contents are heated in a steam cooker to 121°C for 9 to 30 min depending upon maturity and cultivar (Dr. Jerry Dull, personal communication).

Pureed peaches are a popular baby food product. The fruit are handled in the same manner as in canning except the peach halves are pureed following peeling and the puree is cooked for a longer period.

Fruit used for drying must have a high solids content. The fruit are pitted and halved then dried by one of several mechanical measures or by sun-drying. In the case of sun-dried fruit, the fruit halves are treated with sulfur fumes before initiating the drying process.[35] Sun-drying requires approximately 4 to 6 d and is terminated when the halves have a leathery texture and have been dried to a ratio of from 4 to 1 for some cultivars and 8 to 1 for others. Cultivars that yield at least 1 lb of dried peaches per 5 lb of fresh fruit are preferred. Dried fruit must be stored at low temperature (2.8°C) and relative humidity to maintain quality. Although dehydrated peaches and plums are concentrated in nutrients, during the process of dehydration some vitamin potency may be lost. Vitamins A, C, and thiamine can be destroyed by heat and sulfuring while riboflavin is light sensitive.[35] The extent of nutrient loss will depend upon care exercised during the drying process, the type of process used, and storage conditions of the dehydrated product.[35] The demand for dried peaches has decreased during the past 30 years and continues to do so.[17] Currently dried fruit are used for extruded snacks, compressed bars, and etc.[35]

B. PLUMS AND PRUNES

Processing of plums consists of either freezing, canning, juicing, preserves, or drying (prunes). There is a growing demand for frozen plums (*insititia* cultivars) and several hundred tons are frozen each year, but constitute a very small part of the total processing market.[17]

Plums for freezing or canning should be of large size (12/lb bring a premium price and smaller than 20/lb are not accepted), firm, and green or yellow in color. Many of the same procedures for canning and freezing peaches is followed with plums except the fruit are sometimes peeled and are processed whole.

Cultivars of *insititia* are preferred for drying. The fruit are mechanically shaken from the trees and transported to mechanical dehydrators for drying. The fruit are generally pitted before drying. As in the case with peaches, plums must have a high solids content to produce a quality dried product. Approximately 2 1/2 to 3 lb of fresh fruit are required to produce 1 lb of dried prunes. The fruit are dipped in an alkaline solution before initiating the drying process. When mechanical dehydrators are used rather than sun-drying, the natural toughness of the skin of many cultivars presents a problem. Juices made from prunes are very popular and prune purees are used as baby food.[17] Prune juice is made by adding 1 part dried prunes to 4 parts water then heating to 85°C. The resulting juice is added to a second batch of prunes and the process repeated. The same process is repeated a 3rd time then the resulting juice is filtered, diluted to 20% Brix, heated to 93°C then canned or bottled.[95]

REFERENCES

1. **Hedrick, U.P.,** Sturtevant's notes on edible plants, *Rpt. N.Y. Agr. Exp. Sta.,* J.B. Lyon Co., Albany, NY, 1919, 686.
2. **Li, Zai-long,** Peach germplasm and breeding in China, *Hort Science,* 19, 348, 1984.
3. **Wang, Yu-Lin.,** Peach growing and germplasm in China, *Acta Horticulturae,* 173, 1985.
4. **Cullinan, F. P.,** Improvement of stone fruits, *USDA Yearbook of Agriculture,* U.S. Government Printing Office, Washington, D.C., 665, 1937.
5. **Magness, J. R.,** How fruit came to America, *National Geographic,* 100, 325, 1951.
6. **Knight, T. A.,** An account of a peach tree produced from seed of the almond tree with some observations on the origin of the peach tree, *Trans. of the R. Hortic. Soc. of London,* 3, 1, 1820.
7. **Pliny, Secundus,** *A Natural History,* Translated by Bostock, J. and Riley, H. T., 1956, Bohn H.C., Ed., Covent Garden, England, 1956.
8. **Zielinski, Q. B.,** *Modern Systematic Pomology,* Wm. C. Brown Co, Dubuque, IA, 296, 1955.
9. **Hesse, C. O.,** Peaches in *Advances in Fruit Breeding,* Janick, J. and Moore, J.N., Eds., Purdue University Press, W. Lafayette, IN, P 285, 1975.
10. **Weinberger, J. H.,** Plums in *Advances in Fruit Breeding,* Janick J. and Moore, J.N., Eds., Purdue University Press W. Lafayette, IN, 1975.
11. **Savage, E. F. and Prince, V. E.,** Performance of peach cultivars in Georgia, *University of GA Res. Bull.* 114, 1972.
12. **Westwood, M. N.,** *Temperature Zone Pomology,* W. H. Freeman, San Francisco, CA, 1978.
13. **Woodroof, J. G. and Luh, B. S., Eds.,** *Commercial Fruit Processing.* AVI Publishing, Westport, CT, 1975.

14. **Anon.**, *Dried Fruits and Tree Nuts*, Assoc. of California, Santa Clara, CA, 1984.
15. **Drake, C. R.**, Pertinent information concerning nectarines and peaches from thirty-three states, *Rep. Natl. Peach Counc.*, Martinsburg, WV, 1985.
16. **Odell, L. L., Williams, L. S., and Severson, S.**, Fruit and nut acreage, *CA Crop and Livestock Reporting Serv.*, Sacramento, CA, 1985.
17. **Childers, N. F.**, *Modern Fruit Science*, Horticulture Publications Gainesville, FL, 970, 1976.
18. **Dejong, T. M..**, Peach planting/training systems in California in *The Peach*, Childers, N. E., and Sherman, W. B., Eds., Horticultural Publications Gainesville, FL, 1988.
19. **Seelig, R. A.**, Plums-prunes; fruit and vegetable facts and pointers, United Fresh Fruit and Vegetable Association, Washington, D.C., 1969.
20. **Griggs, W. H. and Hesse, C. O.**, Pollination requirements of Japanese plums, *University of CA Leaf.*, 163, 1963.
21. **Latham, A. J. and Norton, J. D.**, Incidence of plum leaf scald in Alabama. *Auburn University Agri. Exp. Stn. Bull.*, 525, 1980.
22. **Norton, J. D. and Rymal, K. S.**, A. U. Producer plum, *HortScience*, 13, (4), 487,.
23. **Anon.**, *Agricultural Statistics*, U.S. Department of Agri., Washington, D.C. 19846.
24. **Ramos, D. E.**, *Prune Orchard Management, University of CA Special Pub.*, 3269, 1981.
25. **Powell, Arlie A.**, Split pit, *Peach Production Handbook*, Myers, S., Ed., *Ga. Ext. Serv. Handbook 1*, University of Georgia Cooperative Extension Service, Athens, GA, 1989.
26. **Ryall, A. L. and Pentzer, W. T.**, *Handling, Transportation, and Storage of Fruits and Vegetables*, Vol II, 2nd ed. AVI Publishing, Westport, CT, 610, 1982.
27. **Hopfinger, Wells, J. M., Bennett, A. H., and Hale, P. W.**, Harvesting and handling peaches, *Peach Production Handbook*, Myers, S., Ed., *Ga. Ext. Serv. Handbook 1*, University of Georgia Cooperative Extension Service, Athens, GA, 213, 1989
28. **Anon.**, Standards of Grades of Fresh Plums and Prunes, USDA Consumer and Marketing Service, Washington, D.C., 7, 1969.
29. **Anderson, R. E. and Penny, R. W.**, Intermittent warming of peaches and nectarines stored in a controlled atmosphere or air, *J. Am. Soc. Hortic. Sci.* 151, 1975.
30. **Pentzer, W. T. and Heinze, H. P.**, Postharvest physiology of fruits and vegetables, *Ann. Rev. Plant. Physiol.*, 5, 205, 1954.
31. **Anderson, R. E.**, The influence of storage temperature and warming during storage on peach and nectarine fruit quality, *J. Am. Soc. Hortic. Sci.*, 104, 459, 1979.
32. **Ben-Arie, R., Lavee, S., and Guelfat-Reich, S.**, Control of wooly breakdown of 'Elberta' peaches in cold storage, by intermittent exposure to room temperature, *J. Am. Soc. Hortic. Sci.*, 95, 801, 1970.
33. **Watada, A. E., Anderson, R., and Aulenbach, B. B.**, Sensory, compositional, and volatile attributes of controlled atmosphere stored peaches, *J. Am. Soc. Hortic. Sci.*, 104, 626, 1979.
34. **Werner, R. A., Hough, L. F., and Frankel, C.** Rehardening of peach fruit in cold storage, *J. Am. Soc. Hortic. Sci.*, 100, 90, 1978.
35. **Salunkhe, D. K.**, *Storage Processing, and Nutritional Quality of Fruits and Vegetables*, CRC Press, Boca Raton, FL, 1976, 166.
36. **Morris, J. R.**, Peach processing in *The Peach*, Childers, N. F., and Sherman, W. B., Eds., Horticultural Publications, Gainesville, FL, 1988.
37. **Anon.**, U.S. Standards for Grades of Fresh Freestone Peaches for Canning, Freezing, or Pulping, U.S. Dept. of Agri., Consumer and Marketing Service, Washington, D.C., 1966.
38. **Anon.**, Administrative rules and regulations for the inspection and clarification and collection of assessments for cling peaches pursuant to the joint marketing order for clingstone peaches (California), U.S. Dept. of Agri., Washington, D.C., 1985a.
39. **Boggess, T. S., Jr., Heaton, E. K., Savage, E. F.**, Canning evaluation of non-melting type clingstone peaches grown in Georgia. *University of GA Res. Rep.*, 87, 1970.
40. **Van Blaricom, L.O., Sefick, H. J., Baumgardner, R. A., and Ferree, R. J.**, Evaluation of peach varieties for processing in South Carolina, *SC Agri. Exp. Stn. Bull.*, 539, 1968.
41. **Boggess, T. S., Jr., Heaton, E. K., and Shewfelt, A. L.**, Processing sweet pickled peaches, *University of GA Res. Rep.*, 184, 1974.
42. **Culpepper, C. W. and Caldwell, H. S.**, The canning quality of certain commercially important eastern peaches, *USDA Tech. Bull.*, 196, 1930.
43. **Boggess, T. S., Jr. and Heaton, E. K.**, Canning southeastern freestone peaches, *University of GA Res. Rep.*, 159, 1973.
44. **Anon.**, Pricing newsletter. Cal-freestone Peach Association. Modesto, CA, Aug. 6, 1985c.
45. **Woodroof, J. G., Shelor, E., Cecil, S. R., and Atkinson, Ida**, Preparation of peaches for freezing, *GA Exp. Stn., Experiment, GA (unnumbered)*, 1947.
46. **Anon.**, U.S. Standards for Grades of Frozen Peaches, U.S. Dept. of Agri., Agri. Mark. Serv., Washington, D.C., 1961.

47. **Tressler, O. K. and Josylin, M. A.** *Fruit and Vegetable Juice Production,* AVI Publishing NY, 1954.
48. **Heaton, E. K., Boggess, Jr., T. S., Shewfelt, A. L., Li, K. C., and Woodroof, J. G.,** The production and utilization of peach pulp, juice drink, and concentrates, *University of GA Agri. Exp. Stn. Res. Bull.,* 136, 1973.
49. **Woodroof, J. G., Heaton, E. K., Boggess, T. S., and Li, K. C.,** *The Peach,* Childers, N. C., Ed., Horticultural Publications, Gainesville, FL, 659, 1975.
50. **Rom, R. C., Dozier, W. A., Knowles, J. W., Carlton, C. C., Arrington, E. H., Wehunt, E. J., Yadava, V. L., Doud, S. L., Ritchie, D. F., Clayton, C. N., Zehr, E. I., Gambrell, C. E., Britton, J. A., and Lockwood, W. W.,** Rootstock effect on peach tree survival, growth and yield, A Regional Research Project Report, *Compact Fruit Tree* 18, 85, 1985.
51. **Couvillon, G. A., Rieger, M., Harrison, R., and Daniell, J.,** Stress-mediated responses of own rooted peach cultivars, *Acta Hortic.,* 243, 221, 1981.
52. **Bertrand, P. F. and Motsinger, R.,** Nematodes in Peach Growers Handbook, Ferree, M. E., and Bertrand, P. F., Eds., *GA Coop. Ext. Service Handbo No.* 1, 1983.
53. **Anon.,** *Georgia Farm Report,* Georgia Crop Reporting Service, Athens, GA, GFR-85-Vol. 9, 1985b.
54. **Ballard, J. K., Proebsting, E. L., and Tukey, R. B.,** Critical temperatures for blossom buds in peaches, *Wash. State University Coop. Ext. Serv., Ext. Cir.,* 373, 1983.
55. **Horton, B. D., Wehunt, E. J., Edwards, J. H., Bruce, R. R., and Chesness, J. L.,** The effects of drip irrigation and soil fumigation on 'Redglobe' peach yields and growth, *J. Am. Soc. Hortic. Sci.,* 106, 4, 438, 1981.
56. **Rogers, B. L.,** Peaches in Maryland: a decade of soil moisture and irrigation studies, *University of MD Agri. Exp. Stn. Bull.,* A-148, 1967.
57. **Morris, J. R., Katham, A. A., and Arrington, E. H.,** Response of 'Elberta' peaches to the interactive effects of irrigation, pruning, and thinning, *Proc. Am. Soc. Hortic. Sci.,* 80, 177, 1962.
58. **Kotze, W. A. G. and Bothma, K.,** Factors that have an effect on the incidence of pitburn in Kakamas peach, *Deciduous Fruit Grower,* 39, 5, 159, 1989.
59. **Weinberger, J. H.,** Prolonged dormancy trouble in peaches in the Southeast in relation to winter temperatures, *Proc. Am. Soc. Hortic. Sci.,* 67, 107, 1956.
60. **Hubbard, E. E. and Purcell, J. C.,** Georgia commercial peach tree survey, *University of GA Res. Rep.,* 463, 1984.
61. **Krewer, G. W., Westberry, G., and Evarts, D.,** Effect of tree height on mid-season peach blush, size, and yield, *ASHS Ann. Mtg. Program and Abstr.,* (Abstr.), 124,s361989.
62. **Bargioni, G., Loreti, F., and Pisani, P. L.,** Performance of peach and nectarine in a high density system in Italy, *Peach Times,* 29, 84, 1984.
63. **Havis, A. L.,** Effects of time of fruit thinning of 'Redhaven' peach, *Proc. Am. Soc. Hortic. Sci.,* 80, 172, 1962.
64. **Oskamp, J.,** The rooting of deciduous fruit on different soils, *Proc. Am. Soc. Hortic. Sci.,* 29,213, 1932.
65. **Johnson, S., Mitchell, G., Yoshikawa, F., DeJong, T., Martino, G., LaRue, J., Beede, B., and Norton, M.,** The effect of cultural practices on fruit size and quality, *CA Fruit Grower,* Vol. 62, 1, 15, 1985.
66. **Ballinger, W. E., Bell, H. K., and Childers, N. F.,** *Peach Nutrition in Nutrition of Fruit Crops,* Childers, N. F., Ed., Horticultural Publications, Gainesville, FL, 1966.
67. **Lilleland, O., Uriu, K., Muraoka, T., and Pearson, J.,** The relationship of K in the peach leaf to fruit growth and size at harvest, *Proc. Am. Soc. Hortic. Sci.,* 81, 162, 1962.
68. **Brittain, J. and Ferree, M. E.,** Nutrition, in *Peach Growers Handbook,* Ferree M. E., and Bertrand, P. F., Eds., *GA Coop. Ext. Serv. Handbook No.* 1, 1983.
69. **Dickey, R. D. and Blackmon, G. H.,** A preliminary report on little leaf in Florida, zinc deficiency, *FL Agri. Exp. Stn. Bull.,* 344, 1940.
70. **McLarty, H. R. and Woodbridge, C. G.,** B in relation to the culture of the peach tree, *Sci. Agri.,* 30, 392, 1950.
71. **Jackson, J. J.,** Vertebrate pests in *Peach Growers Handbook,* GA Coop. Ext. Serv. Handbook No. 1, 1983.
72. **Jones, A. L. and Sutton, T. B.,** *Diseases of Tree Fruits, MI State Univ. Coop. Ext. Serv.,* 1984.
73. **Evert, D. R.,** Phony peach in *Peach Growers Handbook.* Ferree M. E., and Bertrand, P. F., Eds., *GA Coop Ext. Serv. Handbook No.* 1, 1983.
74. **Alden, C. H.,** Peach insects and diseases and how to control them, *GA State Board of Ento. Bull.,* 71, 1930.
75. **Ellis, H. C.,** Plum Curculio, sucking bugs, oriental fruit moth, and peach twig borer in *Peach Growers Handbook,* Ferree M. E., and Bertrand, P. Fl, Eds., *GA Coop. Ext. Serv. Handbook* No. 1, 1983.
76. **Peterson, D. L.** Status of mechanical harvesting of peaches in *The Peach,* Childers N. F., and Sherman, W. B., Eds., Horticultural Publications, Gainesville, FL, 1988.
77. **Mitchell, F. G.,** Stone fruit handling and cooling in *The Peach,* Childers, N. F., and Sherman, W. B., Eds., Horticultural Publications, Gainesville, FL, 1988.

78. **Ernest, J. V., Birth, G. S., Sidwell, A. P., and Golumbic, C.,** Evaluation of light transmittance techniques for maturity measurements of two varieties of prune-type plums, *Food Tech.*, 12, 595, 1958.
79. **Proebsting, E. L. and Mills, H. H.,** Low temperature resistance of developing flower buds of six deciduous fruit species, *J. Am. Soc. Hortic. Sci.*, 103, 2, 192, 1978.
80. **Ballard, J. K., Proebsting, E. L., and Tukey, R. B.,** Critical temperatures for blossom buds in prunes, *Wash. State University Coop. Ext. Serv., Ext. Cir.*, 372, 1973.
81. **Kinman, C. F.,** Plum and prune growing in the pacific states, *U.S. Dept. of Agri. Farmers' Bull.*, 1371, 1938.
82. **Uriu, K., Hansen, C. J., and Smith, J. J.,** The cracking of prunes in relation to irrigation, *Proc. Am. Soc. Hortic. Sci.*, 80, 211, 1962.
83. **Brooks, R. M.,** Climate in relation to deciduous fruit production in California II, Effect of the warm winter of 1940-41 on apricot, plum, and prune varieties in northern California, *Proc. Am. Soc. Hortic. Sci.*, 40, 209, 1942.
84. **Maxine, E. C. and Claypool, L. L.,** Heat injury in prunes, *Proc. Am. Soc. Hortic. Sci.*, 69, 116, 1957.
85. **Hendrickson, A. H. and Veihmeyer, F. J.,** Some factors affecting the quality of dried prunes, *Proc. Am. Soc. Hortic. Sci.*, 44, 205, 1944.
86. **LaRue, J. H. and Gerdts, M.,** Commercial plum growing in California, *University of CA Coop. Ext. Serv. Leaf.*, 2458, 1976.
87. **Hendrickson, A. H. and Veihmeyer, F. J.,** The effects of yields upon the apparent drying ratios of 'French' prunes, *Proc. Am. Soc. Hortic. Sci.*, 37, 178, 1939.
88. **Wells, J. M. and Bukovac, M. J.,** Effect of fruit thinning on size and quality of 'Stanley' plum (*Prunus domestica* L.), *J. Am. Soc. Hortic. Sci.*, 103, 5, 612.
89. **Proebstring, E. L.,** The effect of nitrogen on non-irrigated prunes, *Proc. Am. Soc. Hortic. Sci.*, 53, 49, 1949.
90. **Sprague, H. B.,** *Hunger signs in crops,* McKay Company, New York, NY, 1964.
91. **Wilson, E. E. and Ogawa, J. M.,** Fungal Bacterial and Certain Nonparasitic Diseases of Fruit and Nut Crops in California, University of CA. Ag. Science Pub. 4090, Berkley, CA. 1979.
92. **Starnes, H. N.,** The plum in Georgia, *GA Exp. Stn. Bull.*, 67, 1904.
93. **Starnes, H. N.,** Japan and hybrid plums, *GA Exp. Stn. Bull.*, 67, Experiment, GA, 1905.
94. **Slingerland, M. V. and Crosby, C. R.,** *Manual of fruit insects,* Macmillan, New York, NY, 1930.
95. **Cuess, William,** *Commercial Fruit and Vegetable Processing,* McGraw-Hill, New York, NY, 1948.

Chapter 7

THE GRAPE

Klayton E. Nelson

TABLE OF CONTENTS

I. INTRODUCTION

A. ORGIN AND DISTRIBUTION OF GRAPES

Grapes are the world's largest fruit crop with approximately 40 million tons produced each year. Further, they are the eighth-largest food crop in the world. Almost all this fruit is from one species—*Vitis vinifera* L. This species grows best in a Mediterranean type of climate with long, relatively dry summers and mild winters. These conditions are restricted to areas in geographical belts generally between the 30th and 45th parallels of the northern and southern hemispheres. Although most grapes are made into wine, and another large part dried into raisins, still a significant part are marketed annually as fresh fruit, making table grapes one of the world's prominent fresh fruit crops. This fruit has special significance as it is the only form of the grape that reaches the market with much the same appearance that it had when harvested from the vine. Fascinating to the consumer is the variety of colors and flavors of the fruit as well as the convenient bite size of the berries. As a result, fresh grapes are in demand not only for eating, but also for decorative arrangements. Further, table grapes attract attention by their long history, rich in tradition. The Bible makes numerous references to the fruit of the vine, and frescoes painted on ancient Egyptian tombs depict servants serving clusters of fresh grapes to guests.

V. vinifera probably originated somewhere in the Middle East between India and the Mediterranean Sea. Its use by man antedates recorded history. Grapes were undoubtedly first used as table fruit, that is, consumed directly from the vine. The fresh fruit was so perishable that it was available only when the grapes were ripe and use was restricted to the immediate area of production. The only way that man could preserve this fruit for use later was to allow the grapes to dry into raisins on the vine, or to dry the harvested fruit in the sun much in the same manner as is done today in the San Joaquin Valley. This probably was the first method of preservation of fruit developed by man. Later the phenomenon of fermentation was discovered and man learned to preserve the grapes in the form of wine.

Cultivars of the vine slowly spread eastward across southern Asia and westward around the Mediterranean Sea. In the process, selections with larger berries became preferred for table use. After discovery of the New World, man carried the vine to North and South America, South Africa, and later Australia. A suitable climate was found in those areas of the Americas known today as Mexico, southwestern U.S., Chile, and Argentina. It should be mentioned that man found many native species of *Vitis* in the New World. Cultivars of some of these, or of hybrids with vinifera, are used for table fruit. Notably, the 'Concord' (*V. labrusca*) is popular locally as a table grape in the northeastern U.S. and in the Great Lakes area. It can tolerate more summer rain and harsher winters than the vinifera cultivars. However, the fruit cannot stand either the rigorous postharvest handling or extended storage that many of the vinifera cultivars can.

B. MARKETING GRAPES IN CALIFORNIA

Until the latter part of the 19th century, table grape production in California (*Vitis vinifera* L.) was limited almost exclusively to that for local markets within the state. This market by present standards was very small. However, the situation changed with the completion of the transcontinental railroad in the 1860s followed later by the development of the ice-refrigerated railroad car. These changes made accessible the large markets of eastern North America. Expansion was slow at first, while the technologies of harvesting, packing, refrigerated car design, and railroad schedules improved to the point that grapes of acceptable quality could be delivered to these markets as far as 3000 miles away. The volume shipped east increased to nearly 14,000 tons annually by 1900 and to 10 times this volume by 1916. During the next 8 years the volume increased rapidly to about three quarters of a million tons annually. This dramatic increase was due to several factors; (1) more efficient and complete re-icing services across the United States, (2) faster railroad schedules, (3) advent of Prohibition which caused many grape growers to switch from wine to table greap production, and (4) enactment of standardization laws prescribing minimum standards of quality for the grapes.

Still the market quality of the grapes left much to be desired. Delayed and inadequate cooling often resulted in soft unattractive berries and dry stems that broke readily during handling. Decay was a major problem especially in years when wet weather occurred before harvest. Further, the grapes had to be marketed very soon after harvest which often caused market gluts and low prices for the fruit. Technology was not advanced to the point that the fruit could be held in cold storage for more than a few days without drastic impairment of market quality. To meet these challenges the industry cooperating with the Experiment Station of the University of California and the U.S. Department of Agriculture made substantial contributions. Marked improvements in quality and marketing schedules came about after the 1920s with the application of improved technology in five areas; (1) more complete information on the relation of the chemical composition of the grapes and flavor to consumer preference, (2) more accurate assessment of the decay potential of the grapes at the time of harvest, (3) more precise programs and more options for the utilization of sulfur dioxide to control postharvest decay of the fruit, (4) more critical application of programs for temperature management of the grapes to retard the spread of decay, and (5) better control of air temperature, velocity and relative humidity to reduce the rate of grape deterioration from water loss and decay.

II. ATTRIBUTES OF TABLE GRAPES

As the culture of *V. vinifera* spread, some cultivars emerged as more desirable for table fruit. Significant are the morphological features and chemical composition of these cultivars. The berries are conspicuously larger than those of either wine or raisin grapes. Large berries are not only more attractive, but also more convenient to eat when of ''bite size''. Further, colored grapes are brilliantly prigmented berries—either bright red or jet black—not with intermediate hues of orange, brown, or purple. Although a berry that has a tender skin and detaches easily from the stem is desirable for satisfaction in eating, relatively tough skins and strong berry attachments are essential attributes for table grapes, which must withstand the unfavorable environment of rigorous handling, storage, and transport. Such rough treatment often extends through long periods of storage and over long distances.

Flavor is another important attribute of table grapes. The sweetness of the sugars dominates, but at the same time is complemented by the tartness of the organic acids. The amounts of these major components are usually less than in wine grapes—the sugars, because high sugar levels (although usually desired by the consumer) are associated by the trade with overmaturity and lack of keeping quality; and the acids, because high acid levels accentuate a ''sour'' or ''green fruit'' taste. Prominent varietal flavors such as those of the 'Italia'', ''Muscat of Alexandria,'' or ''Concord'' are usually desirable. However, excessive astringency, caused by the tannins found chiefly in the skins and seeds, is undesirable. Recently, seedlessness has become much preferred over seededness, especially in American markets. Regularly these markets pay a premium for this characteristic. As proof, growers of 'Thompson Seedless' (classified as a raisin variety) go to great expense to girdle, berry-thin, and apply growth regulators to the vines to increase the size of seedless berries to that acceptable as table fruit.

A. VASCULAR ANATOMY OF THE GRAPE BERRY

Botanically, the fruit of the grapevine is a berry. The fruit is borne on clusters consisting of one to several hundred individual berries. The stem structure consists of a rachis (central stem), primary, and secondary laterals or branches. Branching from these laterals are pedicels, often called capstems, each with a berry at its terminal end. The stem structure has special significance with respect to the carrying quality of the fruit. The stems will dry readily and become brittle, especially rapidly when cooling is delayed excessively. Flexing of the stems from rough handling, often causes breakage at the bases of the laterals and pedicels resulting in detached berries—a form of so-called ''dry shatter'' or ''dry drop''. The greatest bending occurs at the junction of these parts of the stems which crushes the tissue and ruptures the cuticle. The result is that the rate of browning and shrinking is accelerated owing to exposure to oxygen and loss of a water vapor barrier that results in rapid loss of water. Rough handling may also tear berries loose leaving broken vascular strands as a wet "brush" still attached to the pedicel. This type of separation constitutes a form of so-called "wet shatter." A weak pedicel attachment, or breaks in the skin around the pedicel attachment caused by stress of handling, aggravates this wet-shatter problem. Another form of wet shatter or "wet berry drop" results when decay organisms such as *Botrytis cinerea* Pers. ex Fr., *Cladiosporium herbarum* (Pers.) Lk. ex Fr., or *Stemphyllium* sp. rot the attachment area and cause the berry to slough off. Symptoms of this type of shatter are the absence of a wet brush adhering to the pedicel, and maceration of the berry tissue in the attachment area.

A serious form of dry shatter often occurs when the pedicel detaches easily and cleanly from the berry during postharvest handling. The problem can be especially acute with the variety 'Thompson Seedless' when as much as 15% of the berries may be left in the container after the clusters have been removed in the market. This happens sporadically, shatter being more severe in some seasons than in others. This type of shatter is definitely aggravated by improper cooling that results in dry stems, by advanced physiological maturity, and by rough handling.

Shatter is not necessarily restricted to clusters that are very ripe as measured by a high degree Brix (% soluble solids in the juice). Advanced physiological maturity may occur when harvest must be delayed because of a heavy crop resulting from inadequate thinning coupled with large berries induced by girdling and gibberellic acid applications. Under this condition maturity, or even senescence, of the vascular tissue of the pedicel attachment continues at a normal rate even if the increase in degree Brix is abnormally slow. As a result, an abscission zone may form that weakens the attachment area causing shatter of berries that may still be relatively low in sugar content. This usually happens when a cork cambium develops from parenchyma cells and as a result intercellular bonding weakens. Cleavage of the tissue in this zone can now take place, often with little stress as when the pedicel is twisted or bent during handling. Separation leaves the attachment area of the berry essentially healed with the central vascular strands still embedded in the berry.

A peripheral vascular system originates at the pedicel attachment area and forms an envelope of strand surrounding the pulp just underneath the subepidermal cell layer. The network arrangement of the strands is the result of the enormous enlargement of the pulp cells during berry growth that stretched these strands laterally causing them to assume a "diamond mesh" pattern. At the same time expansion of the pulp cells isolated this peripheral system from the central system.

The central part of the berry holds zero to four seeds. In a 'Thompson Seedless' berry rudimentary seeds can usually be detected indicating that the embryos aborted at a very early stage of development.

The pulp tissue extends outward to the peripheral vascular envelope. It is composed of relatively large isodiametrically shaped cells with large vacuoles and thin walls. Most of the soluble solids are in in these vacuoles. Just outside the peripheral vascular envelope are four to six layers of much smaller cells that constitute the subepidermal envelope. The shape of these cells ranges from isodiametrical to laminated. The depth of the cells is much less than the width because the cells are stretched in a lateral direction during expansion of the pulp cells. These cells have thicker walls than those of the pulp and are heavily impregnated with pectic compounds, which form the middle lamella or intercellular bonding material. These cells contain the anthocyanin pigments of colored grapes (which consists of six to eight cell layers).

B. EPIDERMAL ANATOMY OF THE GRAPE BERRY

The epidermis is a well-defined layer of cells outside the outer subepidermal layer. The cells are even more distinctly laminated than those of the subepidermal layers. Closely bound to the epidermal layer is a relatively impervious cuticle, about 2 to 4 μm thick. This cuticle is composed of a complex matrix of wax and waxlike compounds, highly resistant to the transfer of water or water vapor.[1] This property of the cuticle is particularly effective in retarding the rate of water loss from the berry after harvest and thus keeping the berry longer in a fresh crisp condition. However, water loss is still a formidable problem in maintaining a crisp fresh condition. The taut cuticle gives the berry a shine or luster. Progressively the berries become duller in appearance as the cuticle loses tension becuase the berries shrink from loss of water. This change in appearance might be likened to that of an inflated balloon—becoming duller in appearance and finally developing wrinkles as it is deflated.

Water loss also causes the skin to pull away from the pedicel attachment and form a depression around the base of the stem. The more pronounced shrinkage in the attachment area is due largely to the "wicking" action of the stem. Stems lose water much more rapidly on a weight/unit volume basis than do the berries because of the high surface-to-volume ratio and because they have a high concentration of stomates and lenticels through which water vapor can escape readily. Further, the rate of water loss is drastically increased through breaks in the cuticle at the stem—injuries commonly caused by the stress of handling.

The "bloom" on the surface of the cuticle is the effect of light reflected and diffused by the overlapping wax platelets.[2] Altering the orientation of these platelets by rubbing destroys this bloom giving the cuticle a shine rather than the desirable luster effect. It is this bloom, or its absence, that makes it such an important quality factor (sales appeal). To the experienced grape merchandizer the condition of the bloom is also an informative symptom of the amount and roughness of handling that the fruit has received.

C. EFFECT OF SULFUR DIOXIDE

Lenticels are present on the berry as very small and widely scattered mounds of suberized cells. They are located on the pedicel also, but are much larger and closer together than they are on the berry surface. Sometimes these lenticels are incompletely suberized and therefore are pervious to chemicals such as sulfur dioxide. This fumigant can penetrate such lenticels and cause localized bleaching of the skin tissue.

Sulfur dioxide bleaching is more conspicuous in red or black grapes than in white grapes because the gas bleaches the red anthocyanin pigments as well as any carotene or chlorophyll pigments present.

There are normally no functional stomata in the cuticle of the mature grape berry; however, they are present in the pedicel. Their absence in the berry is one reason why the cuticle of grapes is relatively impervious; the fruit can tolerate sulfur dioxide to a much greater extent than most other fruits.

D. EFFECT OF AMMONIA

Ammonia as well as sulfur dioxide have on occasion injured grapes. Ammonia is not used deliberately, as is sulfur dioxide, but reaches the fruit as a result of accidental leaks in the refrigeration system.[3] On many occasions storage grapes have been damaged, sometimes so severely as to be worthless. The mode of penetration is similar to that for sulfur dioxide, but the symptoms of injury are quite different. Any opening in the cuticle can be an avenue for penetration, so symptoms of mild injury are spots in the skin. However, these spots are brown in color for white grapes with no sign of bleaching. Also, the stems of the injured fruit are deep brown to black whereas the stems of sulfur-dioxide-treated grapes would still be some shade of green. If the stems are still fairly plump, when exposed to the gas, they will have a watersoaked appearance. If the injury is severe, the berries will be deep brown to almost black. Red or black grapes do not show the dramatic color change when injured, but the symptoms are still very characteristic. A distinct bluish color is imparted to the anthocyanin pigments because the strong basic reaction of ammonia with water raises the pH of the juice to, or above, neutrality.

E. EFFECT OF FREEZING

Table grapes, being living tissue, are damaged by freezing.[3] On rare occasions this may occur in California vineyards before harvest. If vineyard temperatures drop to −2 to −3°C (26 to 28°F) for several hours, freezing of the stems can be expected; if the temperature drops further, the berries will probably freeze, too. When frozen stems are thawed, they first become water-soaked in appearance, then turn brown and, finally black. They will shrivel rapidly and become brittle if drying conditions prevail. The pulp of the berries has a water-soaked translucent appearance upon thawing. This pulp starts to turn brown, especially when exposed to the air. Berries that have been frozen exuded juice like those in storage that had been injured by sulfur dioxide. After freezing, browning is associated with this ''leaking'', but not bleaching as with sulfur dioxide injury.

Occasionally, grapes have been frozen accidentally after harvest. This is most likely to occur in fruit most exposed to the air coming from the refrigeration surface in the cooler, storage room, or carrier. As in the vineyard, if the air temperatures stated persist for several hours, the symptoms of freezing described and illustrated for fruit in the vineyard will likely appear. However, they may not be evident for two to four days following injury because of the low prevailing temperatures of storage and transit.

F. EFFECT OF PATHOGENIC ORGANISMS

Several fungi will attack grapes and cause decay or scarring of the berries. The most important postharvest decay organism is *Botrytis cinerea* Pers. It is expecially serious because it grows vigorously at ordinary vineyard temperatures of 50 to 77°F (10 to 15°C). Even in storage it will continue to grow, although slowly, at a temperature of 30°F (−1°C). Further, it can infect grapes by direct penetration—not even requiring a wound to gain entrance.[4] The mycelium produces enzymes including pectinase that hydrolyze the pectic compounds of the middle lamella between the cells, causing these cells to separate. Such lesions are easily seen on white grapes as brownish areas. On red grapes they can sometimes be seen as a faint grayish area. The term ''slip-skin'' is often applied to this type of infection which is characteristic of an early *Botrytis* infection. Often the lesion may encompass the entire berry with the result that all the skin will slip from the berry leaving the pulp tissue intact. Only the epidermal and subepidermal cell layers are macerated until advanced stages of decay are reached. Turgor pressure of the berry often breaks the fragile skin over a field infection of *Botrytis* with the result that the fungus grows out through the crack forming a ridge of ash-colored conidiophores and conidia. This phase of *Botrytis* infection is often termed ''gray mold'' rot.

If, as often happens, a *Botrytis*-infected berry is not detected and is packed, the fungus will continue to grow in the berry. The fungus cannot be eradicated with sulfur dioxide, even at severe dosage levels.[5] However, if storage fruit is retreated every week with the proper dosage of the fungicide, the infection can be contained within the berry. At storage temperatures the fungus produces few spores so the mycelium is white as contrasted with the gray color of vineyard infections. Without effective sulfur dioxide treatment the fungus ''nests'' and may infect a whole cluster. If unchecked, the entire pack may become a mass of moldy fruit.

Occasionally *Botrytis* may invade the stem from an infected berry and spread to other berries by this route. The fungus can invade the stem from the infected berry, then subsequently spread throughout the stem structure in storage. Although the fungus can be contained within a berry, it is difficult, if not impossible, to prevent its spread in stems, hence the importance of trimming out all infected parts during packing.

The foregoing suggests the insidious nature of *Botrytis* as a postharvest mold. The threat of undetected field infections in storage grapes is the primary reason why sulfur dioxide must be applied weekly as long as the grapes are stored. Even complete containment of the infections with the fungicide may not be adequate if too many infected berries are present to begin with. For example, a tolerance limit of 0.5% decay by weight represents about 12 normal sized 'Emperor' berries per box. Obviously, if the number of infected berries packed in each container exceeds this number, the pack could not pass a condition inspection later in storage, no matter how near-perfect the decay control program had been.

Fortunately some precautions can alleviate the problem of *Botrytis* mold in storage. The fungus is most apt to infect grapes after they are ripe, and even then moisture on the berries or a high relative humidity must persist for a day or two to allow time for conidia to germinate and establish infections.[6] Such conditions are usually brought about by rain, dew, or fog. Heavy cover crops and rank vine growth may aggravate the problem by causing the moist conditions to prevail until infections can be established. If infections are suspected under these conditions it is wise to suspend harvesting for at least 2 or 3 d. By this time infections that would otherwise be undetected will now be more apparent and can be trimmed out before the grapes are packed.[4]

Botrytis may cause decay of storage grapes that have never been exposed to wet conditions before harvest. This decay may result from infections that have been latent in the berry. This fungus has been reported capable of infecting the berry through the style during bloom, then remaining dormant during the growing season.[7] As maturity approaches, the fungus may reactivate soon enough to cause rot easily discernible at harvest, or it may not cause symptoms of rot until the grapes have been packed and stored.

At least three fungi cause a form of storage decay called "black spot". Most troublesome in some years is that caused by *Cladiosporium Pherbarum* (Pres.) Lk.[8] It can infect uninjured grapes much in the same manner as *Botrytis*,[7] which makes it particularly significant. The incidence of decay by the fungus varies widely from year to year, but does seem to be most severe following early fall rains.[9] It is most troublesome in 'Emperors', especially those harvested late in October or early in November.

Whether it infects through the uninjured skin or through wounds, the fungus grows slowly in storage causing characteristic black spots. These spots may not appear for several weeks. The decay is not confined to the epidermal layers as is that caused by *Botrytis*, but macerates the pulp tissue, often to the seeds. The decayed tissue has some integrity and can sometimes be lifted from the berry leaving a deep hole, as contrasted with the decay caused by some other fungi that turns the pulp tissue to a watery consistency. If not contained by weekly sulfur dioxide treatments, the fungus can spread slowly to other berries by contact infection. Eventually the mycelium may erupt through the skin, white in color at first, then appearing as an olive-colored felt covering over the berry. The fungal infections are usually not detectable at harvest, hence the storage fruit should be closely monitored, expecially if there has been wet weather before harvest.

Black-spot decay is also caused by species of *Alternaris* and *Stemphylium*.[10] Growth patterns and appearance of the lesions of these fungi are similar to those for *C. herbarum*. However, these fungi can infect the berry only through wounds. As a result, it is commonly these fungi that are associated with the black-spot decay of the stem-attachment area. These fungi, as well as *Cladiosporium*, if it does infect the pedicel area, will usually rot the vascular strands so that if the berry is detached or sloughs off there is no wet "brush" left on the pedicel. Early fall rains increase the incidence of *Alternaria* and *Stemphylium* rots just as it does that caused by *Cladiosporium*. Wet conditions probably encourage growth of these fungi on dead organic matter or injured fruit which builds up a spore load in the vineyard. With more spores present, infections through wounds caused by postharvest handling are more likely to increase.

Species of *Penicillium* (blue mold) can infect grapes, but are rarely a postharvest problem. They require a wound for entrance into the berry, which limits their infection capability. Moldy harvesting equipment such as field lugs can be a source of inoculum, and if berries rub against the bottom and sides of the containers, infection can result. Subsequent rough handling causing further injuries increases the possibility of blue-mold infection. *Penicillium*-infected berries can be detected by the watery consistency of the contents and a pronounced moldy odor. If the fungus grows on the surface it will appear as tufts of white mycelium which rapidly turn blue or green as a dense mat of spores is produced.

Aspergillus niger v. Tiegh. is frequently encountered in the vineyard as a part of a hot-weather bunch-rot complex. Although not a true smut fungus it is frequently termed ''smut'' because of the black sooty appearance of infected berries caused by the black conidia of the fungus. The rot, like that produced by *Penicillium* sp., reduces the pulp tissue of the berry to a watery consistency. The fungus grows little or not at all at normal transit and storage temperatures, therefore the presence of *Aspergillus* infections in harvested table grapes indicates one of two things; (1) poor trimming so that field infections of the fungus were not eliminated from the pack; (2) gross temperature mismanagement so that high temperatures prevailed long enough for the fungus to become established. Under normal circumstances, then, this fungus is not a postharvest problem. However , in the market, if displayed fruit is allowed to remain at or above 68°F (20°C) for a day or more, this fungus may appear. If sporulation takes place, the spores have the added nuisance effect of ''soiling'' sound berries as they are smeared about during handling.

What has been said for *Aspergillus* sp. applies generally to *Rhizopus* sp. This fungus is a common hot-weather bunch-rot organism which if trimmed from harvested fruit does not become a postharvest problem with normal handling. This fungus grows rapidly under warm, moist conditions producing a coarse gray mat of mycelium. Distriubuted in this mat are numerous sporangia (spore balls), which gives the mycelium the appearance of having been sprinkled with fine black pepper.

The fungus *Uncinula nectrix* does not produce rot of grapes, but can seriously detract from their quality. During the growing season, if not controlled, it invades the epidermal cells of the green surfaces of the vine including the cluster stems and berries. It does not spread further on the stems and berries after harvest, expecially after the first treatment with sulfur dioxide. However, it leaves a characteristic ''lacy'' scarring pattern on the berries. Heavy infections, expecially later in the season, may show the presence of the external mycelium and conidia as a powdery covering on the berry; hence the name powdery mildew. The berries may even be deformed or cracked as a result of severe infection leading to invasion by mold organisms that cause decay. The stems may also be affected showing the same appearance. Especially serious is that these infected stems shrivel and break readily causing shatter of the berries when the clusters are removed from the lug.

A type of scarring similar to that caused by powdery mildew is caused by thrips, a very small insect that scars the surface of the berry early in its development. Commonly the insects will remain under dead floral parts at the calyx end of the berry shortly after set and their feeding on the berry in this area will cause rather coarse scarring not having the delicate lacy pattern of that caused by mildew.[11]

III. CHEMICAL COMPOSITION OF TABLE GRAPES

The principal components in the ripe grape are carbohydrates — chiefly sugars. Glucose and fructose are present in about equal amounts. Normally, there is less than 1% sucrose and practically no starch. The total sugar content in most table-grape varieties, when considered commercially mature, ranges from about 14 to 18%. The degree Brix (% soluble solids of the juice) is often called the sugar percentage but, as will be explained later, other compounds affect the degree Brix value, although usually to a small extent. These sugars accumulate rapidly only during the 21 to 60 d before harvest. Heavy cropping, disease, or inadequate sunshine will slow the rate of this increase, sometimes drastically.[12]

The organic acids, although present in small amounts compared with the sugars, contribute markedly to the overall taste. They do not ordinarily exceed 1% by weight of the juice and may be as low as 0.4%. Tartaric is usually the dominant acid, although malic may be present in significant quantities. Citric, succinic, and other acids may also be present, but only in very small amounts.

The acid concentration increases rapidly early in the season, often to as much as 3% shortly after berry set, then declines until harvest. However, the *total amount* of acid *per berry* increases until the berry is nearly full size.[13]

The phenolic compounds in grapes are significant in many ways. The anthocyanins in the subepidermal cells of colored grapes are synthesized in the berry as it reaches maturity. The simpler complexes of malvidin, cyanidin, petunidin, and delphidin-3-monoglycosides impart the bright red color characteristic of 'Flame Tokay' and 'Emperor' grapes. Additional compounds such as these pigments acylated with *p*-coumaric and caffeic acids and peonidin-3-monoglucoside impart more intense pigmentation ranging from deep red to jet black as in 'Ribier'.[13]

Tannins, the condensation products of phenolic compounds, are found chiefly in the epidermal cell layers and the seeds. They cause the astringent taste. Further, they impart the brown color to injured or senescent cells. 'Italia' berries bruised by lid pressure commonly show this symptom. Sometimes this browning develops in the area of the rudimentary seeds of 'Thompson Seedless,' commonly referred to as "internal browning". Tannins are formed when the enzyme polyphenoloxidase is released into the vacuoles of the cells from injured or senescent cytoplasm, there complexing the phenolic compounds into the brown tannin end product.

Characteristic varietal flavors appear to be produced by many compounds, present for the most part in very small amounts. Only two have been identified as being at least the dominant flavor component—terpene linalool causing the characteristic flavor of muscat grapes, and methyl anthranilate the "foxy" flavor of the 'Concord' variety.[13]

IV. PHYSIOLOGY OF GRAPES

Grapes being living tissue respire, although the rate is low compared with that of most other fruits.[14] Very small amounts of sugar and organic acids are slowly converted into CO_2, H_2O, and heat. There are few other significant chemical changes. There are no starches to convert to sugar, hence there is no increase in sweetness. Any softening of the tissue is due to flaccidity from water loss as there is little, if any hydrolysis of the intercellular pectic compounds. The grape, then, can live a relatively long time after harvest if protected from water loss, decay by microorganisms, and injury from rough handling. This life can be extended to as long as six months or more (depending on the variety) if the temperature is kept as low as possible without freezing. The low temperature minimizes the respiration rate, which prolongs the normal metabolism of the fruit and thereby its useful postharvest life.

V. TABLE GRAPE VINEYARD

A. TRELLISING AND TRAINING

Trellising and training is more important for a table-grape vine than for either raisin or wine-grape production, because greater emphasis is placed on *appearance* as well as total amount of the fruit. A well-trellised vine can produce more usable table fruit partly because more of the clusters are attractive and of uniform maturity. This is brought about because the clusters hang isolated, free of being rubbed by leaves, shoots, or trellis wires. Highest production can be obtained only if the leaf canopy is spread out so that a maximum number of leaves "see" the sun. Only in this green chlorophyll tissue are the sugars synthesized, and only if exposed to sunlight. A green leaf traps about 90% of the radiant energy of the sun leaving only about 10% for the leaf in its shadow.[15] In fact, densely shaded leaves are a liability to the vine, consuming more energy in growth than they provide in the trapped energy of synthesized sugars.

Leaves should "see" the sun, but the fruit should not. Exposed berries frequently sunburn in the severe temperature and light-intensity conditions of California vineyards. At the same time, pigmented varieties such as 'Emperor' and 'Tokay' grapes will not color satisfactorily in deep shade. Such fruit will color though if the leaf canopy is thin or if a reasonable amount of reflected light reaches the berries from the sky. How extensive a trellis can be to provide the foregoing conditions often is limited by cost or cultural operations. Most California grape growers favor a crossarm type. This system provides good exposure of the leaf canopy to the sun, yet the fruit is well shaded to minimize sunburn. Also, the clusters are isolated from the leaves and easily accessible for harvest. However, if the crossarm is offset and slopes rather steeply, the fruit of a cordon-trained vine will frequently be excessively crowded under the lower end. Further, deep shade in this area often results in poorly colored fruit of red and black varieties. Less slope and less offset (or none) will alleviate these problems. Favored by many table grape growers is a horizontal crossarm system of either one or two crossarms. This trellis provides easier access to the fruit from both sides of the vine, and the distances from the head of the vine to the trellis wires are more uniform to facilitate tying the canes of 'Thompson Seedless' vines to the wires. Also, for a cordon-trained vine there is more uniform spacing of clusters, and coloring of red and black varieties is more uniform than with a sloping offset trellis.

The number of crossarms, their length, height from the ground, and angle of slope varies depending on several factors: (1) vigor of vine growth, (2) whether the vines are cane or cordon pruned, (3) shoot and leaf removal needed so clusters hang free, (4) berry thinning needed, (5) growth regulator and pest applications of sprays or dusts that are required, (6) irrigation system used, and (7) cover cropping program followed. The vines are usually spaced 10 to 12 ft between rows and 6 to 8 ft between vines in the row. Rows are seldom longer than 220 yd, so irrigation runs are not excessively long and harvesting crews do not have to haul fruit long distances to the avenues.

More elaborate trellis systems are used in other table-grape districts of the world, especially where labor costs are lower. In South Africa the sloping crossarm may be so long that the higher ends are supported from the next row giving a "factory roof" appearance to the vineyard. In Spain and Chile, a horizontal overhead trellis of beams or wires is commonly used. Such systems facilitate more intricate hand-thinning and uniform applications of growth regulators, which further improve the appearance of the fruit. For example, in South Africa meticulous hand thinning of the normally tight clusters of the 'Barlinka' makes this variety feasible to grow whereas in California labor costs would make it prohibitive. In Chile, often clusters of 'Thompson Seedless' are hand-sprayed individually several times with gibberellic acid. Such treatments not only increase berry size, but the increase is more uniform.

B. CULTURAL PRACTICES BEFORE HARVEST

Precautions should be taken in preparing the vineyard for harvest to facilitate the operation and retain the attractive appearance of the fruit. The avenues should be treated to prevent dust clouds created by moving vehicles from settling on the clusters detracting from their appearance. The space between the rows should be prepared to facilitate movement of vehicles between the vines for distributing picking containers and to pick up those that are filled. This preparation necessitates withholding irrigation water sufficiently long that the soil is stable for such traffic. Irrigation furrows may have to be leveled and high cover crops removed. Even some pruning of long shoots or leaf stripping may be necessary so that the pickers will have easy access to the fruit.

VI. TABLE GRAPE VARIETIES OF CALIFORNIA

A. EMPEROR

This variety accounts for about 20% of the table grape production of California. There are about 22,000 acres of vines, largely limited to a belt near the foothills of the Sierra Nevada Mountains in Kern, Tulare, and Fresno Counties. The clusters are large, conical, and well filled with large red to reddish purple berries that have seeds. The berries are elongated obovoid or ellipsoidal. The texture of the berries is moderately firm, neutral in flavor, with thick tough skins. The stems are strong and the berries adhere very strongly. The harvest season extends from September 15th to November 1st. It is the foremost storage variety and is now marketed until January or even March. Gibberellic acid is sometimes applied to the clusters two weeks after fruit set to reduce shriveling of the berries. Ethephon is often applied at the first sign of pigmentation to enhance color.

B. CARDINAL

This variety accounted for about 2% of the table grape production of California in 1982. Production is in the Coachella Valley and from Fresno south in the San Joaquin Valley. The clusters are large to medium in size and loose to compact. The berries are very large, round to short oval, dark red to reddish black in color with advanced maturity, and are seeded. Ethephon is often applied to enhance color and the vines sometimes are girdled to advance maturity. The variety is very early being harvested in the Coachella Valley the first two weeks of June and in the San Joaquin Valley the first two to three weeks of July. The berry is thin-skinned and the stem attachment is weak so the clusters must be handled carefully to avoid injury to the fruit.

C. RIBIER (ALPHONSE LAVALLEE)

This variety accounts for 4 to 6% of the table production of California. It is produced largely in the lower San Joaquin Valley. The clusters are of medium size, short conical varying from loose to compact. The berries are very large, oblate to short ellipsoidal, black in color, seeded and moderately tough-skinned. The stems are tough and the berries firmly attached. A few vines are girdled at fruit set for larger berries and often at 4% color to advance maturity. The berries are prone to cracking from lid pressure if packed when very turgid. Careful packing alleviates this and the polished effect caused by rubbing the berries. It is a mid-season grape and stores well.

D. QUEEN

This variety was introduced in 1954 from a cross between Muscat Hamburg and Sultinina. It now produces about 0.5% of California's table grape production. The clusters are large, loosely winged, and filled with large berries ellipsoidal in shape and firm with seeds. The color is dark red. The fruit matures during September in the San Joaquin Valley. The vines are often girdled at 1 to 5% color to advance maturity and frequently sprayed with ethephon at the same stage to enhance color.

E. CALMERIA

Introduced in 1950 from an open-pollinated seedling of Almeria (Ohanez). It produces over 4% of California's table grape production. The clusters are large, well-filled with large, ovoid elongated seeded greenish berries with tough skins. The stems are tough and berry attachment strong. The variety matures late and is well adapted to extended storage. The harvest season is during the latter part of September and during October and marketing extends until the end of December or even into January. It is replacing the Almeria as a storage white grape.

F. ALMERIA

This variety now accounts for only about 0.5% of California's table grape production, being slowly displaced by the Calmeria. The clusters are medium to large, filled to compact, with white seeded berries with thick, tough skins and strong stem attachments. The berries are very subject to heat injury (Almeria spot). Harvest season extends from mid-September through October.

G. ITALIA

This variety produces less than 0.5% of California's table grape production. The clusters are large to medium in size and well filled with large, oval, white seeded berries that have a slight muscat flavor when very mature. The skin has a pronounced astringent taste. The berries are very subject to brown discoloration, especially from abrasion during handling. The harvest season extends from mid-August through September.

H. FLAME TOKAY

This variety now produces only 2 to 4% of California's table grape production. The clusters are large and well filled to compact with large ovoid, truncate berries, bright red in color and seeded. The berries are very subject to sunburn and often fail to color adequately, especially during very hot weather in Lodi district where it is produced almost exclusively. The harvest season is from mid-August through September.

I. FLAME SEEDLESS

This variety was introduced in 1973 and has increased from no production five years ago to 4.4% of California's production during the 1982 to 83 season. It appears to be displacing the Cardinal. Current production is chiefly in Kern County, starting the first part of July, although there is some production in the Coachella valley a month earlier. The clusters are medium in size with small bright red crisp seedless berries. To achieve larger berries the vines are girdled and treated with gibberellic acid at fruit-set. At the first sign of color the vines are treated with ethephon to advance maturity of the fruit.

J. PERLETTE

This variety produces about 6% of California's table grape production. It is the earliest commercial variety and is grown almost entirely in the Coachella Valley. The harvest season extends form about May 20 to the middle of June. The variety was introduced in 1946. It has large clusters and requires heavy thinning by a combination of "brushing" during the prebloom stage and flower cluster thinning during bloom. Berry size is enhanced by girdling and application of gibberellic acid at the fruit set stage. The berries are medium sized, white, thin-skinned, crisp and seedless.

K. THOMPSON SEEDLESS

This variety, although officially classified as a raisin variety, produces about half of California's table grapes. It is extensively planted in the Coachella Valley where the harvest season extends through the latter half of June. Much larger plantings are in the San Joaquin Valley from Madera southward to Arvin. For successful table grape production the vines are usually treated at bloomtime with gibberellic acid for thinning and again at fruit set to increase berry size. This program is supplemented with girdling for berry size and cluster thinning at the fruit set stage. This program results in a moderately loose cluster with berries ranging from 4 to 6 g in weight, instead of about 2 g for a "normal" berry for raisin production. The berries are oval to elongated, fairly thin-skinned, seedless, greenish, and with a pedicel attachment that sometimes is very weak resulting in a heavy "shatter" of abscissed berries during postharvest handling.

L. RUBY SEEDLESS

This variety produces less than 0.5% of California's table grape production but is increasing slowly. It was introduced in 1968. The clusters are very large, conical and well-filled with berries medium in size, seedless, ovoid, in shape, skin reddish-black to dark red and tender. The variety ripens mid-season to late in the lower San Joaquin Valley. Berry attachment is firm and the clusters resist damage well during postharvest handling.

These 12 varieties account for about 95% of California's table grape production. There are several varieties of minor importance including: Beauty Seedless, Exotic, Red Malaga (Molinera), Malaga (White Malaga), Olivette blanche, Rish Baba, and Kandahar. To these may be added new varieties with varying degrees of promise: Superior Seedless, Centennial Seedless, Dawn Seedless, Blush Seedless, Redglobe, and Christmas Rose. Interesting are two varieties with a mild *labrusca* flavor—Niabel and Early Niabel. There are several varieties of significance in areas outside of California including Concord and Niagara in the Eastern, North-Central, and Northwestern United States.

VII. FACTORS AFFECTING HARVESTING OF TABLE GRAPES

Table grapes should not be harvested until mature. Unlike many fruits, grapes do not "ripen" after harvest, so they should be picked only after they reach the optimum stage of acceptability in appearance, flavor, and texture.

A. COLOR AND MATURITY

Appearance is determined chiefly by color, especially for red and black grapes. In fact, standards for color (white varieties excepted) are specified in the *U.S. Standards for Grades of Table Grapes.*[16] These standard may be supplemented by higher standards set by local marketing groups such as the Tokay Marketing Agreement in Lodi, California. The U.S. Standards Grades for Table Grapes include: U.S. No. 1 Table, U.S. Fancy Table, and U.S. Extra Fancy Table. The U.S. No. 1 grade requires that for red varieties each bunch have at least 60% of the berries showing "characteristic color" for the variety, and for black varieties at least 75%. For U.S. Fancy each bunch of a red variety must have at least 66 2/3 percent of the berries showing "good characteristic color" but 'Flame Tokay' and 'Cardinal' bunches must have at least 75% of the berries showing "good characteristic color". For black varieties each bunch must have at least 85% of the berries showing "good characteristic color". For U.S. Extra Fancy each bunch of a red variety must have at least 75% of the berries showing "good characteristic color", and for black varieties at least 95%. The minimum requirements of the desirable attributes of the grapes, such as color and bunch size are usually progressively higher from U.S. No. 1 to the U.S. Extra Fancy grades. At the same time the permissible limits of most undesirable attributes (defects) such as crushed, scarred, or shot berries, are progressively lower.

Three aspects of color are considered in determining whether a bunch of grapes meets the minimum standard for color: (1) what is the lowest limit of color intensity considered characteristic for the variety (2) what percentage of the berry surface must have at least this minimum intensity to be considered colored (3) what percentage of the berries of the bunch need to fulfill these first two requirements for the bunch to meet the color standard for the grade?

The requirement "characteristic color" specifies that for red varieties the color may be pink to dark red, except that for 'Flame Today' the color may range from light pink to dark red and for 'Cardinal' light pink through purple. For black varieties, the color may range from reddish-purple to black. The requirement "good characteristic color" specifies that for red varieties the color may range from light through dark red, except that for 'Flame Tokay' it may be pink through dark red and for 'Cardinal' light red through purple. For black varieties, the color may range from purple to black (no reddish tint). For both "good characteristic color" and "characteristic color" at least two-thirds of the berry surface must show the kind and intensity of color specified for the berry to be considered "colored". The term "fairly well colored" is used when the bunch meets the color requirements of the U.S. No. 1 grade. For the U.S. Fancy grade the term "reasonably well colored" is used, and for the U.S. Extra Fancy grade the term "well colored" is applied.

B. TOTAL SOLUBLE SOLIDS AND MATURITY

The total soluble solids of the expressed juice of the berry contains acids as well as sugars in solution. However, the acid fraction is usually very small compared with the sugar fraction—so small, in fact, that for practical purposes the soluble solids are considered as the sugar content. It is measured on a degree Brix (or Balling) scale,—a scale based on g of sucrose in 100 g of a sugar-and-water solution.

The minimum standard degree Brix, like that for color, varies depending upon the variety. Degree Brix and the Brix/acid ratio are specified for all varieties whether colored or white. The minimum may vary also depending upon the district in which the grapes are produced. For several varieties this minimum is one degree less if the grapes are produced in the Coachella Valley rather than San Joaquin Valley, the argument being that the higher ripening temperatures of the Coachella Valley depress the acid content more than the less severe temperatures of the San Joaquin Valley. The result is that the Brix/acid ratio may be the same for the fruit from each valley even though the degree Brix may differ by one. The net effect is that both lots may taste the same because taste response correlates more closely with the Brix/acid ratio than with degree Brix alone. This is especially true for the higher-acid grapes early in the season.

C. CLUSTER STEMS AND MATURITY

Cluster stems may vary from those quite woody with advanced maturity (called "cured") to those more succulent. The cured stems have a lower water content than the succulent ones, hence will shrivel less with desiccation. Further, they are tougher, break less easily, and pedicel attachments are usually stronger. Such stems have much better storage potential and therefore are an especially significant quality factor for extended storage grapes such as 'Emperor', 'Calmeria,' 'Almeria,' and 'Ribier.'

D. SAMPLING METHODS TO DETERMINE MATURITY

The table-grape crop, unlike that for raisins or wine, is seldom harvested all at one time. In early districts, the first harvest may start when as little as 10% of the crop is ready in order to take advantage of early-season high prices. Also, starting early and repeating a picking every week or less allows the harvest to be completed without some of the crop becoming over mature. As many as five pickings may be necessary to do this. Light crops, as from severe thinning, and even heavier crops in late districts may be harvested with as few as one or two pickings.

As harvest time approaches, the progress of fruit maturity can be monitored roughly by sampling a few berries from clusters selected at random. The berries (3 to 6) should be taken from the middle area in order to secure an average for the cluster because the least mature berries are on the lower end and at the tips of the laterals and the most mature near the base of the laterals at the top. Each berry should be squeezed between the thumb and finger until a few drops of juice can be deposited on the glass prism of a hand refractometer and the reading taken immediately. The instrument should not be rinsed between readings because residual water will dilute the juice of the next sample. If rinsed, the instrument should be wiped dry before reuse.

A more critical assessment of maturity can be made by crushing entire clusters in a container, straining the juice through cheesecloth, then secure one or more readings from this more representative sample. If there is enough juice to float a Brix hydrometer (200 to 300 ml in a tall slender cylinder), maturity can be determined this way rather than with a refractometer. However, the industry (including inspectors) now use the hand refractometer almost exclusively because of its portability, speed of operation, and durability.

Precautions should be taken that refractometer readings are accurate. The effect of temperature must be considered because the juice expands as temperature increases. With fewer sugar molecules between the prisms, the refractive effect is less and readings will be lower. Conversely, readings increase as the temperature is lowered. Instruments are available with temperature compensation built into the optical system. Others have a small mercury thermometer attached to the instrument which senses the temperature of the instrument (and hence that of the thin film of juice) and indicates the plus or minus correction factor to be applied to the indicated reading. Still other refractometers rely on "zeroing" the instrument with distilled water. Subsequent readings are valid as long as the instrument does not change temperature. To be effective the instrument should have a degree-Brix range no greater than 0 to 30° and be graduated to 0.1°B divisions if possible, or at most 0.2°B.

Although the hand refractometer is relatively durable, it is a precision instrument and should be handled as such. A severe shock such as dropping it on a hard surface can damage the optics—either the prism or eyepiece mechanism. It should be kept clean by rinsing the prism end with clean water and carefully removing dust from the eyepiece with lens paper or a soft cloth. Juice should not be allowed to evaporate from the prism surface because evaporation concentrates the sugars and will result in a high reading. Conversely, residual water from rinsing will dilute subsequent samples and result in low values.

Sometimes it is necessary to determine the total titratable acid (total acid expressed as tartaric acid) of the grape sample. This becomes necessary when minimum standards permit grapes to pass at some degree Brix less than the qualified minimum standard, provided the acid content is low enough so that the Brix/acid ratio is above a specified level. This ratio denotes the number of parts sugar (% soluble solids) to one part acid (% total acid). This value has been shown to be superior to the degree Brix alone in predicting the palatability of grapes because it takes into account both the acid and sugar levels as taste factors.[17] Winkler in early studies showed acceptance more closely correlated with the ratio than with degree Brix. This was confirmed in later studies with taste panels and supermarket customers.[18,19] This determination is most commonly used early in the season when the acid level is still relatively high. The procedure is more laborious and requires more equipment than the soluble-solids determination, hence its application is usually restricted to these early-season grapes to screen out high-acid fruit when ripening temperatures have been relatively cool, yet permit shipment of lower-acid fruit if higher temperatures prevailed before harvest.

Needed for the determination is a supply of distilled water, some sodium-hydroxide solution, standardized preferably to 0.1333 *N*, phenolphthalein indicator, and an assortment of glassware consisting of a 10 ml burette, 10 ml pipette, and one or two 250 ml Erlenmeyer flasks.

In practice, a sample of 10 ml of clear juice is placed in a flask with the pipette. This juice is diluted with about 100 ml of distilled water and 2 to 3 drops of indicator are then added. The sodium hydroxide is added slowly from the burette until a persistent faint pink color is reached. If 0.1333 *N* sodium hydroxide is used, the number of ml needed can be converted directly to grams total acid/100 ml (% total acid) by dividing the value by 10. The Brix/acid ratio can then be derived by dividing the degree Brix by the total acid value.

E. EFFECT OF WEATHER CONDITIONS

Seldom is the grape harvest interrupted by adverse weather conditions until late in the season by fall rains or frost. If rain does wet the clusters thoroughly, it is prudent to suspend harvest for at least 3 d. If the rain does cause fungus infections to take place, this period will allow symptoms of the infection to develop to the extent that pickers can detect and trim out infected berries. Even so, fruit picked this soon after a rain should be sold immediately or if stored kept segregated and monitored closely for any development of decay.

On rare occasions grapes can be frozen late in the season. A light freeze may damage only the stems, but a hard freeze will damage the entire cluster. Freezing damage can first be detected by a blackening and rapid shriveling of stems. If the berries are affected they become mushy and exude (leak) juice. Harvest should be suspended for at least a day for these symptoms to be expressed. Clusters with freezing damage should not be packed, even those with damage restricted to the stems, because blackened stems are unsightly and will break easily if dry, or are subject to decay if wet.

VIII. POSTHARVEST HANDLING AND PACKING OF TABLE GRAPES

A. PACKING

1. House or Shed Packing

Part of the California table-grape crop is packed in sheds or specifically designed packing houses. The operation can have advantages as contrasted with the other primary method of packing in or beside the vineyard. Packing personnel are exposed less to the somewhat severe working conditions of the vineyard. Many sheds are air conditioned adding further to comfort. More advantage can be taken of mechanization in shed packing, which can make handling operations more efficient. Quality control can be close since supervision of packers on a packing line is more centralized. Further, with the shed system it is more feasible to pack more than one quality grade or even to use more than one type of shipping container.

The system consists of distributing empty field lugs under the vines from trucks moving between the rows. The pickers select suitable clusters, cut them from the vine with harvesting shears, then place them in the lugs one layer deep without further trimming. The filled lugs are loaded onto the trucks for transport to the shed. Many grape growers have mechanized this stage of the operation. An elevator pulled behind the truck eliminates the high lift from ground to truck bed and even to the top of palletized stacks of lugs. At the shed, the palletized lugs are handled by fork lift to bring them to the head of the packing line where each lug is placed on a conveyor either mechanically or by hand.

Packers are almost exclusively women. As they take the grapes from the field lug, the clusters are trimmed with shears to remove defective parts, then packed into the shipping container. The packed lugs are moved by conveyer through a lidding machine, stacked on pallets, then strapped for stability in preparation for fumigation and cooling. As the packer empties the field lugs, they are diverted to another conveyor and moved to where they are repalletized and loaded on the vineyard trucks.

2. Field And Vineyard Packing

A large part of the California grape crop is packed in the vineyard. This method has some advantages over the housepack method: (1) less capital investment in packing facilities, (2) only a vineyard crew to manage rather than a picking and packing crew whose production must be coordinated, and (3) no large inventory of field lugs.

There are two types of operations depending on where the packing takes place. A very small part of the crop is packed under the vine with the shipping container on a small stand or propped against the vine trunk. Here the picker is also the trimmer and packer. Supervision is difficult, and frequently quality control is a problem. The method is usually used by small operators or for small lots of grapes that require special handling.

Far more popular is the avenue pack. An overall view of an avenue-pack operation would show several packing stands, each at the head of a row of vines. In between two rows would be one harvesting crew which services one packing stand. Here the grapes are picked, trimmed, placed in lightweight trays. Four to ten trays are placed on a lightweight hand cart and transported to the avenue. Packing is done on the portable packing stand with enough space for a few of the trays, one or two packing scales, a rack for packaging supplies, and a canopy for shade.

A crew operating as a unit usually consists of four to six men or women. One is the packer at the stand, another the hauler, and the rest pickers. An attractive feature of this method is that the crew commonly operates as a cooperative unit. The packer is the key person of the crew, being responsible for not only the packing, but handling the packaging materials, placing the required markings on the lugs, lidding them with "snap-on" lids, and stacking them at the side of the avenue. The lugs are palletized on trucks as soon as possible and transported to the cooling and fumigation facility.

There may be as many as 20 such crews operating under a foreman. As a crew finishes its row, it moves to the next unoccupied row taking along all the equipment. Thus the crews proceed "leap-frog" fashion along the avenue. Although supervision may not be as close as in a house-pack operation, it is considerably better than with the in-row method, especially considering the volume of fruit packed. A significant breakthrough that popularized this method was the adaption of the "snap-on" lid eliminating hand nailing or moving a bulky nailer about in the vineyard.

3. Precautions Before Packing Grapes

Considering the anatomy of the grape cluster certain techniques of handling will prevent many problems of poor market quality of table grapes. Several phases of handling may be considered, the first being that of selection. Few clusters are so near perfect that they require no trimming before packing. At the same time there are others so imperfect that it is immediately obvious that they should be rejected. The first major decision then is how much trimming is economically feasible to bring a less-than-perfect cluster up to acceptable standards by removing defective parts. Labor costs and market value of the fruit will largely determine the amount of trimming economically acceptable.

Clusters rejected as not feasible to trim are those that are: (1) inadequately colored, (2) so compact that the interior parts cannot be examined for defective berries or stems, (3) so straggly that the exposed stems make them unattractive, (4) so filled with shot berries as to detract seriously from the appearance, (5) too small (at least 1/4 lb for U.S. No. 1 and 1/2 lb for U.S. Fancy), (6) have excessively sunburned, scarred, mildewed, decayed, raisined, cracked, crushed, irregular-shaped, or under-sized berries, and (7) have sunburned, decayed, mildewed, shrunken, or blackened stems. The occasional defective berry or part should be removed with special harvesting shears with curved blades.

The cluster should always be held by the peduncle, avoiding touching the berries as much as possible. Rubbing between berries or with the hands destroys the "bloom" or luster making the berries appear shiny. Further, excessive rolling of the cluster during trimming causes flexing of the stems which can result in breakage of laterals and injury to the pedicel attachments of the berries. Points of greatest stress are at the junctions of the laterals and pedicels. Bending of these parts crushes the stem tissue, which subsequently turns brown, dries, becomes brittle, and breaks easily. It takes little twisting stress during rough handling to rupture the skin at the junction of the pedicel and the berry. Such injuries (usually not visible at the time of packing) are common causes of wet shatter of berries or breakage of lateral stems of clusters. In addition, SO_2 can penetrate readily into the broken pedicel attachment causing bleaching and wetness. The bleaching is good in that it indicates the wound is at least sterilized and will not be a source of decay. However, such injured berries are unattractive and probably inedible because of off-flavors and stickiness.

4. Shipping Containers

Most California table grapes are packed in lug boxes $16^1/_8$ in. long (inside dimension) with widths of either $13^1/_2$ or 14 in. (outside dimensions) and with depths ranging from $4^1/_2$ to $5^1/_2$ in. (i.d.). The net weight of fruit packed without the clusters being individually wrapped is 22 lb (10.5 kg) for Coachella Valley grapes and 23 (11 kg) for fruit packed in the San Joaquin Valley. The container may be tapered up-and-in along the tops of the sides or down-and-in along the bottom sides to facilitate packing and cooling. The lug may be either of all wood construction or only the ends of solid wood and the sides and bottom of a paper-wood laminate (TKV).

The volume of grapes packed in expanded polystyrene containers has been increasing. The inside dimensions are 18 in. long, $11^1/_2$ in. wide and 6 in. deep and they are packed to the same net weight. Other than a bottom cushion pad, no curtains or paper liners are used as the polystyrene does not have the abrasive properties of wood that can scuff or bruise the grapes.

Some grapes are packed in lugs $15^3/_4 \times 19^{11}/_{16}$ in. (outside dimensions) with depths of 4 to $5^1/_5$ in. Each cluster is wrapped completely in tissue paper and packed one-cluster-deep.

Treated corrugated paper has shown considerable promise for shipping containers for table grapes. Extensive tests have shown the feasibility of this material if wax-dipped, curtain-coated with a plastic-wax emulsion, or treated with polyphenolic resin to resist the weakening effect of moisture uptake during storage and transit. Systems of handling have been devised to use full- or half-telescope containers. Conspicuous has been the problem of designing cartons that will withstand 3-pallet stacking. However, pallet corner supports or racks could largely solve this problem. Cooling is generally slower with these containers than with lugs; however, as cooling techniques are improved, especially by the use of "forced air", this problem can be largely eliminated. To date, use of this packaging method has had limited acceptance for table grapes, although the corrugated carton is widely used for tree fruits.

5. Cluster Arrangement in the Package

The clusters are packed in a stems-up arrangement in all the foregoing containers except the wide shallow cluster-wrapped pack. Most common is the so-called "naked pack", but to a significant extent individual clusters are "chimney wrapped". This pack, like the cluster wrap, keeps the stems fresher in appearance than the naked pack because the wraps protect them from drying, especially during long-term storage. Further, the wraps retain berries that may shatter and otherwise be lost when unpacked. However, problems of rapid cooling and SO_2 penetration during gassing are increased. The cluster-wrap pack has the additional disadvantage of making it more difficult to examine the fruit after packing.

When packing the naked pack, the lug is first equipped with a bottom cushion pad which may be part of a one-piece liner that protects the fruit from the ends and even the sides of the lug. Vents in a one-piece liner are provided to register with those along each side at the bottom. The lug is oriented with the label end toward the packer and tilted at an angle of about 15° for ease of packing. The trimmed clusters are placed in the lug starting at the lower end, holding the cluster by the stem in an upright position at the desired height, then inserting other clusters underneath as necessary. Packing continues toward the upper end maintaining a slight crown so that when the lid is fastened the pack is compressed slightly. After placement of the last cluster, the pack is usually covered with a curtain. It may be either a separate unit with ends that tuck down inside of each side of the lug or be a part of the liner. Along each side are vents to facilitate cooling and penetration of SO_2 during fumigation. Some grape producers provide a top cushion pad in lieu of, or in addition to, the curtain. This precaution is especially significant for 'Ribier' and 'Cardinal' packs whose berries crack easily during pressure of lidding — especially early in the day when the berries are extremely turgid.

The chimney-wrap pack is made in the same manner except that the trimmed cluster (or two small ones) is first laid on a pad of tissue wraps, then rolled into the top wrap to form a "cup". When in place in the pack, the edge of the wrap is slightly higher than the top lateral of the cluster. For the cluster-wrap pack the wrap completely encloses the cluster.

6. Palletization

Before World War II nearly all California table grapes were handled either by hand, one lug at a time, or with hand trucks capable of carrying as many as 12 lugs. Cooling and storage rooms were typically only high enough for convenient hand stacking of lugs — 12 ff or less. Field lugs of grapes for house packing and field-packed shipping lugs were hand-stacked onto auto trucks in the vineyard, then unloaded later at the dock with hand trucks. This necessitated that all unloading docks be at truck-bed height. Following World War II, forklifts capable of carrying one or more tons of lugs on a pallet rapidly replaced the hand truck. New facilities were constructed to take advantage of the potential efficiencies of this innovation. Packing-shed floors were often placed at ground level for greater ease of moving traffic of fruit and personnel. Also these floors were made of concrete, partly to sustain the heavy loads of the forklifts. The height of cooling and storage rooms was increased to 30 or more feet to accommodate three- and four-pallet stacking.

Three pallet sizes have evolved, each with some advantages. Dominant have been the $51^1/_2 \times 42$ in. (9-stack) pallet and the 35×42 in. (6-stack) pallet. Both accommodate the standard shipping lugs with no open space on the pallets. These sizes have the advantages that no modifications of dimensions of shipping containers were needed (or the size of the containers could be adapted to these pallet sizes), and the palletized lugs had good stability when loaded solid and strapped. The larger pallet has the added advantage of greater efficiency when handling large quantities of fruit with 50% more lugs per pallet than can be stacked on the smaller one. The 6-stack pallet became the preferred size where fast cooling was critical. Even the clipped corners of the lugs, which provide air channels through the pallet, did not suffice for rapid and uniform cooling of the 9-stack configuration using the conventional parallel flow method of handling the cooling air. The center stack, especially, cooled slowly and was the most difficult to retrieve and examine — the specific lugs in which the fruit was most apt to have problems from inadequate cooling and sulfur dioxide exposure for decay control.

The stacks of lugs on both sizes of the pallets required some restraining system to maintain stability during movement of the pallets. In the field-packing system, the palletized lugs on the trucks were simply cinched in place with overhead ropes and restraining bars. At the shed, the house- and field-packed lugs were held in place for movement in and out of cooling and storage rooms with one or more tapes or straps which encircled the palletized unit. These could be removed quickly later for hand loading into vans and reefers.

B. COOLING

Table grapes are available in most major markets of the world for most of the year. This can be ascribed largely to the application of modern technology of postharvest handling of the fruit. Production of this fruit is highly seasonal and limited to relatively small areas of the world. As a result, the grapes must often be transported over long distances and in seasons several months after they have been harvested. To meet the demands of distant markets much of the grape crop must be kept fresh and attractive for several weeks or even months after harvest. No phase in the postharvest handling procedure to maintain quality is more critical than that of cooling — removal of the sensible or field heat from the fruit after harvest.

Development of effective methods of temperature management for table grapes has been especially significant in production areas distant from major markets. A large share of the California crop is marketed in the northeastern part of the U.S. and Canada 3000 miles away, with a smaller share exported to Europe, South America, and Southeast Asia. The Republic of South Africa markets the major part of its crop in the northern countries of Europe, and much of the Chilean crop is exported to other South American countries and the U.S.

There are three compelling reasons why table grapes should be cooled promptly and thoroughly after harvest to maintain satisfactory quality: (1) minimize water loss from the fruit, (2) retard development of decay caused by fungi, and (3) reduce the rate of respiration of the fruit. Probably the most urgent reason to cool the grapes promptly is to reduce the rate of water loss.[20] This phenomenon is strictly a physical factor related to the evaporative potential of the surrounding air. It may be expressed directly as vapor pressure deficit (VPD), a term indicating the combined influence of temperature and relative humidity, and is the one factor related directly to the rate of water loss from the fruit. The equation may be expressed as follows:

$$VPD = VP \times \frac{100 - RH}{100}$$

where VPD = vapor pressure deficit (mm or in. of mercury (Hg)), VP = vapor pressure at a given temperature, (mm or in. of Hg), and RH = relative humidity (%).

It is apparent from the equation that the VPD increases as the VP increases (which would occur with a rise of temperature of the grapes, since VP is related directly to temperature of water and the juice of the berries is largely water). Further, the VPD will increase as the RH is lowered. It is to be expected, then, that the VPD would be especially high during the typically hot, dry conditions that prevail during harvest of California table grapes. Reducing the temperature of this hot fruit promptly and rapidly will therefore drastically reduce the amount of water loss.

1. Effects of Temperature

a. Water Loss

A generalized overview of fruit temperatures during a typical postharvest handling operation for California table grapes would show a wide range of temperatures. Conspicuous would be the relatively high fruit temperatures that prevail from the time the grapes are harvested until cooling is well under way. Especially critical are temperatures during the summer season. Significant, then, is the length of this phase before cooling when the fruit is deteriorating most rapidly. This phase may range from 1 to 12 h, which can spell the difference between top quality fruit and not acceptable fruit. Once cooling has started, the length of this phase, which may range from 3 to 24 h, can likewise have a significant effect on the market quality of the fruit.

The transit phase, too, can be an important factor in the postharvest handling chain, mainly because of the length of the period. It may be only 2 to 8 d for domestic shipments, but as long as a month for export grapes. Usual is a typical transit temperature of 36 to 38°F (2 to 2.5°C), which may be acceptable for the period of a week or less, but for a month this transit temperature should be nearer 32°F (0°C). Often there could be a storage period of as many as six months starting at the end of the cooling phase. Considering the great length of this period, the temperature should be near ideal (30°F or − 1.0°C).

Finally, the temperature environment during the marketing phase can be of utmost importance. It is by nature a severe phase as the fruit normally warms up during distribution handling. It is the "end of the line" — all of the deterioration of the preceding phases is additive, so market quality may appear to drop quickly, especially if good temperature management is not maintained and the marketing period is extended excessively.

Some techniques can be used to avoid the higher fruit temperatures during harvest. Fruit picked early in the morning will normally be considerably cooler than fruit picked later in the day. Cooler fruit means less deterioration and less cooling is needed. Further, exposing the picked fruit to the sun can cause a drastic rise in fruit temperature. Harvested grapes, then, must be kept shaded until cooling can be started. Also, the earlier in the day the fruit is harvested and cooling started the lower the initial temperature, which shortens accordingly the length of the period required to bring the fruit to transit temperatures. In the Coachella Valley where these data were obtained, it is common practice to start harvesting in the morning as early as visibility will permit, then suspend picking operations at noon.[20]

There are at least three symptoms of water loss from grapes. First to appear are shriveled stems that usually become brittle and break easily when handled. With grapes, unlike other fruit, the stems serve as a handle to move the fruit. When this handle is broken, the fruit is lost for all practical purposes, even though the shattered berries themselves still look and taste good.

The second symptom of water loss to appear is browning of the stem. Such stems detract seriously from the appearance of the grapes. The brown-stem syndrome follows closely that for dry stems and shows the same response to temperature.[20] In fact, the rate of stem browning increases more rapidly with temperature then does the rate of stem drying. From 70 to 100°F (21 to 38°C) there is a threefold increase in stem drying after 8 h whereas for stem browning it is more than a fourfold increase.

The third symptom of water loss is shrinkage of the berries. Grape berries do not show symptoms of water loss until shrinkage is quite evident on the stems. However, at about 3% loss in weight the berries start to appear dull as the taut condition of the skin slackens. At 4 to 5% loss the berries feel definitely soft, and above a 5% loss fine wrinkles start to appear radiating out from the pedicel. Market quality is significantly impaired when the berries feel soft, and the berries border on being unacceptable when they show distinct wrinkles. As in the case of the stems, the rate of berry softening is related directly to temperature before cooling.[20] Grapes held 8 h at 100°F (38°C) had 75% of the berries rates "soft", whereas the lot held at 70°F (21°C) had only 45% soft berries. Weights were determined for these lots of grapes during the delay period. The lot at 100°F (38°C) lost 3.0% of the initial weight during the 8 h period whereas that held at 70°F (21°C) lost 0.3% — only 1/10 as much.

That these high temperatures before cooling are relatively severe may be illustrated by comparing the rate of deterioration of the fruit with that after cooling. The general conclusion can be made that harvested grapes will deteriorate more in 1 h at 90°F (32°C) than they will in 1 d at a common transit temperature of 39°F (4°C), or in 1 week under storage conditions of 32°F (0°C). Good temperature management then is essential, considering the appalling effect the relatively short period before cooling can have on the postharvest life of the grape.

b. Decay

Decay is an ever-present hazard in the postharvest handling of table grapes. Surface contamination by fungus spores and even field infections should be assumed to be present in the berries. Two effective tools are available to minimize this danger. First is prompt and complete cooling of the fruit; and second, timely and thorough fumigation with sulfur dioxide—supplemented with weekly treatments if the fruit is stored.

Several fungi can incite postharvest decay of grapes, the most important being *B. cinerea*.

This fungus grows across nearly the entire temperature range in which grapes are handled. However, growth is slowed as the temperature is lowered until it is almost stopped at the optimum storage temperature of 31°F (−0.5°C). It is important, then, that the fruit temperature be lowered promptly and as rapidly as possible to this temperature. In addition to fumigation, an effective supplement to effective temperature management in controlling decay is careful handling of the fruit at all times to minimize injury. Broken berries are not only unsightly but are favorite courts of infection by fungi.

c. Respiration

Table grapes like all other fresh fruit are alive and remain so during postharvest handling and marketing. Being alive, they respire using stored chemical constituents, primarily sugar and oxygen from the air to produce carbon dioxide, water, and energy (heat). Respiration is a deteriorative process; the heat produced must be removed to prevent a rise in temperature that would accelerate deterioration. Relative to most fruits, grapes have a very low respiration rate. Although low, the process requires significant quantities of refrigeration to remove this heat as well as the sensible heat during cooling. Further, after transit or storage temperatures are attained, heat is still being evolved that must be removed in order to maintain the desired temperature. This amount can be significant. If the grapes have been cooled to 40°F (4°C) before being stacked such that no heat enters or leaves the fruit, the rise in temperature in 8 d is 4°F (2°C). However, if the initial temperature is 68°F (20°C), then the rise is about 18°F (10°C). This is a very substantial increase — certainly sufficient to aggravate problems of decay. Loading inadequately cooled grapes into mechanical refrigerated carriers (either rail or truck) can cause such problems since these carriers cannot adequately cool grapes, partly because of lack of refrigeration capacity and partly because of lack of air circulation throughout the load.

2. History

Until about 1950, most California table grapes destined for distant markets were transported in ice-refrigerated rail cars. These cars had bunkers in each end with a total capacity for about 5.5 t of ice. Properly used, the refrigeration capacity was sufficient to cool adequately the approximately 15 t of grapes. The cards became relatively efficient coolers when fan systems were installed during the 1930s and 1940s. These fans reversed the natural convection causing the air to rise in the bunkers, flow out over the load, down between the braced packages, and back under the floor racks to the bunkers. Greater efficiency of cooling was attained when provision was made to power these fans with electricity at the loading dock, starting them immediately after the car was loaded with warm fruit and then running them continuously for a day before releasing the car. The alternative was to release the car when loaded, with the result that the fans would run only when the car was in motion. In practice, during this first day or two of the transit period, the cars stood still in re - icing stations or switching yards as trains of these cars were assembled for the transcontinental trip. Unfortunately, these long delays occurred precisely when the need for uninterrupted cooling was most critical.

After 1950 more and more of the fruit was transported by carriers equipped with mechanical refrigeration. Refrigerated highway vans carried an increasing share of the grapes. At the same time the railroads shifted gradually to mechanically refrigerated cars (reefers), until by 1970 the ice-refrigerated car was essentially phase out. These changes had a marked effect on methods and location of grape cooling. It become evident that these cars could not adequately cool grapes because of the limited capacity of the refrigeration unit and the much heavier loads being hauled. The units in these carriers were able to hold satisfactory transit temperatures *provided* the fruit had been brought to transit temperature *before* loading. Even for shipments to points within California only very limited quantities of table grapes are moved without cooling. Such movements are usually overnight hauls from the Coachella or San Joaquin valleys into Los Angeles and San Francisco Bay markets. Unless marketing is very prompt, the practice can be hazardous because of rapid deterioration of this warm fruit.

3. Methods

There are four general methods of house-cooling grapes based on how the cooling air is brought to the fruit: (1) *parallel flow* when the cooling air is delivered by fans on the side of the room to the palletized fruit with two sides of each pallet exposed to the air flow, (2) *parallel flow* when the air is delivered from ceiling jets downward between pallets with four sides exposed, (3) *forced air* when the air is delivered directly to the fruit by establishing a pressure gradient across the lugs in the pallet, and (4) *conduction* when the air is delivered to the *unvented lug*. The parallel-flow methods may be regarded as approaching natural-convection cooling — the velocity of the air along the sides of the containers causing turbulance that results in air exchange through the vents of the package. The forced-air method may be considered as simply forced-convection cooling. The conduction method is just that. The berries in contact with the cold unvented liner are cooled by conduction and they in turn extract the heat from those deeper in the pack by the same process with no air movement involved. Actually all the methods involve at least some degree of conduction in heat transfer to the cooling air. Each method has advantages and disadvantages; however, they do differ widely in the rate of cooling the grapes. There is a close relation between the cooling rate and the accessibility of the fruit to the cooling air. Exposing the containers reduces the cooling time, but when the fruit itself is brought into close contact with the air (forced air) the cooling time is drastically reduced.

a. Parallel-Flow Cooling

Fans direct the air across the top of the room after which it flows downward and back through the channels between pallet stacks. The air is then pulled up through the refrigeration bunker by the fan to start the cycle over again.[21] Several means are used to make the fruit as accessible to the air stream as possible. Channels through the stacks of TKV lugs in the center stack help in cooling the center of the pallet, especially if the pallets are spaced with 1 or 2 in. between pallets in the row. Channeling within pallets not only shortens the total cooling time, but, more significant, it reduces the range in temperature between the least and most exposed fruit in the pallet.[22] Effective are thick cleats on both ends of the lids and/or bottoms of the lugs. These provide channeling between the layers of lugs in the pallet. Lack of such channels nearly doubled the time required to cool a pallet of TKV lugs in a 9-stack pallet.[22] Interior channeling becomes even more significant if packing materials within the lugs block air movement. Vented curtains will increase the cooling time three- to four-fold as compared with no liners. Chimney wraps and wraps that completely enclose the clusters have as much or more effect on blocking the cooling air. The outer lugs of the pallet are cooled by the ram-jet effect of the turbulant channel air, which forces cold air through the side vents along the top and bottom of the lugs. The velocity of the air in the channels should be at least 200 ft/min to assure satisfactory cooling. A conventional room should have a fan capacity to deliver 6 to 8000 ft^3/min of air per 1000 lugs of grapes (3). Also, since about $3^1/_2$ t of refrigeration are required to cool 1000 lugs (22 lb net) of grapes 40°F (22°C) in 24 h, the capacity should be adjusted not only to the amount of fruit to be cooled but also to the time required for cooling. For example, if the room holds 4000 lugs, the amount of refrigeration would be 14 t for a 24-h cooling period. However, if the cooling is to completed in 12 h, the refrigeration capacity of the room must be twice that or 28 t. A ton of refrigeration is that required to freeze 1 t of water at 32°F (0°C) into ice at the same temperature in 24 h.

Some table-grape operators place the fruit to be cooled in the same room with storage grapes. This practice is not as desirable as that using a separate room just for cooling because the rapid air movement required for good cooling dries excessively that fruit which has already attained the storage temperature. The objection to this practice can be overcome somewhat by designing the cooling room into "bays," each separated by a wall that extends part way across the room.[23] If each bay is served by a fan or damper system that can be controlled independently, the air environment of this bay is semi-isolated from that of other bays. As long as the temperature of the fruit is being lowered, high-velocity air conditions can be maintained without affecting materially the environment of adjacent bays, such as passing warm air over fruit that has already been cooled. For a cooling bay to operate properly, the cold air must be delivered to the back of the bay where it then can move through the stacks of fruit to the center of the room. Once storage temperature is attained, either the rate of air movement may be reduced or the fruit moved to a "storage" bay.

Cooling rates can be increased in a parallel flow system if the air is brought to the pallets of fruit through jets in the ceiling. Faster rates are possible, largely because the air is brought directly to all four sides of the pallet directly beneath the jet, rather than to only two sides in a "channel flow" situation. However, the faster cooling benefit of this system is offset somewhat by the elaborate systems of plenums needed and problems of balancing the air distribution pattern. Further, since cross as well as longitudinal channels must be provided, the pallets must be positioned so that each intersection of the channels is precisely under a ceiling jet.

b. Forced-Air Cooling

The feasibility of forced air cooling (pressure cooling) as a rapid method of removal of field heat from grapes was first demonstrated at the Earl Fruit Company in Lodi, CA. The rate of cooling with this system is greatly increased over that of the parallel flow system because the cooling air is brought directly to the *fruit* in the package rather than just to the *package*. By setting up a pressure gradient *across* the package there is a positive flow of cooling air *through* the container from one side to the other providing direct contact with the packed fruit.

Maintaining high efficiency of a forced-air system requires close monitoring of several factors. The room temperature should be checked frequently during the day to verify that it does not rise above the permissible level. This is especially important during the afternoon and evening when demands for refrigeration usually peak from cooling larger and larger quantities of fruit being brought in that is progressively warmer. Forced-air systems can make enormous demands on the refrigeration capacity because they extract heat from the fruit at a rate many times that of a parallel flow system. As the temperature of the cooling air rises, owing to inadequate refrigeration capacity, the cooling rate of the fruit is slowed down correspondingly. This prolongs the cooling period and thereby increases the amount of dehydration of the grapes.

In a well operated forced-air system both the static pressure and the air volume parameters are critical. Operating either beyond certain limits results in slower cooling and/or excessive use of power for the air handling system. Generally the system should be designed to deliver about one cubic foot of air per pound of fruit per minute (ft^3/lb/min) or (0.128 m^3/g/min). However, this volume will vary depending on the length of the cooling period desired. The fan capacity required to deliver a given volume of air depends on the static pressure difference across the packed fruit. In turn, the static pressure difference required to deliver a given volume of air depends on: (1) length of the channels, e.g., width and number of containers, (2) size, number, and placement of vents in the containers, (3) resistance of packaging materials such as liners and cluster wraps to air flow, and (4) area of the channels that allow the air to bypass the packed fruit — gaps between containers, pallets, and baffles. Close monitoring of the pressure differential will indicate whether there is excessive air leakage due to faulty placement of pallets or baffles. Effective for checking pressures is a manometer calibrated in hundredths of an inch water pressure.

As channel length is increased, the pressure differential must be increased greatly to keep the air volume constant. This means correspondingly more fan power. For a pallet where the air must traverse three container widths, a differential of 0.25 in. of water was found sufficient to cool naked packs of grapes covered with vented curtains.[23]

The cooling rate within the pallet may be uneven, especially when the air channels are long. It takes 1 to 2 h longer to cool the downstream fruit in a pallet three packages wide than it does to cool the upstream fruit. This problem may be overcome somewhat by allowing some of the room air to bypass the upstream packages.[23] The mixing of cold room air with that air which has traversed the upstream packages permits more uniform cooling of the entire pallet of fruit. Careful monitoring of the temperature of upstream and downstream fruit will indicate the best width of channels to use for mixing the air streams to obtain the most uniform cooling rate. In all instances the warmest location in the pallet would be monitored and the temperature of the fruit at this location should be used to determine when cooling is adequate.

The size, number, and placement of vents in the packages may have a significant effect on the length of the cooling period. For example, with a static pressure of 0.25 in. and with the grapes packed without cluster wraps in vented liners, increasing the vented area in the sides of the package from 1 to 3% shortened the cooling period by about 1 h. With cluster wrapped grapes and no liner, the cooling period was shortened about 2 h with 3% as compared with 1% venting. Two to 5% of the area of the sides of the average package should be vented for effective forced air cooling. The precise percentage will depend on the length of the cooling period desired and on the presence of packaging materials in the packages that may be a factor limiting the rate of air flow through the fruit pack.

Orientation of polystyrene packages in the pallet can have a significant effect on the length of the cooling period because of the extent of vented area in the sides as contrasted to that in the ends of the packages. Commonly these packages are stacked in an interlocking pattern in the pallet and as a result in a forced air system the air flows lengthwise through some packages and crosswise through others. Cooling takes about 2 h longer for the lengthwise than for the crosswise packages because the percentage of vented area is less in the ends than in the sides. Increasing the number and/or size of the end vents in the bottom as well as in the lid of this package would correct this deficiency.

Vents are especially critical in the polystyrene package since the field heat in the fruit must be extracted almost entirely through them. The walls of this package have such effective insulative properties that practically no heat is transferred through the material by conduction. This is one reason why the forced air system for fast cooling of grapes in this package is preferred so much over the much slower parallel-flow system. The same could be said for corrugated cartons which are usually stacked with only the sides exposed to the cooling air, leaving very little surface through which conduction cooling can take place. Here too, size and placement of the side vents are critical regardless of whether the cooling system is with parallel-flow or forced air.

The resistance to air flow and insulative properties of packaging materials in the packages have a marked effect on the length of the cooling period. A vented liner with curtain in the TKV package lengthens the cooling period about one hour as compared to that for cluster-wrapped grapes in the same package. Incorporating both cluster wraps and a liner about doubles the length of the cooling period of that for cluster wraps only.

Air that bypasses the fruit pack does little cooling and therefore has no significant effect on the length of the cooling period. However, even relatively small openings around the packages can increase significantly the fan capacity required to maintain a given static pressure difference. For example, whether or not TKV containers have cleats on the bottom makes no significant difference in the length of the cooling period; however, one fourth inch cleats about doubles the volume of air required to maintain a given static pressure difference. One half inch cleats about triples the volume needed. This large increase in volume requires a great increase in fan capacity to cool a given quantity of fruit. Therefore much more power is needed and consequently is much more costly. Bottom cleats (and top cleats also) may increase the rate of cooling in a parallel-flow system by exposing more of the package surface to the cooling air for conduction, but they are a detrimental factor in a forced air system because of the waste of energy for fan power. For maximum efficiency in a forced air system the only air that should be permitted to pass through the pallet is that which comes into direct contact with the fruit.

Forced air cooling has the advantage that the short length of the cooling period makes it possible to cool and ship fruit the same day that it is harvested and packed. Until a reliable cooling period could be reduced to 9 h or less, it was necessary to carry over the last fruit harvested until the next day so cooling could be completed during the night. This situation aggravated problems of congestion in the cooling facility and also delayed by at least one day the fruit reaching the market.

Schedules usually become exacting in a well-run forced air cooling facility owing to the rapid turnover as warm fruit is placed "on line" and cooled fruit removed. As a result it is often helpful to predict when a given lot of fruit will be cooled adequately and can be removed for storage or shipment. For this, the half-cooling method of calculating cooling rates is useful. This method may be defined as the time it takes to reduce the temperature of the commodity half way from its initial temperature to that of the cooling medium (air in this case). The utility of this method may be demonstrated as follows. If, for example, at 2 p.m. after 2 h of cooling, the temperature of grapes on the forced-air cooler is 56°F (13°C), using 32°F (0°C) air for cooling, and the initial temperature of the fruit was 80°F (27°C), it can be predicted that this fruit will reach 38°F (35°C) in 4 more hours or at 6 p.m. In this way the arrival of refrigerated carriers can be closely scheduled to the time the fruit is adequately cooled and ready for loading.

For practical purposes the length of the cooling period may be expressed in terms of the "7/8ths cooling time". A typical cooling curve for grapes would show that when they were packed at a temperature of 80°F (26°C) they would be 7/8ths cooled at 38°F (3°C) which is within the range for satisfactory transport or for transfer to storage. However, when only "half-cooled" the temperature would be 56°F (13°C) which is well *above* a suitable temperature for shipment or storage.

c. Conduction Cooling

A third method of cooling is by conduction, that is, transfer of heat only through the walls and liner of the package. This method is widely used by the Chilean and South African table grape industries, but certain packaging requirements must be met. The cold air is circulated around the *unvented* (or nearly unvented) packages. Cooling takes two to three times longer than good room cooling does, yet can be done with practically no drying of the stems or berries. However, it is necessary to provide a water vapor barrier completely around the fruit to retain this freshness. Further, a cushion pad (preferably of excelsior) must be included, not only for physical protection of the fruit, but also to absorb that moisture which condenses on the inside of the unvented vapor barrier during cooling and would otherwise collect on the bottom of the liner, soaking the bottom berries. Indispensable to control decay is an in-package sulfur dioxide release device that must be placed — one just under and one over the fruit inside of the vapor barrier. Following is the sequence of packing a lug of grapes equipped with top and bottom two-stage sulfur dioxide generators: (1) the lug box is first lined with an unvented liner, preferably including a water vapor film, followed by a cushion pad, (2) a two-stage generator is placed on the cushion pad with the first stage uppermost next to the fruit, (3) the fruit is packed in the usual way, either as a naked pack or with chimney wraps, (4) the top generator is placed on top of the fruit with the first stage downward next to the fruit, (5) an unvented curtain, preferably with a water vapor barrier film is installed over the top generator, and (6) the unvented liner is completed and the end guards are bent in for lidding.

4. Relative Humidity Factor

Maintaining a low temperature is the primary consideration in securing fast cooling of grapes, but maintaining a high relative humidity during the process is important and often neglected. Considerable water loss can be caused by a low relative humidity during cooling, even though the period may be relatively short. Such water loss frequently happens when the coil temperature is lowered in order to increase the cooling capacity of the unit. Unfortunately, when this is done the colder coil condenses more moisture from the air, lowering its relative humidity. Further, the source of this moisture is largely from the fruit itself. Coils for coolers in California are usually designed to operate at about 20°F (11°C) below the temperature of the return air. This temperature difference will result in a relative humidity of about 70% — low enough to cause heavy condensation on the coil surfaces. The temperature difference can be reduced by increasing the coil surface. However, to raise the relative humidity to 90% (which would reduce condensation by half) would require coil surfaces three to six times larger. The cost of doing this can be reduced by using a packed column as the heat exchanger with the air. Enormous heat exchange surfaces can be built up relatively cheaply if plastic filaments or other material for the column are used. This large surface compensates somewhat for the relatively inefficient heat-exchange characteristics of air. At the same time, the amount of the more expensive expansion coil can be kept small by submerging it in a tank as an ice accumulator — water being a more efficient heat-exchange medium than air.

The relative humidity of the cooler can also be raised by a system of fog-spray nozzles. The fog spray increases condensation on the coils, which adds to the refrigeration load; however, improvement in fruit quality usually will more than offset this disadvantage. Certainly most grape containers (excluding polystyrene and treated corrugated) and the packaging materials within, absorb considerable moisture during cooling and storage. In fact, the gain in weight of each container may be as much as 0.5 lb after a month in storage. The most rapid rate of absorption is during cooling when the containers are driest. During this period the fog-spray supplies at least part of this moisture, which would otherwise transpire from the fruit.

5. Fumigation with Sulfur Dioxide

Sulfur dioxide was first used in California to prevent decay and fermentation of wine grapes.[24] However, it was several years later before a satisfactory fumigation program was developed for table grapes. The reason for this delay was, in part, that the high concentrations of sulfur dioxide used for wine grapes caused unacceptable levels of injury in table grapes, and adversely affected flavor and appearance. An initial gas treatment was developed, applied before cooling, which effectively controlled decay during the 8 to 10 d required to transport the refrigerated fruit to eastern markets.[3] Later the treatment schedule was expanded to include periodic retreatments for grapes held in storage for as long as 6 months.

Currently it is standard practice in california to apply the initial fumigation the same day that the grapes are packed. This treatment surface sterilizes the berries, especially wounds made during handling. The treatment may be applied in a special gassing room before cooling is started. A fan system must include a purge system so the residual gas can be exhausted to the roof or otherwise disposed of. It should be possible to activate this purge system by remote control, and then open the door slightly to allow fresh air to enter the room as the gas laden air is exhausted. More usual is to accumulate the packed fruit in the cooler during the daily packing and fumigate it at the end of the day. In this way cooling is not delayed and the treatment is applied when most of the working crew has left and will not be exposed to the irritating vapor of the gas. However, when grapes are accumulated in the cooler during the day, fumigation of the fruit should not be delayed longer than *12 h* after harvest. Periods longer than this before application of the gas increase the danger that fungus spores on the surface of the berries will have time to germinate and infect the fruit, especially if cooling is slow and, as a result, the air remains warm and humid around the berries.

Other commodities should not be treated with the grapes, or even be where the fumigant can reach them, because most of them are easily injured by the gas. Because grapes also can be injured, they should be exposed to only the minimum quantity of the gas necessary. This amount will depend upon (1) the decay potential and condition of the fruit, (2) the amount of fruit to be treated, (3) the type of containers and paçkaging materials, (4) the air velocity and uniformity of air distribution, (5) the size of the room, and (6) the losses through leakage or sorption on room surfaces. Under favorable conditions, a basic sulfur dioxide concentration of 0.5% by volume applied for 20 min is adequate. For cluster-wrapped packs this time should be extended to 30 min. To keep the concentration at this level, the absorptive capacity of the lugs and fruit as well as their volume must be taken into consideration.[75] The dose can be calculated from the following formula:

$$\text{weight } SO_2 = \frac{A \times B}{C} + (D \times E)$$

where A = the concentration of sulfur dioxide to maintain (%), B = the unoccupied space in the room; this value would be the total volume of the room minus that occupied by the lugs (ft^3 or m^3), C = volume occupied by one lb or kg of sulfur dioxide gas; at 32°F (0°C) this would be 5.5 ft^3 or 0.156 m^3 and at 68°F (20°C) this would be 6.0 ft^3 or 0.17 m^3, D = number of lugs in the room to be treated, and E = weight of sulfur dioxide absorbed by each lug at the concentration desired (lb or g).

Adequate for factor E for TKV or all wood lugs is 0.0011 lb (0.5 g) when (1) the fruit is sound, (2) air velocities assure good penetration, (3) the room is relatively gas tight and (4) the gas has little opportunity to be absorbed on walls and refrigeration surfaces. Allow 0.0006 lb (0.25 g) of sulfur dioxide per lug for polystyrene and waterproof corrugated containers. If chimney wraps or a top pad are used in these last two types of containers, the absorption factor should be increased to 0.007 lb (0.33 g) per lug.

The gas must be distributed quickly and evenly to all parts of the room. This can be done by spacing special nozzles 6 ft apart along the ceiling in the room. If the outlet is placed in front of a fan there should be one for each fan, or the air from the single fan should be distributed evenly across the room through a plenum system so that the gas is well diluted by the air before it reaches the fruit.

The *hot*-gas method of delivery may be used if the room requires 10 lb or less of gas. The steel cylinder containing the liquid sulfur dioxide is first connected to the gas inlet, and the valve is then opened. The cylinder should then be placed in a pot of boiling water to vaporize the fumigant as rapidly as possible. Only about 1 lb/min can be delivered this way. With some precautions the *cold*-gas method may be used by inverting the cylinder when the valve is opened. It is faster and requires no supplementary heating equipment. A "riser" extends to the bottom of the larger steel cylinders so that, when the valve is opened, liquid sulfur dioxide flows into the plumbing line and does not change into a gas until it escapes from the nozzles in the room. Every precaution should be taken that air volume and velocity are adequate to vaporize and mix the gas thoroughly with the air before it reaches the fruit. Up to 100 lb of gas can be released in 2 to 3 min with this method.

IX. STORAGE OF TABLE GRAPES

Before the 1930s table grapes could generally be stored for extended periods only if embedded in some dry medium. Since ancient times grapes in the Middle East have been embedded in packing materials such as barley to preserve their "fresh" state. In California during the 19th century 'Mission' grapes were occasionally packed in sawdust and transported within the state for use after the vintage season had ended.[26] Toward the end of that century 'Almeria' grapes from Spain began to arrive in eastern seaports each fall. These grapes were packed in barrels with ground cork, each holding nearly 50 lb. of fruit. The volume of this imported fruit increased to 20,000 t annually before 1913. At the same time, the volume of California grapes being shipped to eastern markets was increasing. However, although the 'Ohanez' grapes, packed in cork, could be marketed during the fall and even into the winter, the California fruit, packed in plain packs (no embedding material), had to be disposed of within 6 weeks of harvest.[27]

The California industry attempted to compete for the late-season market by storing grapes in modified versions of the Spanish barrel. About 1912, California grapes were first packed in ground cork imported from Spain. This attempt and others to store grapes in cork dust were frustrated by the high cost of the material, hence extensive studies were made to develop a cheaper packing material. Redwood sawdust, properly screened and dried, was found superior, keeping grapes longer and in better condition than the more expensive cork product.[28]

The sawdust keg was for many years the principal storage and export container. It held 32 to 34 lb (14.5 to 15.5 kg) of grapes and about 14 lb (6.4 kg) of sawdust. A straight-sided drum of the same capacity was used to some extent, as well as a 22 lb (10 kg) sawdust chest (actually a paper-lined lug). During the 1920s the so-called "export chest" came into use. It held the same amount of fruit as the keg, and since it occupied less shipping space it gained favor, particularly as an export container. Also, with the "stem up" pack, the chest was easier to inspect. The keg, however, remained the main domestic storage container because it held the grapes in the best condition. Redwood sawdust was gradually replaced by a lighter-colored material made from spruce, because of better color contrast with red 'Emperor' grapes and because it was less dusty and provided a better cushion than the redwood.[29]

The volume of sawdust-packed grapes reached a peak in 1928 at about 2000 carloads, and then declined to 500 to 600 by 1932. This was the beginning of the trend eliminating the sawdust pack as a domestic storage container. Grapes could be stored for several weeks in the cheap plain pack as cooling and handling methods were improved, especially when sodium bisulfite was placed in the cushion pad to control the spread of decay. Finally, when a program was developed for retreating the grapes with sulfur dioxide in storage, plain packs could be stored longer or as long as those that had been stored in sawdust. The fruit was fresher without the drying effect of the sawdust and has less "woody" flavor. Further, the cost of the plain pack was less than that of the sawdust chest.[29]

A. CURRENT VOLUME AND MARKETING

The movement of grapes into storage usually builds up in the latter part of August as the production of 'Thompson Seedless' peaks supplemented by such mid-season varieties as 'Ribier,' 'Flame Tokay,' 'Italia,' and several minor ones. By this time, harvest in the early districts (Coachella and upper or lower San Joaquin valleys) is largely complete. Little or not attempt is made to store the early varieties from these districts such as 'Perlette,'

'Thompson Seedless,' 'Cardinal,' and even 'Ribier.' The storages are fairly well filled by the middle of September. During the following 8 to 10 weeks this stored fruit is moved to market and replaced by the late season varieties of 'Emperor,' 'Calmeria,' 'Almeria,' and even late-harvested 'Ribier.' Toward the end of October storages normally reach their capacity and may contain as many as 8 million lugs. These storages are concentrated largely in a belt extending from Fresno to Bakersfield along the eastern side of the San Joaquin Valley.[30]

The harvest season closes during the first half of November, and from then on the market is supplied entirely from these storages. The volume is heavy until after the Christmas-New Year holiday season, then tapers off gradually during January and February. Some storages continue to supply grapes into March and even April.[30] The length of the storage life of the major varieties varies and falls into the following ranges depending upon the initial quality of the fruit and degree of excellence of postharvest handling:

'Cardinal'	1 to 2 months
'Thompson Seedless,' 'Italia,' 'Tokay'	$1^1/_2$ to $2^1/_2$ months
'Ribier,' 'Almeria,' 'Calmeria'	2 to 4 months
'Emperor'	3 to 5 months

B. STORAGE ENVIRONMENT

The best temperature to attain the full storage potential of table grapes is 30°F (−1°C). The fruit can tolerate temperatures slightly lower than this but the margin of safety above freezing the grapes becomes so narrow that for practical purposes this temperature should be considered the minimum.[30] Even so, careful monitoring of the temperature of the air coming from the refrigeration surfaces is essential, because freezing of the commodity can be disastrous. At 30°F, the vapor pressure deficit (VPD) of the air is very low, yet significant, considering the long duration of the storage period.

The relative humidity of the air should be as high as practicable — 95% if possible. Limiting factors are: dripping from walls and ceiling and wet slippery floors. A combination of this high relative humidity and the lowest practicable temperature reduces the VPD to the smallest value possible and thereby the lowest possible rate of water loss from the fruit. At the same time the rate of growth of mold organisms is slowed as much as possible, and also the rate of respiration — all to reduce the rate of deterioration.

Methods of maintaining high relative humidity levels, discussed under cooling, apply especially here for storage. Refrigeration surfaces (dry-coil, packed-column, or brine-spray) should be as large as possible to minimize the temperature difference between the surfaces and the storage air. Supplementary moisture in the form of fog spray is helpful in maintaining a high relative humidity, especially during the first month of storage when the dry wood and paper of the packages are absorbing moisture to reach equilibrium with the storage air. Each TKV or wooden lug will absorb 0.33 to 0.67 lb of water during the first month of storage.[31] Every pound of water supplied from a supplementary source saves an equal amount that would otherwise come from the fruit.

Air velocity, so important to be high during cooling to bring the grapes to a low VPD as quickly as possible, becomes a liability when the storage temperature is attained. The rate of moisture loss is related directly to air velocity, hence in grape storages the velocity should be reduced to that needed only to maintain the desired fruit temperature. Careful monitoring of grape temperatures within pallets will dictate how much velocity is needed, which may be as little as 10 to 20 ft/min in the channels between the pallets.[32]

To market good-quality grapes after several months of storage will require a relative humidity approaching 95% and an air velocity approaching 6 ft/min, if we consider fruit that has lost more than 6% of its weight as being unacceptable.[30]

Uniform container alignment is just as important in the storage room as it is in the cooler. Misalignment will block the slow air movement with the result that "hot spots" may develop in the room where the air is still. Further, when the velocity should be high for the periodic sulfur dioxide treatment, poor penetration of the gas may result with blocked channels. Pallets should be spaced so that there are channels on each side (2 to 4 in.) and also between the ends (1 to 2 in.). Avoid removing an entire line of pallets across the room because this leaves a wide "hole" through which the air will move and bypass the channels between the remaining pallets. Also, it is advisable not to have the lines of pallets excessively long, because the concentration of sulfur dioxide drops rapidly from absorption by the containers and fruit as the air traverses the channels between the pallets. It is better to have the air moving across the narrow dimension of the room with more fans (and gas outlets) than moving it along the longer dimension with fewer fans.

C. STORAGE DISORDERS

Decay is the chief disorder problem of table grapes. Fortunately, sulfur dioxide is effective in retarding the spread of decay fungi. Even so, a heavy mold potential may result in an unmanageable problem, particularly if any facet of the fumigation program is less than perfect. The undetected infection during packing poses the greatest hazard. Often it takes several weeks at storage temperature for such an infection to develop to the extent that it can be seen easily. By then, spread may have taken place to the extent that the fruit is ruined.[33] Under these circumstances it is important to be able to predict early in the storage period the level of decay potential present in the grapes. This knowledge makes it possible to market those lots soonest that have the highest potential. Harvey, in 1955, established the feasibility of a method to achieve this. Samples of berries taken during packing were subjected to the same conditions as the packed fruit except that they were held at room temperature. As a result, infections could be detected in a week at this elevated temperature that would otherwise take several weeks at storage temperature to become evident.[35] Marketing schedules could then be arranged based on the results. Details of applying this method with techniques and materials needed were subsequently published.[36]

Supplementing the use of this method for predicting decay should be thorough and frequent inspection of the stored fruit, particularly grapes in packs deep within the pallets—those least likely to receive adequate sulfur dioxide for decay control. Such inspections will not only help in forestalling decay, but also show whether the sulfur dioxide used weekly is being evenly distributed—whether it is too much in some areas causing unnecessary bleaching, and too little in others allowing decay to spread. If both problems exist, they are probably caused by two factors, or a combination of them: (1) inadequate volume of air being moved during fumigation, and (2) excessive barrier effects of packaging materials (cluster wraps, curtains, or inadequate venting of the container). Further, a consistent pattern of bleaching indicates an excessive dosage of sulfur dioxide and conversely, a consistent pattern of spread of decay indicates an inadequate dosage consistent with the decay potential of the fruit.[30] Calculation of the correct storage dosage to use will be discussed later.

Physiological aging of grapes is a normal phenomenon. Eventually the pulp tissue turns brown, which gives white berries a dull grayish appearance and red or black grapes a brownish tint. However, in some years this browning may appear sooner than usual in storage and show no consistent pattern of occurrence. It can be particularly troublesome in 'Thompson Seedless', sometimes after only a month of storage. On close inspection berries have dull gray color, and when cut open show a brown color largely restricted to the rudimentary seed area.[30] There is no clear explanation for the sporadic occurrence of this disorder, even though the onset of browning can be accelerated by physiological stress such as bruising and elevated temperatures. Browning has been correlated directly with storage temperature in the range of 32 to 39°F (0 to 4°C). Further, berries flattened by lid pressure and those areas subjected to abrasion between berries during transit vibration, showed a greater tendency to brown than those not subjected to stresses.

Frequent and close inspection of the grapes during storage is the best assurance of detecting the disorder early enough to dispose of the fruit before the market quality is impaired appreciably. At the onset of browning, it is often difficult to detect; the berries should therefore be examined under strong light — light transmitted through the berry as well as that reflected from it. Early detection is essential because several days are required to market affected fruit. During this period, stress conditions of handling and higher transit temperatures favor more rapid development of the disorder.

1. Berry Cracking

Infrequently, 'Thompson Seedless' berries have cracked during storage for no apparent reason, and the incidence can be severe. This type of cracking is not to be confused with that caused by excessive pressure on the fruit during lidding. Typically the cracks appear as "hair line" fractures of the skin with no consistent pattern of distribution, either on the berry surface or among berries in the cluster. They do appear most frequently on fruit that is less mature—berries that are thin-skinned and greenest in color. This is the type of fruit typically from heavily overcropped vines. The condition does not appear to be related either to rate of cooling or to high relative humidity during storage. This condition has been observed occasionally in Chilean 'Thompson Seedless' grapes during and shortly after cooling. A possible cause could have been unusually heavy irrigation to within a day of harvest. An abundant water supply for the roots, coupled with a high relative humidity under the overhead trellis (which would retard evaporation from the leaves), may have resulted in sufficient turgor pressure of the berries to have caused the cracking. This theory is supported by the observation that cracking was severest in fruit picked and packed in the morning (when turgor pressure would be highest), less in that picked and packed in the afternoon, and least in that carried over on the packing shed floor at room temperature and packed the following day.

As for browning, the most effective treatment is frequent and thorough inspection of the fruit. These inspections should be started immediately after cooling and continue during storage. Besides, if cracks appear, they should be checked closely for signs of bleaching along the edges. Absence of bleaching indicates that insufficient sulfur dioxide has reached these injuries to sterilize them and prevent decay. Prompt action should be taken to treat the fruit thoroughly with the gas and dispose of the fruit as soon as possible.

D. SULFUR DIOXIDE TREATMENT OF GRAPES DURING STORAGE

In extended storage of table grapes, control of decay is a critical operation. For practical purposes, this operation becomes the limiting factor for maintaining quality when satisfactory temperature, relative humidity, and air-movement conditions are provided. Only sulfur dioxide is currently acceptable as a chemical agent to accomplish this control of decay.[30] Its successful use is confronted with two formidable problems: (1) established mold infections at the time of packing — and they should be assumed to be present — cannot be eradicated by the initial treatment, however well it is applied; and (2) the retreatments during storage to prevent these infections from spreading and to resterilize the surface of the berries cause chemical injury to the fruit that is accumulative. As a result, the margin of safety between controlling decay and keeping injury to the fruit within acceptable limits becomes increasingly narrow as the storage period is extended.[33] Therefore, the correct dosage of sulfur dioxide for these weekly treatments becomes progressively more critical with storage time. A long storage period of 140 d compounds the problem. 'Emperor' grapes were stored at 35°F (2°C) and treated weekly with sulfur dioxide. A decay potential was established by placing nine *Botrytis*-infected berries in each pack, each infected berry being surrounded by sound fruit.

The spread of decay was measured by calculating the average number of infected berries at each innoculated site. Decay increased at an accelerating rate with time in storage because as the fungus spread out in all directions from the original infected berry it made contact with more and more sound berries. As a result, infection proceeded at an exponential rather than linear rate. The 0.2% treatment allowed practically no spread of decay, even after 140 d of storage. However, the amount of bleaching at this time was rated "severe". A concentration of 0.1% allowed the fungus to infect two sound berries, but bleaching was rated as only "moderate". At 0.05% $3^1/_2$ berries were infected, at an average, at each site in addition to the original infected berry. Bleaching in this case was rated as only "slight" (hardly detectable).[33] A good decay-forecasting method would have indicated what concentration would have been needed to keep decay within the required limits. For example, if the method showed only two infected berries per packed lug, there would have been only nine infected berries after 140 d if a 0.05% concentration was used — a level of decay still within the 0.5% by weight limitation for U.S. No. 1 on a condition inspection. At the same time, bleaching would have been only slight. However, if the method showed three or four infected berries, a concentration of 0.1% would have been necessary to keep decay within this limit, even if the bleaching level would be increased to moderate.

The most effective dose will vary depending on the specific set of factors prevailing at the time of treatment. These factors, important as they may be during the initial treatment, become more significant during storage. Containers, especially those of untreated wood or paper packaging materials, absorb moisture rapidly during the first three weeks of storage and this moisture in turn considerably increases their absorptive capacity for sulfur dioxide during fumigation. Consequently, the concentration of the gas during the treatment period is depleted more rapidly with damp containers and, therefore, the fruit is actually exposed to less of the fumigant on a time-concentration basis than would be expected.[25] As a result, the rates and amounts of depletion are related directly to the level of the storage relative humidity (resultant moisture content of the packaging materials) and the number of containers in the storage room.

When sulfur dioxide is used, greatest protection from decay with least injury is obtained over extended storage periods by retreating the fruit at weekly intervals. Longer intervals between treatments to provide equal protection require substantially higher concentrations of the gas, with the net result that the total exposure to sulfur dioxide is appreciably higher. This interval of one week is based on the storage temperature being not more than 32°F (0°C).[34] Higher temperatures would require shorter intervals between treatments or higher concentrations of the gas, because higher temperatures allow decay fungi to grow faster. A weekly treatment with sulfur dioxide at a concentration of 0.1% allows virtually no spread of decay during 140 d if the storage temperature was maintained at 31°F (−0.5°C). However, at 35°F (2°C) decay triples and 39°F (4°C) decay increases nearly sixfold. From these data it is apparent that a weekly treatment of 0.1% at 31°F (−0.5°C) will control decay, as well as a concentration of 0.2% at 35°F (2°C).[33] The lower storage temperature, then, not only minimizes problems of decay control but bleaching as well, if an appropriate concentration of sulfur dioxide is used for the weekly treatments.

The formula for calculating the correct dose of sulfur dioxide is the same as that discussed for the initial treatment except that the values for the concentration and absorption factors are lower. A concentration of 0.1% on a volume basis is adequate if the grapes have a low decay potential when packed, for example, if they have not been exposed to rain or other wet conditions that would encourage field infections of *B. cinerea*. Further, a treatment period of 30 min is adequate unless excessive packaging materials, such as cluster wraps, delay penetration of the gas; or there is poor circulation of air in the room which would delay distribution of the fumigant.[30] Under these conditions a packed TKV or wood lug with 23 lb (10 kg) of fruit would absorb about 0.001 lb (0.45 g) of sulfur dioxide. Under the same conditions, a packed polystyrene lug would absorb 0.0006 lb (0.225 g) because the plastic absorbs essentially no sulfur dioxide. No adjustments need be made in dosage whether or not cluster wraps are used. because when they are used the containers are usually packed to 21 instead of 23 lb net weight (9 to 10 kg). The amount of the gas absorbed by the wraps about offsets that which would otherwise have been absorbed by two additional pounds of fruit.

Dosages may be adjusted above or below the 0.1% concentration depending on decay potential of fruit. It may be reduced to as little as 0.05% if the fruit has essentially no decay potential. The absorption factor should then be reduced accordingly — by about one third from the 0.1% factor. This program should cause no discernable bleaching of the fruit even after five months of storage. However, frequent and close inspections should be made at these low concentrations should any decay be present. If the grapes have been exposed to persistent and heavy fog or dew before harvest, and especially if they have been exposed to considerable rain, the dosage should be increased. If the decay prediction data indicate a high potential, the concentration should be raised to 0.15 or even 0.2% with a corresponding increase in the absorption factor. At 0.2% concentration, field infections should be completely contained (no spread of decay to unaffected berries). However, significant bleaching of the berries can be expected, especially after 2 months of storage.[30]

A small correction factor should also be considered in calculating the dosage, to allow for the effect of storage relative humidity on the absorptive capacity of the containers. The values shown are based on a relative humidity of 90%. If the relative humidity has been 95%, the absorptive factor should be increased by about 10%; if it has been 85%, the factor should be reduced by about 10%.[25]

1. Removal of Residual Sulfur Dioxide

Usually there are significant concentrations of SO_2 left in fumigation rooms at the end of the treatment period. This gas is usually purged from the room through the roof to quickly reduce the concentration to levels that will not injure the fruit or personnel who must re-enter the room. Unfortunately, SO_2 even at very low concentrations is very irritating to the mucous membrane of the nasal passages and lungs as well as the eyes. Effluent gas wafted back down where personnel are located can be very objectionable. Further, drawing large volumes of warm air into a storage room to replace the gas laden air usually causes a rise in temperature of the fruit, sometimes accompanied by undesirable condensation of water on the berries.

Varying amounts of SO_2 are removed from the atmosphere during fumigation by absorption on the fruit and lugs. These amounts reduce the quantity that must be removed by purging on coil sprays after the treatment is completed. An example might be cited of a typical commercial storage room that was treated on two different occasions with the same dosage of SO_2 to give a concentration of about 900 ppm. In one instance the room was nearly filled with 21,000 lugs of grapes and in the other nearly empty with only 1000 lugs. Even with the room nearly full, the fruit and lugs had absorbed the gas to the extent that 200 ppm still remained in the room at the end of the 30 min treatment period. In the nearly empty room the concentration was still close to 600 ppm.[30] In both instances, considerable SO_2 remained that needed to be removed in order to reduce the concentration to a safe level. The maximum concentration permitted by OSHA is 10 ppm for a maximum 15 min exposure per d or 5 ppm for an 8 hr exposure per d for personnel without protective equipment. Even if personnel need not re-enter the room, exposure of the fruit repeatedly to 100 ppm SO_2 for several hours will cause bleaching of berries, hence the need for an internal scrubbing system that will rapidly remove the residual gas to required levels.

One method of removing SO_2 from the room without purging is by circulating the gas laden air through the refrigeration coils of a wall bunker with defrosting sprays turned on.

The fumigant is very soluble in cold water, but unfortunately, a large spray surface is needed to reduce the concentration to tolerable levels at an even reasonably acceptable rate. Further, most fumigation facilities are not equipped to use such a scrubbing system without drastic modifications. Also, this system exposes the defrosted (unprotected) coils to large amounts of the gas which, with water, forms sulfurous acid which can be very corrosive to many metals.

Considerable attention has been directed recently towards the feasibility of a portable scrubbing facility—one that can be moved from room to room and is self-contained except for an electrical power source.[37] Two systems have evolved, both using cold water as the primary absorber of the fumigant. The "Filocel" system involves pumping water from a self - contained reservoir over a packed column through which the gas laden air of the room is forced by a fan. The other system involves forcing the water from the reservoir through Venturi nozzles at pressures sufficient to draw the room air through the unit and the spray. De - mister pads are usually needed for both systems to prevent excessive amounts of suspended water being blown in the room. The efficiency of gas removal of both systems can be increased considerably by adding NaOH to the water. Neutralization of the sulfurous acid appears to increase the capacity of the water to remove the gas and do it more quickly. Unfortunately, the salt Na_2SO_3 resulting from the reaction between the H_2SO_3 and the NaOH tends to clog the de - mister pads, probably because of its low solubility in water at storage temperatures.

Very slow removal of the gas has often been experienced with these units, especially in very large storage rooms. This may be due to one or a combination of several factors: (1) high ratio of volume of the room air to the volume of air handled by the scrubber, (2) inadequate mixing action of the air handler of the room, or (3) inadequate capacity of the water reservoir of the scrubber. Grape storage rooms commonly have volumes of several hundred thousand cubic feet while these scrubbing units have a capacity of 10 to 20 thousand ft/min. If it is assumed that all SO_2 is removed from the air as it passes through the scrubber each time and that the air in the room is kept mixed thoroughly by its fan system, it would be expected that the concentration of the gas would be reduced by half for each complete air change of the room. On this basis the "half-scrubbing time" for a 10,000 ft^3/min unit in a 100,000 ft^3 room should be 10 min. It would require six half-scrubbing times or 1 h to reduce a concentration of 600 ppm to below 10 ppm (600 to 300 to 150 to 75 to 38 to 19 to $<$10 ppm).

Several factors must be considered in choosing a scrubbing unit for removal of SO_2 from the fumigation atmosphere. The air capacity needed for the unit will be determined mainly by three factors: (1) volume of the room to be scrubbed, (2) the difference in the concentration of SO_2 from the initial level to that required, and (3) the length of the period in which scrubbing must be completed. For this calculation, two assumptions should first be made: that the unit operates at 100% efficiency (complete removal of SO_2 from the air as it passes through the scrubber each time), and that the air handling system of the room keeps the SO_2 concentration uniform throughout the room during scrubbing.

Many grape storage rooms are very large—200,000 ft^3 or even more. For practical purposes, the scrubbing period should not be longer than 1 h. Under these conditions, a 200,000 ft^3 room would require a scrubber with a capacity of 20,000 ft^3/min to effect one air change in 10 min or six air changes within the hour. This combination would reduce the concentration of SO_2 from 600 to 10 ppm—a level permissible to work in without protective equipment for not longer than 15 min/d. Precooling rooms are often used to give the grapes the initial treatment of SO_2 which may reach concentrations of 5000 ppm or more.[37] A 1 h scrubbing period would require nine to ten half-scrubbing times of about 6 min each. The same scrubbing unit of 20,000 ft^3/min could accomplish this if the room was no larger than 130,000 ft^3.

The efficiency of the scrubbing unit will be determined by: (1) the amount of water surface to which the gas laden air is exposed as it passes through the scrubber, (2) volume of water in the reservoir, if a closed system, (3) volume of flow-through water, if an open system, (4) temperature of the water, and (5) whether the pH of the water is kept high as with NaOH.

Extrapolation of the data relative to the volume of water needed to provide near 100% efficiency of the scrubber indicates an impractical solution for a portable scrubber. Required would be about 20,000 gal of water for a 200,000 ft^3 room—a highly impractical volume.

Encouraging was the effectiveness of NaOH in reducing the amount of water needed for near 100% efficient scrubbing. Twenty gallons of water with NaOH was just as efficient as 100 gal without it.[37] Likely the lowest practical volume of water that could be used would be determined by the solubility of NaOH and/or Na_2SO_3 in water. At some volume when NaOH is used as a neutralizing agent in lieu of water as an absorbing agent, the saturation point of the base (about 3.7 lb/gal at 32°F) will limit the amount of NaOH feasible to use. At this point, no more NaOH can be introduced effectively into the system. Further, the low solubility of the Na_2SO_3 end product (about 1.1 lb/gal at 32°F) may impose even more severe restriction on the minimum amount of water that can be used. The precipitated salt will form crystals in the system blocking air flow through the demister pads or even cause plugging of spray nozzles and pumps.

A 200,000 ft^3 room with 1000 ppm of residual SO_2 will require about 35 lb of NaOH to neutralize the gas. An NaOH saturated solution of about 10 gal at 32°F will be required for this condition. On the other hand, the end product Na_2SO_3 at saturation will require about 53 gal of water to keep the salt in solution and thereby avoid problems of crystallization on the scrubbing components.

A feasible solution to the problem would be to connect the scrubber to an outside source of water so that a flow-through system would provide fresh water during scrubbing. A flow of about 80 gal/min through a 20,000 ft^3/min air handler would be adequate to degass a 200,000 ft^3 room from 1,000 to less than 10 ppm SO_2 in 1 h. This volume of water could be handled by hoses and pumps of modest size. In fact, a pump may not be necessary if the pressure of the water supply is adequate for the drip or spray nozzles. The flow may have to be increased to some extent if the water is not cold, since water at room temperature is about half as efficient in absorbing SO_2 as it is near 32°F.

X. SHIPMENT OF STORED GRAPES FOR THE DOMESTIC MARKET

Movement of palletized grapes directly into the carrier (as discussed under palletization) has largely eliminated the need for sulfur dioxide fumigation in the carriers.[30] The rigorous treatment of hand loading and bracing the lugs in the carriers formerly made it advisable to then fumigate the fruit to sterilize fresh injuries. Current handling of the fruit as palletized units has largely eliminated this need. The value of the treatment is especially questionable, if there is no circulation of air during fumigation. This happens in nonrefrigerated containers and even in those with refrigeration if the fan and the refrigeration unit are activated by a thermostat set at some transit temperature above normal storage temperatures. When storage fruit at 32°F (0°C) is loaded into such a van, there is little opportunity for air movement. Hence, if sulfur dioxide is injected over the load of fruit in the van, the more exposed fruit is overexposed to the gas and the fruit deep in the load is underexposed or may receive none at all. To overcome this problem the grapes should be given their final sulfur dioxide treatment not longer than 2 d before loading, and the carrier treatment should then be eliminated. If the treatment must be applied in the van, temporary fans should be installed to assure circulation of the air during the treatment period. Methods for doing this have been reported in detail.[30] Unfortunately, the technique requires some time to install the fan system. Further, the carrier must remain at the loading dock for the entire treatment period so that the fan equipment can be recovered from the load. Largely as a result of these inconveniences the system has not gained acceptance.

Good temperature management is the most important factor in preserving the market quality of the fruit during transit. It is to the best interest of the shipper as well as the carrier that the pulp temperature of the fruit when loaded be at least as low as the lowest practicable transit temperature that can be maintained in the carrier. Failure to accomplish this has been the cause of many expensive lawsuits to assess the blame when poor market quality at destination results in a financial loss. A further precaution is to have a temperature record of the van or reefer during the transit period. The air temperature over the load can be ascertained by installing a recording thermometer at time of loading. This record is especially valuable to assess the operational efficiency of the refrigeration system—whether there were interruptions in transit, stoppages, or limit points were exceeded for the maximum and minimum temperatures. As a further precaution, the shipper can install recording thermometers in the fruit packs. One monitored pack should be placed in the part of the load that is subject to the widest fluctuations in temperature. This would most likely be in the top layer most exposed to the air blast from the refrigeration coils. Here the danger of freezing would be greatest, if the refrigeration system should operate at too low a temperature, and also here the fruit would warm up fastest if refrigeration should cease. The other pack should be buried deep in the load where air circulation would be least, and therefore there would be greatest danger of a rise in temperature of the fruit from the heat generated by respiration. Suitable recording instruments to provide this information can usually be rented from service agencies.

It is important that the grape containers be well braced in the carrier so that there will be no shifting of the load. Three directions of movement must be considered: (1) Lateral shifts that are caused by the sway of "whip" of the carrier as it changes direction; these forces are usually most severe in the top layer of containers—those farthest from the roadbed where the amplitude of the swaying is greatest. (2) Longitudinal shifts brought about by sudden acceleration or deceleration forces on the load; stress can be severe with sudden braking, or, in a rail reefer, as the slack is taken up in a long train starting to move, the jolt can cause severe shifting of an unrestrained load; also, in switching operations recoupling of rail cars can sometimes cause very sudden changes in velocity of the reefer. (3) Vertical shifting caused by an uneven roadbed coupled with the action of the suspension system of the carrier that moves the load up and down. If these movements are in the natural frequency range of the load, the danger of shifting of containers is increased considerably because shifting movements may gain momentum with each impulse.

Proper strapping of pallets is the first precaution to be taken to prevent shifting of loads in transit.[30] With vertical as well as horizontal straps in position, and well tightened, the containers should be well protected from shifting *within* the *pallet*. However, this strapping will not prevent the *entire* pallet from shifting as a unit in the carrier. It is necessary, then, that these units which may weigh a ton, be securely braced against each other and against the walls of the carrier during loading. Dunnage strips should be nailed firmly across the top of the pallets so that the outer ends press firmly against the walls of the carrier to prevent lateral shifting. Alternatively, the pallets may be wedged tightly with fold-out expandable filler materials that can be cut to desired widths. These materials are available from suppliers. Longitudinal shifting can be prevented by placing the pallets tightly together, starting at the forward end of a van and from each end of a reefer. Heavy bracing must be cut to take up all the space between the back of the load and the doors of the van when closed. In a rail reefer, there is a heavy steel-bulkhead door on each side of the middle doorway. When loading is complete, these doors can be swung into place and locked—one jammed firmly against the containers of one end of the car and the other securing those in the other end. Care must be exercised that all slackness is eliminated before the bulkheads are finally locked.

XI. EXPORT OF GRAPES USING THE IN-PACKAGE SULFUR DIOXIDE$_2$ GENERATOR

All California export tables grapes are now packed in the same kinds of shipping packages as those used for domestic shipments. Use of the sawdust chest was discontinued about 1980. This means that there has been a complete changeover from the standard sawdust export chest of 30 years ago. This has come about largely because of palletization and containerization that has made it possible to handle the more fragile domestic packages across the docks without excessive damage to both the packages and the fruit. At the same time, extended protection of the fruit from decay during long export transit periods has been integrated into the handling system.[38,39]

The wooden lug, TKV lug, and the polystyrene package do not require special modifications for export. They are adequate to protect the fruit during the rigorous dock handling procedures as long as they are within the protection of the strapped pallet which must be handled by a fork-lift truck. This is especially true if the pallets are in sea containers that can be transferred directly from a highway truck to a berth on a container ship. This latter system of handling not only eliminates the physical handling of individual pallets across the docks, but also isolates the fruit from unfavorable temperatures and pilferage. These are significant considerations when traffic in the dock area is commonly congested, delays in transfer of pallets often excessive, and little or no security provided to protect the fruit from injury or loss.

Preparation of grapes for export is essentially the same as that discussed for domestic markets. However, in view of the longer transit period required (sometimes as long as a month), some aspects of the quality of the fruit intended for export and precautions in handling need special emphasis. In the first place, the quality of the grapes should be especially good. It costs no more to ship a lug of top quality grapes to a distant market, for example, than one which on arrival contains inferior fruit. Yet, the profit or loss factor will be vastly different in each case. It is especially important, then, that the fruit be thoroughly inspected before loading to ascertain whether it will indeed withstand the rigors of the extended export transit period and still be of acceptable quality on arrival at destination. Usually this is done with the standard Federal State Inspection Service inspection to determine the quality grade. Special attention should be given to the decay potential of the fruit, because this is usually the main quality factor that will change (adversely) during transit and marketing. If the grapes are to be exported from storage, a "condition" inspection should be made, even though an initial inspection at time of packing may have shown the quality of the fruit to be excellent. As during the transit period, the decay factor is the one that usually changes most during storage.

An additional precaution to reduce the danger of decay is effective timing of the final sulfur dioxide treatment before loading. First, the treatment should be done in the shipping facility to be sure that *all* of the fruit is exposed uniformly to the gas. This cannot be done with assurance in the carrier! Second, the treatment should be applied not more than one day before loading to be assured of the longest possible protection period during transit.

Effective in-transit protection from decay can be obtained with an in-package slow-release sulfur dioxide generator. As attractive as this method might appear for providing the decay protection for several weeks, there may be disadvantages in using this method directly for export. The limit of 2 to 3 months of protection with this method means that the grapes could be stored for only 1 to 2 months if 1 month transit time must be added to that for an export shipment. Often, shippers do not know at packing time when or how many lugs they will eventually export. The shipper may be locked into a packaging system at packing time that might prove unworkable later. However, the system can be modified to overcome this disadvantage, yet retain much of the decay-control capability. No modification of packages or packaging materials is necessary. The grapes can be packed in the usual way, fumigated initially, and subsequently at weekly intervals in storage. At export time a slow- or second-stage generator can be inserted into the pack over the fruit and under the curtain. The final storage gas treatment is then applied, after which a polyethylene shroud is installed completely enclosing the pallet to retain the sulfur dioxide produced by the generators. All these operations must be done at storage temperature because it is almost impossible to cool a shrouded pallet of grapes back down to the desired transit temperature.

The in-package method provides effective protection for even the longer transit periods. Also, it has the advantages of keeping the stems and berries fresher because the relative humidity within the shroud is near saturation and because there is no air movement to accelerate water loss from the fruit. Further, the sulfur dioxide released by the generators is retained within the shroud to the extent that little, if any, of the gas can escape. It is advisable that as soon as the pallets reach destination, and especially when they are removed form the carrier to warmer temperatures, that the shroud be removed. Otherwise, residual sodium bisulfite in the generators will break down, rapidly releasing concentrations of sulfur dioxide that may cause appreciable bleaching of the fruit. This precaution becomes particularly significant with transit periods shorter than 3 weeks when considerable quantities of the generating material are still in the inserts.

An alternative method is to install a two-stage instead of the slow-stage-only generator. In this case, the pallet should be shrouded *immediately* so that none of the fast-stage gas is lost. Also, no conventional storage application is needed because the fast- and first-stage generator supplies this need. The feasibility of this method has been demonstrated for a transit period of 4 weeks.

The two-stage sulfur dioxide generator simulates to some extent the initial and weekly sulfur dioxide applications generally used in California except that the fumigant is generated *inside* the unvented container.[40] The first-stage component (functioning as the initial application) starts generating sulfur dioxide within minutes after packing and is exhausted after 3 to 5 d. The second-stage component (functioning as the storage application) starts generating the gas after 3 d and maintains a low concentration for extended periods. This type of pack can be stored for as long as 2 to 3 months without further treatment except that of keeping the temperature at or near 32°F (0°C). The storage period should not be extended, though, beyond the time needed to market the fruit within the 2 month time limitation because there is little assurance that in-package sulfur dioxide generators currently available will afford protection from decay for longer than this. When the generators cease to envolve the fumigant, the fruit becomes vulnerable to attack by decay organisms in the nearly saturated atmosphere of the unvented container.

The first-stage of the two-stage generator consists of a sheet of Kraft paper 8 × 6 in. (20 × 40 cm) to which is affixed 1.5 g of sodium bisulfite ($NaHSO_3$). The second-stage is composed of two sheets of Kraft paper, each faced with a very thin film of polyethylene. These sheets are placed with the polyethylene film of each in contact, then heat-sealed together forming 12 pockets in the generator unit, each containing $NaHSO_3$ in powder form. The 12 pockets contain a total of 5.5 g of the chemical. The two stages may be loosely affixed together to be used as one unit or can be obtained separately and used independently. Either or both stages may be obtained from Uvas Quality Packaging, Inc., P.O. Box 369, Antioch, CA 94509.[41]

Various combinations of the generators can be used depending on several factors: (1) length of the protection period needed, (2) the temperatures that will prevail during cooling, storage, transit, and distribution, (3) the decay potential in the grapes, (4) the depth and quantity of grapes in the package, and (5) the amount of venting required for the gas barrier liner of the package, e.g., vents through which methyl bromide can penetrate to the fruit to fulfill insect quarantine requirements.

The maximum protection period of 2 to 3 months can be expected only if the other factors are favorable. Cooling must be started promptly after the container is closed and the fruit temperature brought down steadily to 32 or even 31°F (0 or −0.5°C), and kept there. The distribution phase, when the temperature of the grapes usually rises, must be short—not over 3 d. The decay potential must be low, that is, the grapes must be relatively free of field infections of *B. cinerea* that develop as a result of damp rainy weather before harvest. The quantity of grapes in the package must not exceed the capacity of the generator. The two-stage generator described will protect up to 23 lb (11 kg) of grapes if packed, one underneath and one above the fruit. No fruit should be further than 4 in. (10 cm) from a generator surface since the concentration gradient of the SO_2 decreases about one half for every inch (2.5 cm) of distance from the generator. If the pack is deeper than 4 in. (10 cm) both top and bottom generators should be used. If the pack is shallower, but wider and longer, both generators may be used side by side on top of the grapes. It is usually in the corners of the pack at the greatest distance from the generator that decay is most apt to develop.

The first stage alone will protect the grapes for as long as 10 to 15 d if all factors are favorable. Compromises on temperature or venting will shorten this period. With no cooling the grapes can be protected from decay for up to 6 d. Any venting in the gas barrier shortens this period.

The second stage may be used alone *only* if the grapes can be fumigated with SO_2 applied from outside the containers in a fumigation room. This requires that the containers be vented to admit the gas to the grapes. Further, the vented containers must be promptly and thoroughly cooled after packing with either a parallel-flow or a forced-air system. When the fruit is at or near 32°F (0°C) the containers must be converted to the unvented mode to confine the SO_2 produced by the second-stage generators. If the containers are palletized, the entire pallet may be enclosed in a polyethylene film cover to confine the gas. Otherwise each container must be converted to the unvented mode utilizing packaging components designed with this capability.

Venting of the gas barrier (as when required for methyl bromide treatment to fulfill insect quarantine requirements) reduces significantly the efficiency of the in-package SO_2 generator. Even the minimum venting of the $^1/_4$ in. diameter (6 mm) on a 4 in. (112 mm) spacing pattern allows enough loss of SO_2 to increase decay slightly. Further, loss of water vapor through these vents is enough to cause measurable drying of stems and berries.[40] Cooling should be started immediately after packing, otherwise excessive amounts of SO_2 will be lost from the first-stage generator. Allowing the packed containers to remain in warm still air accelerates the evolution of the gas from the first-stage and since there is some aeration, the concentration of the SO_2 does not build up to levels that are effective in controlling decay.

REFERENCES

1. **Radler, J. and Horn, D. H. S.,** The composition of grape cuticle wax, *Aust. J. Chem.*, 18, 1059, 1965.
2. **Chambers, T. C. and Possingham, J. V.,** Studies of the fine structure of the wax layer of Sultana grapes, *Aust. J. Biol. Sci.*, 16, 818, 1963.
3. **Ryall, A. L. and Harvey, J. M.,** The cold storage of vinifera table grapes, *USDA Agr. Mark. Serv. Handb. 159*, 46, 1959.

4. **Nelson, K. E.**, The effect of Botrytis infection on the tissue of Tokay grapes, *Phytopathology*, 46, 223, 1956.
5. **Nelson, K. E.**, Some studies of the action of sulfur dioxide in the control of Botrytis rot of Tokay grapes, *Proc. Amer. Soc. Hortic. Sci.*, 71, 183, 1958.
6. **Nelson, K. E.**, Effect of humidity on infection of table grapes by *Botrytis cinerea*, *Phytopathology*, 41, 859, 1951.
7. **Hewitt, W. B.**, Rots and bunch rots of grapes, *Calif. Agr. Exp. Sta. Bull.*, 868, 52, 1974.
8. **Dufrenoy, J. and Genevois, L.**, Developpement du Cladosporium herbarum sur des raisins à basse température, *Soc. de Biol. (Paris) Compt. Rend.*, 118, 708, 1935.
9. **Delp, C. J., Hewitt, W. B., and Nelson, K. E.**, Cladosporium rot of grapes in storage, *Phytopathology*, 41, 937, 1951.
10. **Harvey, J. M. and Pentzer, W. T.**, Market diseases of grapes and other small fruits, *USDA Agr. Mark. Serv. Agr. Handb.*, 189, 37, 1960.
11. **Jensen, F. and Luvisi, D.**, Flower thrips nymphs involved in scarring of Thompson Seedless grapes, *Calif. Agr.*, 27, 10, 8, 1973.
12. **Winkler, A. J.**, Maturity tests for table grapes, *Calif. Agr. Exp. Sta. Bull.*, 529, 35, 1932.
13. **Winkler, A. J., Cook, J. A., Kliewer, W. M., and Lider, L. A.**, *General Viticultures*, University of Calif. Press, Berkeley, CA, 410, 1974.
14. **Lutz, J. M. and Hardenburg, R. E.**, The commercial storage of fruits, vegetables, and florist and nursery stocks, *USDA Agr. Handb.*, 66, 94, 1968.
15. **Kliewer, W. M., Lider, L. A., and Schultz, H. B.**, Influence of artificial shading of vineyards on the concentration of sugar and organic acids in grapes, *Amer. J. Enol. Viticult.*, 18, 78, 1967.
16. USDA Agr. Marketing Serv. U.S. Standards for Grades of Table Grapes (European or Vinifera type), USDA Consumer and Mark. Serv., Washington, D.C., 1971.
17. **Winkler, A. J.**, Maturity tests for table grapes—the relation of heat summation to time of maturing and palatability, *Proc. Am. Soc. Hortic. Sci.*, 51, 295, 1948.
18. **Nelson, K. E., Allen, J. W., and Schutz, H. G.**, Effect of grape maturity, sample order, and sex of the taster on the flavor response of supermarket customers, *Am. J. Enol. Viticult.*, 23, 86, 1972.
19. **Nelson, K. E., Schutz, H. G., Ahmedullah, M., and McPherson, J.**, Flavor preferences of supermarket customers for 'Thompson Seedless' Grapes, *Amer. J. Enol. Viticult.*, 24, 31, 1973.
20. **Nelson, K. E.**, High picking temperatures and rough handling can reduce consumer acceptability of California fresh table grapes, *Blue Anchor* 32, 2, 6, 1955.
21. **Guillou, R.**, Coolers for fruit and vegetables, *Calif. Agr. Exp. Sta. Bull.*, 773, 65, 1960.
22. **Richardson, H. B., Nelson, K. E., and Meredith, H.**, Cooling tests of palletized units of table grapes using a portable and versatile research facility, *Proc. Int. Cong. Refrig.*, Wash., D.C., 3, 147, 1971.
23. **Mitchell, F. G., Guillou, R., and Parsons, R. A.**, Commercial cooling of fruits and vegetables, *Calif. Agr. Exp. Sta. Ext. Serv. Man.*, 43, 44, 1972.
24. **Winkler, A. J. and Jacob, H. E.**, The utilization of sulfur dioxide in the marketing of grapes, *Hilgardia*, 1, 107, 1925.
25. **Nelson, K. E. and Baker, G. A.**, Studies on the sulfur dioxide fumigation of table grapes, *Am. J. Enol. Viticult.*, 14, 13, 1963.
26. **Jacob, H. E. and Herman, J. R.**, Harvesting and packing grapes in California, *Calif. Agr. Exp. Sta. Bull.*, 390, 44, 1925.
27. **Stubenrauch, A. V.**, Factors governing the successful storage of California table grapes, *USDA Bull.*, 35, 31, 1913.
28. **Haight, L. G.**, The keg pack of Emperor grapes, *Blue Anchor* 1, 6, 8, 17, 1924.
29. **Nelson, K. E., Curley, R., and Jensen, F.**, The influence of sawdust handling methods on the quality of table grapes for export, *Am. J. Enol. Viticult.*, 14, 94, 1963.
30. **Nelson, K. E.**, Harvesting and handling California table grapes for market, *University of Calif. Agr. Sci. Pub.*, 4095, 67, 1979.
31. **Nelson, K. E., and Guillou, R.**, Highly marketable grapes maintained by controlled humidity, *Am. Soc. Heating Refrig. Air-Cond. Engineers J.*, 2, 4, 57, 1960.
32. **Nelson, K. E., Baker, G. A., and Gentry, J. P.**, Relation of decay and bleaching injury of table grapes to storage air velocity and relative humidity and to sulfur dioxide treatment before and during storage, *Am. J. Enol. Viticult.*, 15, 93, 1964.
33. **Nelson, K. E. and Richardson, H. B.**, Storage temperature and sulfur dioxide treatment in relation to decay and bleaching of stored table grapes, *Phytopathology*, 57, 950, 1967.
34. **Nelson, K. E. and Richardson, H. B.**, Further studies on factors affecting the concentration of sulfur dioxide in fumigation atmospheres for table grapes, *Proc. Am. Soc. Hortic. Sci.* 77, 337, 1961.
35. **Harvey, J. M.**, A method of forecasting decay in California storage grapes, *Phytopathology*, 45, 229, 1955.

36. **Harvey, J. M.,** Instructions for forecasting decay in table grapes for storage, *USDA Agr. Mark. Serv.*, 392, 6, 1960.
37. **Nelson, K. E.,** Factors affecting removal of sulfur from atmospheres of table grapes, *Am. J. Enol. Viticult.*, 33, 61, 1982.
38. **Nelson, K. E.,** Packaging and handling trials on export grapes, *Blue Anchor*, 47, 9, 11, 1970.
39. **Nelson, K. E. and Ahmedullah, M.,** Packaging and decay-control systems for storage and transit of table grapes for export, *Am. J. Enol. Viticult.*, 27, 74, 1976.
40. **Nelson, K. E.,** Effects of in-package sulfur dioxide generators, package liners, and temperature on decay and desiccation of table grapes, *Am. J. Enol. Viticult.*, 34, 10, 1983.
41. **Nelson, K. E. and Ahmedullah, M.,** Effect of type of in-package sulfur dioxide generator and packaging materials on quality of stored table grapes, *Am. J. Enol. Viticult.*, 23, 78, 1972.

Chapter 8

THE CHERRY

Stephen R. Drake

TABLE OF CONTENTS

I. INTRODUCTION

A. HISTORY

The cherry is one of the most widely cultivated of all fruits and is commonly found in temperature climate zones of the world. The genus *Prunus* represents many stone-fruits (peaches, plums, apricots, almonds, and cherries) with a myriad of species. There are two commonly cultivated species of cherries, *Prunus avium,* the sweet cherry, and *Prunus cerasus* the sour cherry. There are several other species of cherries, (*padus, virginiana, pseudocerasus* to name a few), but they are of little commercial importance.

Cultivated cherries probably had their origin in southeastern Europe or Asia. From this point the cherry was spread by animals and man to all the temperate climates of the world. Early records indicate that the cherry was first domesticated in Greece, but it was probably domesticated before that in Asia Minor. In 300 B.C., Theophrastus described the cherry and in 37 B.C. Marcus Terentius Varro described the procedure for the grafting of cherries. The cultivation of the cherry in early Greece was for timber and not fruit. Records indicate that both the sweet and sour cherries were cultivated in early Greece.

Pliny described ten distinct cultivars of cherries grown in Italy in the first century A.D. and mentioned Portugal, Belgium, and England as other countries where cherries were cultivated. Pliny stated that the Romans brought cherries to England prior to the first century A.D., and cherries became firmly established in Kent where the Romans settled. During the Dark Ages agriculture became a lost art in England. With the introduction of Christianity and the establishment of monasteries an agricultural revival was initiated in England. By the 16th century cherry culture was well established in England. Turner, in 1538, Gerarde in 1596, and Johnson, in 1633, all discussed cherry culture during their time.

The early French planted cherries in North America in and around the areas of Nova Scotia, Prince Edward Island, and along the St. Lawrence River. These cherries planted by the French were probably from the area of Normandy. Cherry cultivation was first mentioned in written records of the Massachusetts Bay colony in 1629. Early records of the New England area state that cherry trees were found in abundance in fields, woods, and fence rows, and were probably progeny from early cultivation. The early colonists also planted cherries in Pennsylvania, New Jersey, Delaware, Virginia, and North Carolina. A nursery was established on Long Island in 1730 and in 1804 listed 20 named cultivars of cherries for sale.[10]

The western movement of cherry cultivation in America occurred as early French settlers moved to such places as Detroit and St. Louis. Franciscan monks probably introduced cherries into California in the late 1700s. Employees of the Hudson Bay Company initiated cherry culture in Oregon prior to 1847, but they used seedling cherry trees rather than grafted cultivars. In the fall of 1847 Henderson Lewelling arrived in Oregon and brought with him grafted trees of known cultivars from the East. In the confusion of transportation, one of the trees transported by Lewelling lost its label and was renamed 'Royal Ann', a misconception in that it was actually a cultivar called 'Napoleon', but the name 'Royal Ann' has persisted. Some of the more notable cultivars developed early in Oregon were 'Bing', 'Lincoln', 'Republican', 'Lambert', and 'Willamette'.[13]

The cherry was widely cultivated in early American but, until recently was grown almost entirely in the backyard orchard. Starting in the early 1900s commercial plantings of cherries began to develop and distant markets were created. This was due primarily to improved shipping and holding facilities. Today much research effort is expended on shipping and holding technology to enhance fresh market quality of cherries.

Despite the fact that the cherry has been a favored home-grown garden fruit and has been planted in every state in the country it is not well-adapted to most areas in the U.S. Washington, Oregon, California, and Michigan account for most of the approximately 320 million lb of sweet cherries produced annually. The states of Washington and Oregon account for over half of the U.S. sweet cherry production. Michigan alone produces over 70% of the U.S. sour cherry production. New York, Wisconsin, Pennsylvania, Utah, and Oregon produce small amounts of sour cherries. Sweet and sour cherries combined have an average annual farm value of $141 million. The value of the crop can vary tremendously depending on many factors such as size of the crop, availability at a given time, and weather conditions.[9] Cherries are either marketed as a fresh or canned product with small amounts being frozen. The amount of sweet cherries marketed as a fresh or processed product depends on the quality of the fresh product. The processing of sweet cherries is generally considered a salvage operation. Sour cherries are almost totally marketed as a processed product.

II. HARVESTING

Color is the most common indicator of maturity for proper harvest and is used as an indicator of harvest time for both fresh and processed cherries. Soluble solids, flesh firmness, and time from full bloom have also been used to some degree as indicators of maturity. Color, used in combination with soluble solids, is probably the best indicator of maturity to produce a product with acceptable consumer quality.[8] Color standards are generally used throughout the industry to establish harvest date. In the past few years there has been research using chemicals such as daminozide [Butanedioic acid mono-(2,2-dimethylhydrazide)] to hasten color development and ethylene [(2-chlorethyl)phosphonic acid] to enhance abscission of the fruit, both to be used as harvest aids.[3,15]

Fresh market sweet cherries are picked with the stems attached. The condition of the attached stem on a cherry is often used as an index of quality by individuals in the fresh produce area. Mechanical harvesting of sweet cherries for the fresh market is not feasible under present conditions. Some sweet cherries that are brined have been harvested successfully by mechanical methods. Cherries that are mechanically harvested tend to be severely bruised and to deteriorate in quality during storage and transit.

Sour cherries are picked by pulling or stripping the fruits from the stems with the stems left on the tree. Stemless cherries are preferred by most processing plants because of the time and expense necessary to remove the stems. After picking, sour cherries are placed in large lugs or tanks containing water which serves as a cushion for transporting hundreds of pounds at a time.[2,8,13]

III. POSTHARVEST HANDLING

After harvest sweet cherries are immediately transported to a warehouse. The field heat (precooling) is generally removed by cold water immersion (hydrocooling) or spray. The cherries are generally placed in refrigerated storage. Fresh market cherries are belt-sorted and sized mechanically. Water containing a fungicide is sprayed over the fruit for washing and decay prevention. Sweet cherries are packed into 12 or 20 lb fiberboard boxes with a polyethylene liner. The polyethylene liner helps establish a modified atmosphere, where oxygen is consumed and carbon dixoide liberated by the respiring fruit. This modified atmosphere retards decay, and the high humidity maintains brightness and stem freshness. Sweet cherry size is expressed in terms of the number of rows of cherries that fit into the 12 or 20 lb box.[2,14]

Mechanically refrigerated rail cars and trucks are used to ship sweet cherries to distant markets. To maintain quality shipment immediately after harvest is best. Since the harvest season for fresh cherries is very short a large volume of fruit must be harvested and shipped in a short period. To prevent market gluts storage is a must for proper marketing of the fruit. With the expanding use of controlled atmosphere storage cherries can be stored for extended periods of time beyond the 14 to 18 d average for refrigerated storage.[2] Recent research with sweet cherries has indicated that this storage period for cherries may be extended for several days or weeks with the use of controlled atmosphere storage.[16]

IV. STORAGE

Storage life of sweet cherries is directly related to temperature and humidity. In general the lower the storage temperature of cherries (down to 0°C) the better the keeping quality of the fruit. A difference of 15.6°C in storage temperature resulted in a 15% or 2 d reduction in storage-life.[13] Cherries refrigerated immediately after harvest maintain quality up to 5 d longer than fruit refrigerated the day following harvest. Respiration rate, the rate at which cherries liberate carbon dioxide, is the generally accepted index of metabolism. Storage temperature directly influences respiration rate. At 15.6°C, respiration rate is approximately eight times faster than at 0°C. In addition lower temperatures reduce mold growth and flavor loss.[8,13]

Water loss through transpiration is the cause of almost all post-harvest weight loss of cherries in storage. Weight loss is a major factor in determining the storage life of cherries. As cherries transpire the fruit shrivels and loses its brightness of color, firmness, and acceptable appearance. In addition the attached stem takes on an unacceptable color.

Transpiration and water loss are directly related to the relative humidity (RH) and temperature of the storage environment. A warm atmosphere has a greater moisture holding capacity than a cold atmosphere. At 15.6°C and 80% RH the air contains more moisture than at 4.4°C and 80% RH. Cherries in storage with 50% RH will lose up to 50% in weight, while cherries stored at 80% RH will lose only 10%.[8,13]

Other chemical or physiological change occurs in cherry fruits after harvest in addition to weight loss due to dehydration. Monosaccharides increase during storage as a result of sucrose and polysaccharide metabolism. Color intensity increases and total acids decrease. All of these changes during storage are slight except for acidity which continues to decrease as storage time increases.[13]

V. COMPOSITION

A. VITAMINS

Only five of the known vitamins occur in cherries to any appreciable concentrations. The vitamins that have been found in cherries are carotene, thiamine, riboflavin, niacin, and ascorbic acid. Research into the nutritional properties of cherries has been very limited, and little is known of the vitamin content of cherries during maturation. The amount of any particular vitamin occurring in cherries varies widely with the cultivar, growing location, and cultural practices. The sour cherry contains substantially more vitamins than the sweet cherry, particularly vitamin A or carotene content and ascorbic acid or vitamin C.[18] Reports indicate that the cv. 'Napoleon' has a higher concentration of carotene than 'Bing', but has a much lower ascorbic acid concentration than 'Bing'.[13] The ascorbic acid concentration in 'Bing' and 'Rainier' cherries can be increased by the use of gibberellic acid, a growth regulator. Other research has indicated that sweet cherries from the Pacific Northwest have a greater vitamin content than sweet cherries from other areas.[4,5]

B. MINERALS

Mineral content does not vary significantly among cultivars. However, ash content varies greatly with fruit maturity. The ash content is directly related to maturity. As cherries mature the ash content increases to about 0.6% of the fruit. The minerals found in cherries, in descending order of concentration, are: Phosphorus, calcium, magnesium, sodium, sulfur, iron, chlorine, silicon, manganese, aluminum, copper, iodine, bromine, and zinc. When compared to other fruits cherries are good sources of phosphorus (31 mg/100 g approximately) and sulfur.[13,18]

C. SUGARS

Total solids of cherries gradually increase as the fruit advances towards maturity. This increase can be very dramatic with as much as a 1% increase every 2 d. This rise is due to the rapid increase of insoluble solids to as much as 20%. As the fruit ripens the insoluble solids decrease to less than 2%.

Soluble solids, expressed as percent soluble solids or degrees Brix, also rise rapidly during the later stages of maturity. Soluble solid concentration remains rather constant at a low level until the fruit begins to develop color. Soluble solids increase very rapidly as color develops in cherries and continues to increase until the fruit is harvested. After harvest there is a little change in soluble solids concentration.

There is a good correlation between color and soluble solids development. These two indices (color and soluble solids) are probably the best indicators of maturity for good fresh market and canning quality.[7,8,11,13] The rate of development and the final concentration of soluble solids varies with the season and cultivar in question. Growth regulators, such as daminozide, have been reported to increase soluble solids, but other reports have refuted this.[6,17] Seasonal rainfall or irrigation can influence soluble solids and size development. Shortages of moisture to the cherry trees prior to harvest will generally produce small fruit with greater than normal soluble solids.

More than 90% of the total sugars in cherries are reducing sugars with some sucrose, fructose, and glucose. About 2% or less of the sugars are sucrose with negligible amounts of fructose and glucose. There is a very close correlation between soluble solids and sugar concentration. As cherries reach full maturity an increasing proportion of the soluble solids are sugars. The sugar concentration can double in less than 3 weeks.[13]

D. ACIDS

Malic, citric, succinic, lactic, and oxalic acids are found in cherries. The predominant acid is malic and the concentration can vary from 0.5 to 2%. The remaining acids are present in such small quantities of little consequence, and only malic acid need be used to indicate total acidity. Total acidity increases in cherry fruit until color begins to develop. As full maturity approaches, the total acidity of the fruit decreases. This decrease in total acidity continues after harvest and during storage.[13]

VI. FACTORS INFLUENCING THE QUALITY OF CHERRIES FOR PROCESSING

Cherries are well adapted for processing. 'Montmorency' is the major processed cultivar of red tart cherries. 'Napoleon' ('Royal Ann') and 'Rainier' are popular light sweet cherries for processing. 'Bing', 'Lambert', and 'Van' are the more popular dark sweet cherries for processing. Sour cherries to be processed are harvested without stems if possible. Sweet cherries are harvested with or without stems, depending on the end use for which the processed product is intended. Stems may be removed mechanically or by hand at the processing plant.[8,13,24] Most of the U.S. production of sour cherries and a significant portion of the sweet cherry production is processed. Processing is an indispensable outlet for the cherry industry. In the sweet cherry industry processing is generally used to market smaller sized fruit, damaged fruit, and generally to salvage fruit that is surplus to the fresh market.

Cherries for processing must be of acceptable firmness, soluble solids concentration, and color. Prematurely harvested cherries will yield a processed product that is soft, has no flavor, and is subject to excessive loss in drained weight. Well-ripened cherries produce a processed product with good texture, flavor, and overall quality. Maturity of the cherries has a very noticeable effect on firmness of the processed product. Cherries picked too early will have a less firm texture than those cherries harvested at a later date. Cherries that are over-ripe when processed tend to produce a product with an unacceptable texture.[8]

Stage of maturity will substantially influence the soluble solids and acid content of the fruit. As harvest season progresses and cherries ripen, the soluble solids concentration or Brix readings will increase and the acid concentration will decrease. The ratio of soluble solids to acids will directly influence the flavor of the processed product. In general, the greater the soluble solids and the less the acid content the more acceptable the overall eating quality of the cherries.

Color of the processed cherries has a direct influence on quality of the product. When fully ripe dark sweet cherries become purple or even black and quality is good. Development of red color in dark sweet cherries is an index of maturity and can be used to predict processing grades of sweet cherries. Light-colored cultivars of sweet cherries develop a red blush, but intensity of the red color is not a good index of maturity for light-colored cherries.[7,11]

VII. PROCESSING

A. INSPECTION

Removal of defective cherries prior to processing is a major cost factor in processed cherries. To remove defective cherries, the cherries are passed over an inspection belt, and defective cherries are removed by hand. If high percentages of fruit are defective the belt speed must be reduced or additional sorters employed. Excessive defective cherries that slip by the inspection process will reduce the ultimate grade for the processed product and thus the monetary return to the processor and grower.[8,13,24]

B. WASHING AND SOAKING

Cherries should be transferred from shipping containers to the processing plant in a manner that minimizes bruising of the fruit. Most transport is accomplished by the use of water flumes or similar types of conveyances. Cherries should be washed by a cold water spray or immersion in water. Washing cherries removes spray residues, foreign matter, and replaces lost moisture. Red tart cherries may or may not be soaked in cold water for several hours prior to pitting to increase the firmness of the cherry. This firming process allows for pitting of the cherries without excessive destruction of the cherry. Red tart cherries are often held in water for one or more days at the plant. Soaking cherries for extended periods of time may result in increased cullage, loss of soluble solids and color, and increased toughness. The major function of soaking the fruit is that of storage. The lower the temperature of the water used to soak the fruit, the better quality of the fruit.[12]

C. GRADING, SORTING, AND PITTING

After washing, the cherries are mechanically graded for size diameter. Size grading allows for the packing of cherries of uniform size. Most of these mechanical size graders operate with the use of screens of various sizes from 20/32 to 33/32 in. Some graders remove large fruit first, while other graders are designed to remove the small fruit first.

Sorting cherries for defects such as lack of color, scar tissue, disease, insect damage, and other types of injury is accomplished by hand. As the cherries pass over an inspection belt human sorters remove the defective fruit. The speed of the inspection belt is regulated by the amount of defective fruit, number of sorters, and ultimate grade desired.

In the past sweet cherries were processed without pitting. Retention of the pit in sweet cherries adds to the appearance of the fruits and imparts a particular flavor to the canned product. The present trend for sweet cherries is to remove the pits prior to processing. Almost all red tart cherries are pitted prior to processing. Pitting is accomplished by mechanical means that punch the pit from the cherry. Pitted cherries are inspected and pits and cherries that were not pitted are removed. The cherry pits are disposed of in a variety of ways but mainly as fuel together with other combustible material for the plant boilers. Pits can constitute from 6 to 17% of the total weight of the cherries and are the major loss when cherry fruits are processed.[8,13,24]

D. CANNING

Cherries are well adapted for either commercial or home canning. Most commercial sweet cherries are canned in the states of Washington, Oregon, and California, with a limited quantity canned in Michigan. Red tart cherries are canned almost exclusively in Michigan with some packed in New York, Pennsylvania, and Wisconsin.

Cherries can be packed in either cans or glass containers. Both red tart and sweet cherries corrode tin plate rapidly. To alleviate this problem, cherries must be packed in cans with enamel linings specifically developed for high acid products. In canning cherries the cans are filled to provide for a minimum drained weight after processing. The cans are filled as full as possible either by hand or by semi-automatic hand pack fillers to meet minimum drained weight standards. Minimum drained weight standards are promulgated by the Federal Food and Drug Administration. The drained weights for processed cherries obtained vary with the size of fruit, total solids, and other factors such as growth regulators applied to the cherry trees. Initial fill weight prior to processing must be established by the individual packer based on local cherries.

Sweet cherries are packed in sugar syrups ranging in concentration from 0° to 56° Brix. Red tart cherries are packed in water or sugar syrups of lighter densities than sweet cherries. The put-in syrup densities used to obtain required cut-out syrup densities that meet grade vary depending on the sweetness of the fruit. Syrup densities greater than 40° Brix plus will result in loss of drained weight and possibly, pitting of the processed fruit. Cherries packed in low syrup densities or water may gain in drained weight or volume. There is a direct relationship between syrup concentration and the weight or volume of the fruit due to osmotic gradient. Syrups should be heated to 100°C to remove oxygen and cooled before being added to the cherries in the cans. The addition of hot syrup to the cans of cherries may cause undue splitting of the cherries. Syrup may also be added to cans by vacuum syrupers. This system allows for a constant fill and, when used with a vacuum closing machine eliminates thermal exhausting.

If thermal exhausting is employed there are two objectives to be achieved: (1) removal of oxygen to reduce corrosion, and (2) production of a vacuum. Cherries are either exhausted in steam or hot water for 4 to 17 min at 98° to 93°C following the syruping operation. The length of the exhaust or time to accomplish the objectives depends on filling temperature and can size. The exhaust time should be long enough to result in a mean internal temperature of 71°C or higher. The exact temperature of the exhaust will depend upon can size and local conditions.

Adequate vacuum can be obtained with atmospheric closure if exhaust temperature is adequate. If the mean internal temperature of the cherries prior to closing is less than 71°C a steam closing device can be used to insure adequate vacuum. If a vacuum syruper is used for canned cherries, it should be used in conjunction with a steam or mechanical vacuum closing device to insure adequate vacuum in the can.

Cans of cherries are closed immediately after exhausting and processed at once. By doing this the heat of the exhausting procedure is not lost before processing begins, and thus cook times can be reduced. Cherries, both red tart and sweet, should be processed long enough to provide a temperature at the center of the can of not less than 91°C. This processing temperature can be achieved by a boiling water bath or by still or agitating retorts. Cherries packed in water will reach a minimum center temperature of 91°C 5 min or more before cherries packed in sugar syrup. Pitted cherries require a longer period of time to reach the desired center temperature than unpitted cherries. An increase of 5 to 10 min may be required for packs using mechanical means of obtaining vacuums, depending upon can size. No specific processing time can be given that meets a variety of conditions.

Immediately after processing canned cherries should be water cooled to 38°C or less. If water cooling is not thorough damage to the cherries may result from heat causing breakdown and discoloration of the product.[8,13,24]

U.S. Standards for grades of red tart cherries and sweet cherries have been established by the Agricultural Marketing Service of the U.S. Department of Agriculture. These standards are not mandatory but are a basis for uniform quality. Color, uniformity of size, absence of defects, and character are quality standards for establishment of canned cherry grades. Color refers to a characteristic cultivar color and to the intensity and brightness of the characteristic color. Uniformity of size refers to the uniformity of diameters of pitted and unpitted cherries and the variation in weight of unpitted cherries. Absence of defects refers to the degree of harmless extraneous material found in the product, including stems, pits, or portions thereof, blemished fruit, and other defects which detract from appearance. Character refers to the tenderness and texture in relation to maturity and to the presence of serious processing cracks in unpitted cherries.

Recommended minimum drained weights are not part of the grade since drained weight is not a factor of quality. Compliance with recommended drained weights is required by the Food and Drug Administration. The liquid media and brix measurements for canned cherries are not part of the grades, but cut-out or Brix measurements must meet certain labelling standards. The liquid ingredients that can be added to canned cherries are cherry juice and water or a combination of the two. The sweetening ingredients from which syrups can be prepared are sucrose, glucose, or corn syrup in specific combinations.[21,22]

E. FREEZING

Frozen cherries have several advantages over canned cherries. Frozen cherries retain more of the original flavor and appearance than the canned cherries. Color retention in frozen cherries is particularly noticeable. When frozen properly cherries have a characteristic bright red or purple color depending upon the type of cherry (red tart or sweet). The characteristic color and flavor of frozen cherries provides an attractive product for hotels, restaurants, bakeries, candy manufacturers, etc.

'Montmorency' is the leading cultivar of red tart cherry used for freezing. Some 'English Morello' and 'Early Richmond' red tart cherries are also frozen, but neither has the quality of 'Montmorency'. Both dark sweet cherries ('Bing', 'Lambert', 'Black Republican') and light sweet cherries ('Napoleon', 'Rainier') can be frozen. Sweet cherries, however, tend to oxidize and darken during freezing and thawing. This darkening of color is particularly evident in light sweet cherries. The darkening color of red tart cherries is generally associated with scald or discoloration as a result of bruising during handling. The extent of this type of discoloration depends upon the initial handling of the fruit. Oxidation also changes the flavor of the cherries. Cherries for freezing must be handled rapidly prior to freezing to prevent oxidation since pitting aggravates the problem.

The preparation of cherries for freezing is similar to the procedures used for canning. In fact, some processors use the same lines for both products. Sweet cherries are almost always pitted prior to being frozen. Pitting of sweet cherries before freezing facilitates penetration of the sugar syrup. When unpitted sweet cherries are frozen the skin must be scored or broken in some manner to allow for sugar penetration. Removal of defective cherries, pits, and other material is of high priority. Defective fruit is readily visible in a frozen pack.

Dry or liquid sugar is added to cherries prior to freezing for several reasons. Sugar helps preserve the color, flavor, and texture of frozen cherries. In addition sugar reduces drained weight or drip loss of cherries after thawing. Dry sugar is usually added to commercial packs (30 lb tins) and liquid sugar to consumer packs (1 to 2 lb packages). Cherries packed in liquid sugar have better color and texture than cherries packed in dry sugar. Liquid sugar is more conveniently added to cherries than dry sugar, and less mechanical damage is evident, thus producing a more attractive pack. Sugar in high concentrations (65° Brix) may cause product shrinkage and toughening. Dry sugar is usually added in a ratio of 5 to 1 or 4 to 1, fruit to sugar. When liquid sugar is added to cherries before freezing a 50° to 65° Brix solution is used.

The oxidation and browning of sweet cherries during freezing and thawing can be reduced. The browning is accelerated by the interaction of oxidative enzymes with oxygen. Oxidation of phenolic compounds is primarily responsible for the brown color. Freezing temperatures retard enzymatic activity. The exclusion of oxygen by the addition of sugar syrup helps to prevent darkening, and sugar can also act as an enzyme inhibitor. Freezing temperatures and sugar retard browning but do not prevent oxidation. Antioxidants such as ascorbic or citric acid alone or in combination should be used to prevent oxidation and browning. Two hundred to two hundred fifty mg/lb of pitted cherries is the general range recommended for use for these food grade acids to prevent browning.

Cherries are packed to 90% of the container capacity to allow for expansion during freezing. Cherries frozen prior to packaging can be filled to the capacity of the container.

Commercial packs of frozen cherries require long periods of time for the center temperature to reach an acceptable level of 4°C or less. Precooling the fruit and/or promptly placing in the freezer aids in alleviating high temperature problems. Cherries should be frozen at as low a temperature as practicably possible, or −23.3 to −29°C. After freezing cherries should be stored at temperatures in the range of −17.5 to −21°C. Fluctuating temperatures that allow for defrosting and refreezing should be avoided. Fluctuating temperatures will reduce desirable texture and enhance dehydration of the cherry fruit. Cherries should be transported in refrigerated containers operating at temperatures to prevent thawing of the product.[8,13,24]

U.S. standards for grades of frozen red tart and sweet cherries were established by the Agricultural Marketing Service of the U.S. Department of Agriculture. Frozen cherries are prepared from the clean, sound, properly matured fresh fruit of any cultivar, which are sorted, washed, and drained, packed with or without the addition of nutritive sweetening ingredients, and frozen in accordance with good commercial practice and maintained at temperatures necessary for the preservation of the product. The grades of frozen red tart or sweet cherries are determined immediately after thawing.[20,23]

Factors of quality for frozen sweet cherries include color, size, symmetry, defects, and character. Color is described as a reasonably uniform color that is bright and typical of well ripened sweet cherries. The percent by count of oxidized or discolored cherries reduces the grade. Size and symmetry refer to uniformity of shape and diameter of the thawed cherries. Diameter is determined on the basis of a rigid hole through which the cherry will pass without the use of force. In pitted cherries the diameter is approximated to the apparent original size. Defects refer to freedom from harmless extraneous material. Character refers to the degree of ripeness, texture, tenderness, and relative thickness of the flesh of the frozen product. Factors of quality for frozen red tart cherries include color, freedom from pits, defects, and character. The standards of quality for red tart cherries follow closely the standards for sweet cherries except size and symmetry are not considered.[20]

F. BRINING

A large quantity of sweet cherries, are preserved by brining. In brining mature whole cherries are soaked in solution of sulfur dioxide of sufficient strength to preserve the cherries. Brined cherries are usually converted into maraschino cherries and used in a variety of products. 'Bing', 'Lambert', 'Napoleon' ('Royal Ann') and 'Rainier' cherries are the cultivars most commonly used for brining. Cherries may be brined at any stage of maturity but quality is optimum with mature cherries. Soft cherries will often be the result of using overripe fruit. In addition, very mature fruit will usually have a high percentage of defects, leading to a reduced grade. When immature fruit is pitted more tearing of the flesh is likely to occur due to thinner flesh and greater adherence of the flesh to the pit.

In general cherries to be brined should have the soluble solids (° Brix) 2° lower than required for fresh shipment. The color is more easily bleached from less mature cherries than from fully ripe cherries. Less mature cherries are firmer and do not bruise as readily as the more mature fruit. Cherries that are bruised before brining will discolor and the discoloration will remain after brining. Cherries to be brined are usually picked with the stems. This avoids injury and browning of the stem cavity. If the cherries are harvested without the stems rapid handling will prevent discoloration.

Cherries to be brined are washed with or without detergents and placed in the brining solution. Brine formulations consist of sulfur dioxide or sulfites and a firming agent, generally lime or calcium or a combination of the two in the form of hydrated lime or calcium carbonate. These solutions will generally contain 1 to 2% SO_2 and 0.5 to 1% calcium. Additional calcium to insure firm cherries may be added in the form of calcium chloride or calcium hydroxide. Calcium concentration should be carefully monitored. Excess calcium may cause the pits to cling to the flesh causing difficulty in pitting the fruit.

The brining solution blanches the fruit to a uniform yellowish-white color and firms the fruit. Cherries should always be added to brining tanks containing brining solution to minimize bruising. Tanks used in the brining operation should be made of wood, concrete or stainless steel. The use of iron equipment should be avoided, since iron will cause discoloration of the fruit during brining. Cherries in brining tanks should be covered with brine at all times. Large brining tanks should be only half-filled until the cherries become firm. By only half-filling the tanks damaged and distorted fruit can be avoided. After 1 or 2 d in the brine, additional cherries may be added to fill the tanks.

Brine tanks should be covered to prevent loss of sulfur dioxide or dilution by rain water (brining tanks are stored outside). Food grade polyethylene film may be used to cover the tanks. Bulk boxes are also used for holding and transporting brined cherries. These boxes are constructed to be staked and palleted, and are generally lined and covered with polyethylene film. A bung hole is provided in each box to replenish or check the brine.

Standard analytical procedures for SO_2 determination are sufficient to determine brine concentrations. The SO_2 content of the brine should be checked constantly and maintained at approximately 0.75% during storage and transportation of the product. There must always be a quantity of free SO_2 or the product will spoil. Excess calcium may tie-up free SO_2, and care should be exercised to avoid this problem. Additional decolorization of dark-fleshed fruit cultivars or oxidation bleaching may be desired. This can be accomplished in a 15 to 20 h period with 0.75% sodium hydrochlorite and 1.0 *N* acetic acid.

After a storage period of a few weeks the brined cherries are ready to be processed into maraschino cherries. The cherries are rinsed in water and then pitted. The pitted fruit is leached in running water to remove the SO_2. Prior to pitting some styles of maraschino cherries are stemmed. Maraschino cherries with stems are called cocktail style. After leaching SO_2, the cherries are dyed with a food grade red dye. The cherries are heated to 100°C for 15 to 25 min in the dye and cooled and allowed to stand for 24 h. During dying a high pH (4.5) is maintained, and the pH is lowered to precipitate the dye in the fruit tissue so that the color will not leach out when the cherries are mixed with other fruit. Adjustments can be made with sodium bicarbonate to raise the pH and with citric acid to lower the pH after the dying process. The heating and standing procedure may be repeated to set the dye. After dying the cherries are rinsed to remove dye. The cherries are then heated in a 30° Brix sugar solution for 5 to 20 min and allowed to stand for several h. After the syrup has been absorbed by the cherries additional syrup at 40° Brix and artificial flavorings are added. The syrup and cherries are again heated and packed hot into glass jars. Preservatives may be used if necessary. Cherries to be used in canned fruit salad are not syruped. The flavors, colors, citric acid, and preservatives added must be listed on the label.[1,8,13,24]

U.S. Standards for grades of sulfured cherries were established by the Agricultural Marketing Service of the U.S. Department of Agriculture. Sulfured cherries are prepare from properly matured whole red tart or sweet cherries of similar cultivar characteristics, packed with or without a firming agent in a solution of sulfur dioxide sufficient to preserve the product. There are several styles of sulfured cherries: unstemmed and unpitted, stemmed and unpitted, and stemmed and pitted. Size is a factor in some styles of sulfured cherries, but is not a quality factor. Color, absence of defects, and character are factors of quality for sulfured cherries.

Color grades for sulfured cherries are based on color uniformity. A light or dark discoloration on the surface not typical of the cultivar will reduce the grade. Absence of defects is the degree of freedom from damaged and misshapened cherries. Damage to sulfured cherries can include mechanical injury, surface discoloration, rain cracks, and blemishes. Character is firmness of the cherry and condition of the flesh.[19]

G. CANDIED FRUIT (GLACE)

Candied fruit is produced primarily for the institutional trade and has unique size, flavor and other ingredients. The product is manufactured to specific customer needs. Candied fruits are generally marketed as a retail item in 0.5 to 1 lb sizes. The manufacture of candied fruit is considered a continuation of the maraschino process before flavoring is added.

In maraschino cherries the syrup concentration does not normally exceed 40° Brix, but in candied fruit syrup buildup is continued until the syrup reaches 70° Brix. This buildup of syrup concentration is slow with about 10° Brix increase each day. Sucrose and dextrose in equal parts are used for increasing syrup concentration. Dextrose is used to prevent the cherries from becoming hard. The cherries are held for about 3 weeks in 70° Brix syrup to allow for syrup penetration of the fruit. Finally the cherries are coated with a light layer of syrup. Glazed or candied fruit must be protected from high humidity and temperature. Candied fruit for the retail market is packed by hand. There are many modifications in the preparation of candied fruit. Most of these modifications shorten the syrup penetration process and some of the processes are patented.[8,24]

H. OTHER USES

Red tart and sweet cherries can also be manufactured into dehydrated, pickled, and jellied products, and juice, cider, and wine. Sweet cherries, higher in soluble solid content than sour cherries, do not collapse and consequently make an attractive, plump raisin product. Sour cherries contribute a characteristic flavor to juice products such as cider and wine. In these products sweet cherries should be used as a blend with sour cherries. In the manufacture of juice products the pits should not be removed before processing. The pits of cherries provide a characteristic almond flavor (due to hydrogen cyanide) to the juice which is preferred by many consumers.

Sweet and sour cherries, fresh or processed, are very attractive fruits enjoyed by many. Cherries are a very perishable commodity, well adapted to preservation by processing. Preservation by processing allows for the enjoyment of this high quality fruit on a yearly rather than a seasonal basis. Manufactured products from cherries are used in many of our confectionary and salad dishes to enhance a meal or snack.[24]

REFERENCES

1. **Bullis, D. E. and Wiegard, E. H.,** Bleaching and dyeing Royal Ann cherries for maraschino or fruit salad use, *Ore. Agric. Exp. Sta. Bull.* 275, 1931.
2. **Childers, N. F., Ed.,** *Modern Fruit Science,* Somerset Press, Inc., Somerville, NJ, 1954.
3. **Couey, H. M., Williams, M. W., Berlage, A. C., and Larson, R. P.,** Effects of ethephon and harvest method on quality of stemless sweet cherries harvested for the fresh market, *HortScience,* 9, 587, 1974.
4. **Drake, S. R., Proebsting, E. L., Jr., Carter, G. H., and Nelson, J. W.,** Effect of growth regulators on ascorbic acid content, drained weight, and color of fresh and processed 'Rainier' cherries, *J. Am. Soc. Hortic. Sci.,* 103, 162, 1978.
5. **Drake, S. R., Proebsting, E. L., Jr., and Nelson, J. W.,** Influence of growth regulators on the quality of fresh and processed 'Bing' cherries, *J. Food Sci.,* 43, 1695, 1978.
6. **Drake, S. R., Proebsting, E. L., Jr., Thompson, J. B., and Nelson, J. W.,** Effects of daminozide, maturity, and cultivar on the color grade and character of sweet cherries, *J. Am. Soc. Hortic. Sci.,* 105, 668, 1980.
7. **Drake, S. R., Proebsting, E. L., Jr., and Spayd, S. E.,** Maturity index for the color grade of canned dark sweet cherries, *J. Am. Soc. Hortic. Sci.,* 107, 180, 1982.
8. **Fogle, H. W., Snyder, J. C., Baker, H., Cameron, H. R., Cochran, L. C., Schomer, H. A., and Yang, H. Y.,** Sweet Cherries: Production, Marketing and Processing, USDA/ARS, *Agric. Handbook No. 442,* 1973.
9. **Greig, Smith,** personal communication, Dept. of Agricultural Economics, Washington State University, Pullman, WA, 1984.
10. **Hedrick, U. P., Howe, G. H., Taylor, O. M., Tubergen, C. B., and Wellington, R., Eds.,** *The Cherries of New York,* J. B. Lyon, Co., Albany, NY, 1915.
11. **Ingalsbe, D. W., Carter, G. H., and Neubert, A. M.,** Anthocyanin pigments as a maturity index for processing dark sweet cherries and purple plums, *Agr. Food Chem.* 13, 580, 1965.
12. **LaBelle, R. L.,** Heat and calcium treatments for firming red tart cherries in a hot fill process, *J. Food Sci.,* 36, 323, 1971.
13. **Marshall, R. E.,** *Cherries and Cherry Products,* Interscience, NY, 1954.
14. **Micke, W. C. and Mitchell, F. G.,** Handling sweet cherries for the fresh market, *Calif. Agric. Exp. Sta. Circ.* 560, 1972.
15. **Patten, K. D.,** Factors affecting the maturation, texture, and mechanical damage of sweet cherries, Ph.D. Dissertation, *Dept. of Hort. and Landscape Architecture,* Washington State University, Pullman, WA, 1984.
16. **Patterson, Max E.,** CA storage of cherries, in *Controlled Atmospheres for Storage and Transport of Perishable Agricultural Commodities,* Richardson, D. G. and Meheriuk, M., Eds., Timber Press, Beaverton, OR, 1982, 149.
17. **Proebsting, E. L., Jr., and Mills, H. H.,** Effect of daminozide on growth, maturity, quality, and yield of sweet cherries, *J. Am. Soc. Hortic. Sci.,* 101, 175, 1976.
18. U.S. Dep. of Agric., *Nutritive Value of American Foods in Common Units,* Agric. Res. Serv. Agric. Handbook 456.
19. U.S. Dep. of Agric., *U. S. Standards for Grades of Sulfured Cherries,* Agric. Mark. Serv., June 12, 1951.
20. U.S. Dep. of Agric., *U.S. Standards for Grades of Frozen Sweet Cherries,* Agric. Mark. Serv., March 1, 1958.
21. U.S. Dep. of Agric., *U.S. Standards for Grades of Canned Red Tart Pitted Cherries,* Agric. Mark. Serv., May 15, 1974.
22. U.S. Dep. of Agric., *U.S. Standards for Grades of Canned Sweet Cherries,* Agric. Mark. Serv., May 12, 1976.
23. U.S. Dep. of Agric., *U.S. Standards for Grades of Frozen Red Tart Cherries,* Agric. Mark. Serv., June 28, 1974.
24. **Woodroof, J. G., and Luh, B. S.,** *Commercial Fruit Processing,* AVI Publishing, Westport, CT, 1975.

Chapter 9

THE STRAWBERRY

Justin R. Morris and W. A. Sistrunk

TABLE OF CONTENTS

I. INTRODUCTION

Strawberries (*Fragaria* × *ananassa* Duch) are one of the most widely grown small fruits in the world. They are adapted to a wide variety of soil and climate conditions and can be found growing from the tropics to subarctic regions.

The strawberry species dates back an estimated 50 millions years[119] and their use by man dates to the Romans and Greeks.[22] The first cultivated strawberries were grown for both ornamental and medicinal purposes and are often depicted in paintings of the 14th century. This strawberry, however, is not the strawberry we know today. The fruit was small and the plants were not productive. The large modern fruit of today was developed in the early 18th century by the French who crossed two species of strawberries which were introduced from the Americas. The cross was made between the wild beach strawberry *Fragaria chiloensis* native to the coast of Chile and the wild meadow strawberry *F. virginiana* native to eastern North America.[13,22] This hybrid was *F. ananassa,* the mother of the modern day strawberry.

This hybrid was returned to North America in the form of the domesticated cultivars in the early 1800s. By the early 1900s, strawberry production had spread from coast to coast with the markets for fresh fruit centered around large cities. By the 1930s large production areas were concentrated in Arkansas, California, Louisiana, Michigan, Oregon, and Washington. However, during the 1950s the commercial processors left the South; this was largely due to the uncertainty of raw product supply and labor shortages.

Today, 75% of the strawberries produced in the U.S. come from California. This is largely due to a 5 month growing season which results in high yields and continuous production. Table 1 shows the West Coast to be the largest producers of fresh market and processing strawberries.

Much of the emphasis on strawberries today is placed on quality. Quality is affected by cultivar, climate, plant nutrition, cultural systems, harvesting and handling techniques, diseases and insects, maturity, transport and storage, and processing methods. With greater consumer demand for quality it becomes increasingly necessary for the growers, shippers, processors, and consumers to be aware of the factors which influence strawberry quality. This chapter will attempt to discuss some of the major factors which affect quality from planting until final consumption.

II. QUALITY CHARACTERISTICS

A. U.S. STANDARDS FOR GRADES

The U.S. Standards for Grades of strawberries were last revised in December of 1981 and provides four grades under which strawberries can be classified: U.S. No. 1, U.S. Combination, U.S. No. 2, and Unclassified.[32] The first three grades list requirements that specify strawberries of one cultivar or similar cultivar characteristics, attachment of the cap (calyx), firmness, maturity, freedom from mold or decay and from damage caused by dirt, moisture, foreign matter, disease, insects, mechanizations, or other means. Minimum requirements for berry size and fruit surface area showing a pink or red color are listed. Tolerances by volume are given for defects and off-sized berries. Unclassified is not really a grade as such, but is used to show that no grade has been applied to a lot.[32]

B. OBJECTIVE TESTS

Many objective tests have been used to define quality parameters in fresh and frozen fruit.[10,76,83,94,95,99,101,102,111,120,122] Color has been measured by many different types of color instruments including the Color Difference Meter.[34,35,89,95,101,102,103] Total anthocyanins have been measured by optical density × dilution after extraction with acidified ethanol.[97] Also,

TABLE 1
Fresh Market and Processing Strawberries: Acres Planted, Harvested, and Yield Per Acre, 1978/82, USDA

State	Acres	Harvested[a] 1000 Cwt	Yield Cwt/A
California[b]	11,000	5,115	465
Oregon	5,300	463	89
Washington	2,900	174	60
Florida	2,500	475	190
Michigan	2,400	124	54
Wisconsin	1,800	57	38
Ohio	1,500	105	70
New York	1,500	87	62
Louisiana	700	39	60
New Jersey	650	27	42
Arkansas	400	26	64
Total 11 States	31,500	6,315	
Total U.S.	34,000	6,790	

[a] Of the harvested berries above, the amount processed in four states was: California—1609; Oregon—420; Washington—128; Michigan—39. New York processed a small amount.

[b] California is by far the leading state in strawberry production from an acreage and volume standpoint. The state accounts for about 75% of the production on 32% of the acreage.[9]

extraction in acidified methanol and separation of pigments by paper chromatography has been used to identify the pigments and define color of strawberries.[98]

C. EFFECTS OF CULTIVAR

New cultivar developments have been rapid since the strawberry was first developed in the 1700s and the change from one cultivar to another has been just as rapid. There are hundreds of cultivars available for production in the U.S. today and many breeding lines are being evaluated for their suitability and adaptability to growing conditions, harvesting methods, market types, and yields.

1. Disease Resistance

Cultivars in use today are rigorously screened for plant and fruit characteristics before they are released from breeding programs. The most important plant characteristic is disease resistance, while vigor, growth habit, and runner development are also important. In terms of disease resistance there are several important diseases that are included in most screenings. Among these are numerous virus diseases, leaf blight *(Dendrophoma obscurans)*, leaf scorch *(Marssonina potentillae)*, leaf spot (Ramularia tulasnei), anthracnose *(Collectotrichum fragariae)*, powdery mildew *(Oidium fragariae)*, red stele *(Phytophthora fragariae)*, and verticillium wilt *Verticillian dehliae)*.

Selection of strawberry cultivars that are the most suitable for fresh and processed fruit may vary from one region to another, depending on the climate, soil conditions, and intended use of the fruit. While environmental conditions determine the adaptability of the plant to a particular region, it is the genetic make-up that governs the development of fruit with an individual set of quality characteristics.

2. Breeding

Although many plant and fruit characteristics such as plant type, disease resistance, productivity, adaptability, etc. should be considered before the assessment of fresh fruit quality, fruit quality should be a major objective of any breeding program. The cultivar selected for processing should have a rich red color throughout the fruit as well as excellent firmness and flavor. Strawberry fruits vary widely in composition of enzymes, pigments, acids, phenols, pectins, cellulose, flavor components, and sugars as well as in size, shape, and cell structure.[101,102] Thus, it is necessary to design a variety of tests appropriate for defining differences in genotypes not only in plant and fruit characteristics but also for quality parameters.

Breeding for fruit quality in strawberries is a challenge because of the complexity of color, character, flavor, and other quality parameters that dictate the final quality of processed products.[66,106] In selecting from large populations of segregating genotypes the ability to choose those genotypes with the most desirable quality for further testing can mean success or failure to any program. Successful breeding and development of superior genotypes usually is best accomplished be a team effort involving a breeder, postharvest physiologist and food technologist. Color, character, flavor, and chemical composition vary widely among cultivars and selections.[37,39,68,83,95,101,102,104,106] Even so, high correlations between quality parameters and chemical and physical characteristics exist over large numbers of genotypes, indicating that a minimum number of these tests can be used to predict quality.

The decline in strawberry production in the U.S. for processing has been due partially to the lack of suitable cultivars that are high in quality.[103] Lack of firmness and uniform red color are the principal factors responsible for low quality. When comparing older cultivars (Table 2) with recent releases, considerable progress has been made in improving quality (Table 3). A few of the selections were much higher in soluble and total solids than the named cultivars except in 'Comet'. The cultivars listed are the most important ones that have consistently performed well in Arkansas. The percent acidity was above 0.80 on all genotypes listed except "A-6665." Selections with low acidity and high pH should be eliminated from a breeding program because of the effect on color stability.[101] Total anthocyanins were shown to be a good measure of color of strawberries, showing a high relationship with sensory color (Table 3).

Color of thawed frozen strawberries was shown to be darker than the color of fresh fruit by measurement of 'L' values by a Gardner Color Difference Meter.[101,102] This color difference was attributed to a decrease in redness, increase in blueness (lower 'b'), and the development of a purplish red color. The magnitude of color differences between cultivars was less in frozen fruit in most instances because of the rapid oxidation after thawing. In cultivars with high pH, brown discoloration masked the L, a, and b values so that they did not correlate with sensory color as well as in fresh fruit among a large number of cultivars.

In the 1960s, the commercial cultivars grown in Arkansas rated poor in quality as compared to new genotypes.[101,102] One of the primary reasons for poor quality was caused by the rapid deterioration in color upon freezing and thawing. Differences between the 'L' and 'a' values of fresh and thawed frozen fruit were a good indicator of the susceptibility of cultivars to pigment breakdown.

Part of this rapid change in color was attributed to pH. Strawberry anthocyanins have been shown to increase in color intensity with a decrease in pH.[83] The correlation coefficient between pH and color was $r = 0.52$; an r value of -0.57 was found between acidity and pH.

Color and flavor are closely related in most pigmented fruits. Strawberries that were rated higher in color by a sensory panel were lower in 'L' and higher in 'a' values.[35,101,104] Color is primarily a characteristic of cultivars, yet other factors such as maturity and delay in harvest influence color of fresh and frozen fruit. Maturity of fruit affected only 'L' value

TABLE 2
Effect of Cultivar on Sensory Quality Ratings of Thawed Frozen Strawberries (4 + 1)[a]

Selections and cultivars	Color	Slice wholeness	Texture	Flavor	General appearance	Mean rating
Albritton	6.1	5.3	5.8	5.9	6.0	5.8
Blakemore	4.8	5.5	4.8	6.1	5.2	5.3
Citation	6.5	6.2	6.4	5.6	6.8	6.3
Dixieland	4.9	6.3	5.7	5.2	5.4	5.5
Earlibelle	7.8	6.4	6.5	5.7	7.2	6.7
Earlidawn	6.2	5.4	5.8	5.2	5.9	5.7
Midland	7.5	5.0	6.5	7.1	6.4	6.5
Midway	6.2	5.6	5.3	6.2	6.1	5.9
Md. U.S. 2289	3.8	5.3	5.4	5.4	4.6	4.9
Md. U.S. 2593	5.8	6.6	6.7	6.1	6.4	6.3
Md. U.S. 2713	8.5	7.3	6.8	6.0	7.9	7.3
Md. U.S. 2819	5.7	6.0	6.3	6.2	6.2	6.1
Md. U.S. 3082	4.4	6.7	6.3	5.2	5.5	5.6
Northwest	8.6	6.4	6.9	6.0	7.0	7.0
N.J. 857	5.1	5.1	5.5	6.0	5.0	5.3
Pocahontas	6.1	6.9	6.2	6.0	6.1	6.2
Redglow	5.9	5.0	5.3	5.1	5.4	5.3
Stelemaster	4.4	5.0	5.7	5.5	4.6	5.1
Sunrise	3.9	4.4	5.0	4.7	4.1	4.4
Surecrop	5.7	5.0	5.3	5.1	5.4	5.3
Tenn. Beauty	6.7	4.9	5.4	5.8	5.9	5.7
LSD at 5% level	1.0	1.1	0.9	1.0	1.1	0.8

[a] Sensory quality rated on thawed frozen fruit (4 + 1) by a panel of 10 members on a scale of 1 = poor to 10 = best.

when measuring color quality on 21 cultivars; 'L' values decreased and 'b' values increased with maturity. Saturation of hue or $(a^2 + b^2)^{1/2}$ was not as useful in defining redness as 'a' values. Correlations between sensory ratings for color on fresh fruit and 'L', 'a', and optical density of extracted pigments were similar to those on thawed frozen fruit.

Darrow et. al.[24] showed that seedlings in progenies from parents having high, medium and low ascorbic acid content attained a higher ascorbic acid content than the level that existed in commercial cultivars.

Sistrunk and Moore[101,102] have shown that ascorbic acid varied significantly among selections and cultivars. "Atlas," "A-6414" and "A-6495" were higher than the others listed (Table 3). Color as rated by a sensory panel appeared to be negatively correlated with 'L' value and not as closely related to 'a' values as in earlier studies.[101] Since both 'a' and 'b' values are a function of 'L' wide differences in 'L' value can affect the correlation of 'a' with sensory color of strawberries.

In a study on 12 cultivars of strawberries Skrede[112] found no relationship between the chemical components and sensory parameters except for color. Also, there was no relationship between drip loss and texture measurements in thawed frozen fruit.

Hansche et al.[38] indicated that fruit firmness is a highly heritable characteristic, yet correlations between firmness and either yield or fruit size were very low. Similar relationships in heritability for softness and firmness were demonstrated by other investigators.[28,68,76] Soluble solids and titratable acidity were found to be consistent from year to year among cultivars.[87] These characters have been shown to be genetically controlled so that low acid genotypes can be eliminated from a development program.[28]

Selection of fruit characteristics depends upon the distance to market. Flesh firmness

TABLE 3
Objective Analyses of Selections and Cultivars of Fresh Strawberries, 1978

	Fresh fruit								Frozen sliced[b]		
							Color difference[a]				
Selections and cultivar	Soluble solids (%)	Total solids (%)	Titratable acidity (%)	pH	Ascorbic acid (mg/100g)	Total anthocyanins OD/gfw at 500 nm	L	a	Color	Wholeness	Flavor
A—6086	5.3	6.7	0.89	3.26	35.6	1.62	27.3	29.9	6.4	7.7	5.7
A—6225	6.2	7.5	1.07	3.26	39.6	2.21	24.4	29.0	9.1	8.8	6.8
A—6385	6.4	7.5	0.83	3.28	40.4	2.01	21.9	26.1	7.8	6.7	7.3
A—6414	6.9	7.0	0.84	3.30	54.4	1.88	23.8	28.7	6.9	8.8	7.1
A—6495	5.9	6.8	0.94	3.20	46.8	2.48	23.8	30.7	7.9	6.2	6.8
A—6568	5.9	7.5	1.13	3.09	38.8	3.20	22.8	30.4	8.0	8.1	7.1
A—6583	6.3	7.5	1.16	3.12	34.8	2.90	19.6	26.9	8.8	8.2	7.4
A—6665	6.7	7.4	0.75	3.30	36.4	2.85	18.2	26.7	8.9	6.9	7.1
A—6673	7.0	7.6	0.85	3.33	32.4	2.36	23.0	28.8	8.3	6.1	6.9
A—6675	5.5	6.6	0.84	3.10	24.0	2.05	21.8	26.3	8.3	6.3	7.7
A—6769	7.5	8.2	0.90	3.18	40.0	3.48	22.5	29.3	8.1	6.5	6.9
A—6791	6.0	7.3	0.81	3.15	21.2	3.32	21.0	24.6	7.7	7.9	7.1
Arking	5.8	6.9	0.89	3.09	20.8	1.44	27.3	27.2	5.8	7.3	6.8
Apollo	4.8	7.7	1.08	3.10	40.0	2.44	24.0	30.5	7.4	6.4	7.3
Atlas	6.5	8.1	1.03	3.21	47.2	1.78	26.2	29.7	7.3	6.9	7.0
Cardinal	5.9	7.1	0.85	3.21	26.8	2.95	18.9	24.7	9.1	7.9	7.4
Comet	8.2	9.8	0.92	3.15	24.8	2.80	19.4	28.1	8.5	6.5	7.0
Delite	5.5	6.3	0.71	3.28	19.6	1.36	28.7	25.3	3.1	5.1	4.6
Earlibelle	5.9	7.2	1.15	3.25	42.0	2.52	23.6	31.5	7.5	6.9	6.7
Sunrise	6.3	7.5	1.08	3.11	42.0	1.24	28.8	28.5	4.6	5.2	6.1
LSD at 5% level	0.3	0.2	0.12	0.06	4.2	0.26	1.2	1.5	1.0	1.1	1.4

[a] Color determined by Gardner Color Difference Meter standardized by a red plaque L = 24.4; a = 25.2; b = 11.8.
[b] Sensory quality rated on thawed frozen fruit (4 + 1) by a panel of 12 members on a scale of 1 = poor to 10 = best.

and skin toughness greatly influence the 'shelf-life' of fresh strawberries and only firm tough skinned cultivars which are adapted to long-distance shipping should be selected.[66] Market acceptance is based primarily on the appearance of the product. For the home or local market, large brightly colored, glossy fruit are desired. This fruit may be excellent in quality but the softer texture restricts its use to a local area as shipment would reduce its acceptability. Some cultivars not adapted to shipping may be used for freezing expecially if the local market becomes overloaded. However, some cultivars if not properly selected for bright and uniform internal color will fail to produce an acceptable frozen product.

Acceptable flavor and texture after freezing and thawing are other important quality attributes to consider in a breeding program. Flavor appears to be correlated with color[101] and with ascorbic acid.[35] However, flavor is influenced by soluble solids and total acidity of the fruit. Flavor differences among cultivars and maturities were more difficult to detect by sensory rating than color and firmness.[101]

D. EFFECT OF CLIMATE

Strawberries are found growing in virtually every part of the world. Different climatic areas have different cultivars which are adapted to that particular area. Within each area, environmental fluctuations such as light, rainfall, and temperature occur which can influence quality parameters.

Photoperiod and temperature have a pronounced effect on quality. Degree day accumulations have been shown to influence the time of fruit maturation,[131,137] and temperature can have a marked affect on strawberry flavor.[23] Sunny days and cool nights tend to produce strawberries of better flavor than cloudy, humid days and warm nights. Reduced sunlight has been shown to decrease pH, ascorbic acid, color, soluble solids, and acidity while increasing the firmness of strawberry fruits.[102] Other factors, such as soil moisture, soil temperature, rainfall, and plant age also contribute to seasonal variations in fruit quality.

The use of supplemental irrigation will often reduce seasonal variations in strawberry production. Supplemental irrigation is necessary for strawberry production in semi-arid regions of the country and in areas where summer rainfall is inadequate or unevenly distributed for best production. Supplemental irrigation produces stronger plants with larger leaf areas, which in turn produce larger more attractive fruit.[85] However, excess irrigation or natural rainfall during the harvesting season can be detrimental to quality by damage to the epidermis, decreasing fruit firmness, and reducing sugar content,[53] and by creating favorable conditions for fruit rot.[130]

The amount of rainfall and humidity as well as the time of harvest influenced the firmness of fresh and frozen strawberry fruit.[20,73] Fruit was softest immediately after a rain. Mold infections were high on strawberries from plants with heavy foliage, especially during periods of rainfall and high humidity. The length of time between pickings had less effect on mold count than rainfall.[8] After some mold has developed in the field or in the storage on berries, it has been shown that some detergents will lower the mold count on the frozen berries.[47] It has not proven to be a practical method commercially of reducing the final mold count in berries.

E. EFFECT OF PLANT NUTRITION AND CULTURAL PRACTICES

Cultural practices include a number of individual operations which the grower must incorporate and time according to the specific cultivar, the climatic conditions, type of soil, fertilization practices, and presence of pests and diseases. Selecting a cultivar which will yield an economical crop of good quality fruit along with a good planting site are two initial factors deserving much thought.

Planting time will vary according to the region and its climate. If planting is done too early for a cultivar, the plants lack vigor and will produce lower yields and fruit quality. Planting too late also results in a low yield but the fruit size and quality are good.

Almost any type of soil containing an adequate supply of organic matter can be used to grow strawberries. Often the best yields are obtained on deep, fertile, well-drained loams with a good moisture holding capacity. Without adequate water drainage, the soil will become saturated thereby killing the old roots, preventing the development of new ones, and increasing the chances of root disease.

1. Plant Nutrition

The application of fertilizers to strawberry plants has a marked influence on plant growth. This is especially true for nitrogen fertilizers. However, it is difficult to demonstrate the effects of most fertilizers on quality except in cases of deficiency or excessive levels. Also, the effects of weather factors and soil types can contribute to quality to a greater extent than commercial ranges in fertilization rates.

The strawberry plant is among the most sensitive fruit plants in response to supply of available soil nitrogen (N).[2,126] Reduction in yields of strawberry plantings have been reported by several researchers when N was applied in early spring.[57,58,126] The most effective time to apply N is in the early fall, immediately prior to flower bud initiation.[42,53,56,57,90,118] The form in which N is applied appears to make little difference provided it is readily available to the plants.[2,50,54,118]

Szymoniak[121] reported that phosphorus was distinctly beneficial in increasing the size and firmness of strawberries and in reducing the proportion of culls. Shoemaker and Greve[57] found strawberries from nitrated plots to be slightly softer than fruits from check plots. Overholser and Claypool[77] reported that nitrogen applications increased the intensity of respiration of strawberries from 10 to 20%, and caused softer berries. Dennison and Hall[27] showed that 100 lb of nitrogen/acre produced significantly more fruit and larger fruits than 0 or 40 lb rates. However, later research showed that yield did not significantly increase when nitrogen was increased from 20 to 100 lb/acre.[15] At high nitrogen levels (up to 100 lb of actual N/acre on alkaline soil) fruit yield, earliness, size, and appearance were adversely affected.[125] In a study on the Florida 90 cultivar of strawberries, in which an annual hill system and black polyethylene mulch was used, the total N content and shear resistance of the fruit increased with an increase in the rate of nitrogen.[84]

The quality of strawberries as influenced by commercial fertilizers has been studied by a number of investigators.[20,45,53,77,96,129] Fruit were lower in soluble solids and total pectins when fertilized with high levels of N, especially in riper fruits. Sistrunk et al.[96] showed significant differences in quality due to fertilizer levels. Percent soluble solids, total sugars, and water-soluble pectins of fresh fruit were higher on the fruit from plots with either a higher level of N or N plus potassium (k)/acre, but shearpress values were lower. In the frozen sliced pack, percent broken slices and syrup flow rates were lower and shearpress values were higher on fruit from plots fertilized with lower rates of fertilizer.

Data on the influence of different harvest dates and fertilizer treatments on quality of fruit indicated that percent moisture and sugar content varied among treatments from one harvest to the next.[53,94] Greater differences in composition occurred in fruit at different harvest dates than between fertilizer treatments. Much larger differences were recorded in firmness in berries from watered versus non-watered plots.[53] Berries were smaller, moisture content lower, and sugar content higher from plots receiving no water and no fertilizer. In similar studies, the total N content of fruit and resistance to shear increased when the rate of N was increased to the plants whereas titratable acidity in the fruit decreased.[84] Increasing the rate of K to plants increased the K and acidity in the fruit while firmness of fruit did not change. The compression rate of the fruit tended to be reduced with an increase in the rate of K to plants while increasing with rates of N. Source of K had little effect on fruit quality.

Overall, the effects of nutrition on strawberry quality can, for the most part, be attributed to indirect effects, except where deficiency levels of an element exist. The strawberry grower must use nitrogen fertilizers cautiously because a stimulation of excess plant growth may contribute to conditions favoring development of field rots, and softer berries especially during wet seasons. An increase in berry size due to fertilization may be offset in value by losses due to shipping quality.[26]

2. Training System

When discussing training systems there is very little information on the effects of the various systems on fruit quality. However, it can be said that without these different systems there would definitely be inherent problems which would have detrimental effects on the final yield and quality.

Training systems change slowly over the years; though at one time the wide matted row of strawberries was generally considered the best training system for the country. This system has gradually changed over the years. There are currently five major training systems in use with modifications being made continually. These training systems are; matted rows, hedge rows, spaced beds, hill rows, and solid beds.[91]

The matted row system is used in most northern regions. The runners are allowed to set at will within the limits of the row. Rows as wide as 4 ft are often used. However, the current trend is toward narrower beds with fewer plants. This will allow plants less competition and in turn produce larger, better quality berries. The larger fruit is particularly desirable for the fresh-market and pick-your-own operations.

The hedgerow system entails definite placing of runners from each plant. All unwanted runners are removed. Fruit quality is usually improved by this system, but high cost of production due to extensive hand labor usually overrides this benefit.

The spaced bed is often used in the East. This system is much like that of the matted row. The main exception is that the number of runners allowed to 'peg down' is limited. Spaced rows often produce larger berries and higher yields. As with the hedgerow system, this system requires intensive hand labor, and thus higher cost input.

Hill rows are used in areas where growth continues for most of the year. There are generally two types, the single and the double hill rows. The single hill beds are used mostly in the Pacific Northwest while double hill beds are used in Florida and California. The beds are raised to about 10 in. and each type ranges in width up to 3 f. The beds are usually covered with black plastic. Drip irrigation and fertilization is provided under the plastic mulch. This type of system will not work in the northern regions because it lowers the temperature of the root zone during cool weather.

Solid bed planting systems are just as the name implies. The beds are completely solid and the runners are allowed to grow at will. Yields are much greater for this system. It requires much less hand labor and may become the system of the future for berries destined for mechanical harvesting and processing. One might expect slightly smaller fruit size due to more plants per unit area with this system, but the solid planting produces extremely high yields. Research in Canada has shown a 40 to 50% increase in yield over the matted-row system.[82]

During summer, spacing and thinning of the runner plants usually increases yield and improves fruit quality. Since the rows are more open, the berries are larger, easier to pick, and less likely to rot since they can dry off more rapidly and can be better reached by fungicides.[49]

Weed control should be started soon after planting. Herbicides are available that are effective in controlling weeds. Weeds compete with the strawberry plant for moisture and nutrients. Excessive weed growth results in reduced yields, and if present during the time of pollination, interfere with proper pollination of the blossoms, resulting in the development of many 'nubbins'. Also, weeds will interfere with harvesting operations.

Commercial strawberry production is largely on raised beds, primarily for facilitating water drainage. In California, bed height is important due to the problem of salt accumulation. Low profile beds are reported to accumulate less salt than higher beds.[124]

3. Irrigation

Irrigation is necessary in most areas for strawberry production. Because of their shallow root system, strawberries require water either in the form of rainfall or irrigation many times during the growing season and especially at planting time, flower initiation, fruit development, and immediately after renovation. Proper control of soil moisture through irrigation will result in better plant stands, increased bud formation and blossom survival and therefore higher yields. Fruit quality can be improved with irrigation by the production of larger, firmer, and better shaped fruit. However, excessive irrigation around harvest time is detrimental to fruit quality as the fruit is softer and more easily bruised during harvest. Excessive moisture at this time also enhances the growth of molds and stimulates fruit rot.

Irrigation is important not only in providing adequate moisture, but also for reducing the detrimental effects of both high and low temperatures. At high temperatures, excessive transpiration will result in reduced fruit size since water is necessary for cell enlargement. The "air conditioning" effect of overhead irrigation helps to lower the leaf temperature and raise the relative humidity of the surrounding air. At low temperatures (frost), part or all of the flower pistils are destroyed resulting in either a complete lack of fruit development or in misshapened fruit. This damage due to frost can be prevented by sprinkle irrigation. Since a severe early frost will eliminate the primary fruits that produce the largest berries, not only will yield be reduced, but the berries remaining for harvest will be smaller.

The major types of irrigation are sprinkle, furrow, and drip. In the eastern U.S., sprinkle irrigation is one of the most useful types as it can provide "air conditioning" at excessively high temperatures, frost protection at low temperatures, and provides a good distribution of water. Furrow irrigation uses a lot of water and leaves the rows excessively muddy. Also, problems with salt accumulation can occur more extensively with furrow irrigation. With trickle irrigation, pipes containing emitters or small orfices are laid on top of the bed thus providing water only to the root zone. With this system, salt accumulation is low, water is used to its maximum, and there is a potential for less disease and rots since the foliage and berries do not get wet.

4. Mulching

Mulching is required in some northern states to protect the plants against the early winter freezes occurring before the plants are completely dormant. In spring, mulching helps reduce injury from frost by delaying bloom. Mulching can also conserve moisture, replace cultivation, reduce the number of dirty and rot infected berries, make picking more pleasant, and may improve fruit color and increase yield.[91] Although the most commonly used mulch today is probably straw or hay, the use of plastic film provides similar advantages with the exception that bloom may be speeded up, exposing the plants to early frost in problem areas. With the film some scalding of berries may also occur on hot days.

F. Pests and Diseases

The strawberry plant and its fruit are susceptible to several types of organisms capable of reducing yield, appearance, and quality of the fruit. Two common causes of serious strawberry losses are due to the presence of viruses and nematodes. These pests are capable of reducing growth and production by half or more.[19] Insects known to damage or destroy floral parts are the spittle bug, tarnished plant bug *(Lygus lineolaris),* mites, and thrips. The damage brought about by these insects result in misshaped and malformed fruit.

Each year fruit rots that are caused by various fungi result in a major postharvest disease

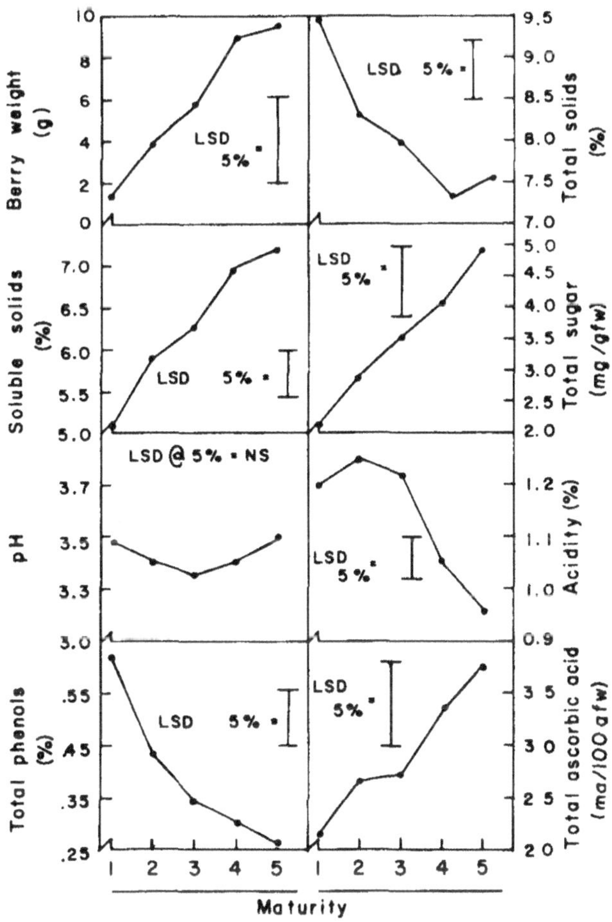

FIGURE 1. Compositional changes with increasing maturity (1, immature green; 2, mature green; 3, inception; 4, firm ripe; 5, processing ripe). Each mean is pooled across 2 cultivars and 2 repetitions. (From Spayd, S. E. and Morris, J. R., *Ark. Farm. Res.*, 30(1),6,1981. With permission.)

loss. The most troublesome fruit rot is caused by *Botrytis cinerea* commonly known as grey mold rot. This fungus causes major losses both in the field and after harvest. Other important fungi resulting in fruit rot are *Colletotrichum*, *Phytophthora* (leather rot), *Rhizoctonia* (hard rot), *Pezizella* (tan brown rot), and *Gnomonia* (stem-end rot). *Rhizopus* is normally associated with the postharvest rot fungi, but can also be found in the field.

III. HARVESTING AND HANDLING

Unlike many fruits, strawberries are harvested when they are essentially full-ripe. Because of this exacting stage of maturity and the very high rate of metabolism, they are one of the most perishable fruits. During maturation, the berry weight, soluble solids, total sugar, total anthocyanin, and ascorbic acid increase, while total solids, acidity, peroxidase and polyphenol oxidase activity decrease with berry maturity.[117] In Figure 1, one of these changes that occur during maturation are shown.

Strawberries are harvested at different stages of ripeness depending on the cultivar grown, harvest temperature, cooling facilities, distance to market, and whether the berries are intended for fresh market shipment or for processing.[67] Cultivars with soft flesh or that soften quickly when ripe should be picked before fully red in color. If the fruit is intended for the local market or for processing, it can be picked when the color is more fully red. Since the strawberry continues to enlarge until fully ripe, the manager of a pick-your-own operation would benefit by encouraging customers to pick only fully ripe berries.

Fresh market berries should be picked by pinching off the stem from each berry since pulling the berry from the plant will usually break the skin and many times sever the cap. Berries are also bruised when the berries are tossed or dropped into the container and during stacking of over-filled crates. Bruised berries and berries without caps will not keep as well during holding for the fresh market. Grading is often performed in the field by the picker as he places the fruit directly into pint containers arranged in fiberboard tray packs. This method reduces amount of handling and speeds the berries to market.

Berries should be precooled immediately after harvest. Strawberries held at 21°C will respire 10 times faster than those at 0°C. This high rate of respiration after harvest accelerates the deterioration.[78] Also, the rate of cooling after harvest affects shelf life. An additional day of shelf life or a greater percentage of marketable fruit may be obtained by rapid cooling bringing temperature of the fruit down below 5°C within 1 h vs. 10 to 12 h.[14,65] Rapid cooling is achieved by forced air methods. In addition to decreasing the rate of respiration, lowering the temperature of the berry has been found to increase the resistance of the epidermis to puncturing, possibly reducing the susceptibility of fruit to mechanical injury and to penetration by molds.[46]

A. MECHANICAL HARVESTING

Mechanical harvesting of strawberries may be a necessity for survival of the U.S. strawberry industry in areas where there is a shortage of harvest labor. A mechanical harvesting and cleaning system developed by the University of Arkansas[69,70,72] has proven to be commercially successful. The newest commercial model of this machine, a self-propelled, two-row unit, has been used in North Michigan and in Oregon.

Overall success of harvest mechanization of strawberries is dependent on developing a total system for the production, harvesting, handling, and utilization of the fruit including all maturity classes from a once-over harvest. Therefore, studies were designed at the University of Arkansas to (1) evaluate the effects of a total systems approach to strawberry harvest mechanization on yield and raw product quality; (2) determine the suitability of the large-size machine-harvested fruit for utilization as a sliced, sugar-packed product; and (3) determine the influence of cultivar, fruit-maturity mixtures, and storage time on quality attributes of jam made from machine-harvested fruit.[69,70,72,100,116]

The commercial model of the strawberry harvester was used in these studies. The harvester, shown in Figure 2, performs three basic functions: (1) Fruit is lifted by air and stripped from the plants, (2) separated from the leaves and other foreign material, and (3) conveyed to transporting containers.

Three strawberry cultivars, 'Cardinal', 'Earlibelle', and 'A-5344', were produced on flat-shaped beds at the Main Experiment Station, Fayetteville. The average percentage of fruit produced in each plot that was picked by machine was 96% for 'Cardinal', 92% for 'Earlibelle', and 94% of 'A-5344'. Details of this production system were described by Morris et al.[69]

Field plots were 4.6 m long, with 4.6 m skips between plots within the row to allow the machine to stop and empty before harvesting the next plot. Twenty-four plots of each cultivar were randomized to allow 4 replications of 2 plant populations and 3 harvest dates. The first harvest (early) was made on May 21, 1975, for 'A-5344' and on May 22 for Cardinal and Earliebelle. Succeeding harvests (middle and late) were made at 2-d intervals for each cultivar.

The fruit from each plot was cleaned, size-graded into small (mostly green) and large

FIGURE 2. Diagram of the 1973 mechanical strawberry harvester. (a) mowing sickle bar, (b) comb-brush picking and conveying system, (c) fan, (d) air-lock valve, and (e) fruit transporting conveyor. (From Morris, J. R., Kattan, A. A., Nelson, A. S., and Cawthon, D. L., *Hort. Science,* 13, 412, 1978. With permission.)

FIGURE 3. A "continuous flow" cleaning and grading line designed at the University of Arkansas to clean and grade machine-harvested strawberries. (From Morris, J. R., Kattan, A. A., Nelson, G. S., and Cawthon, D. L., *Hort. Science,* 13, 413, 1978. With permission.)

(mostly ripe) categories, and weighed. The continuous-flow cleaning and sizing line used in this study is shown in Figure 3. Samples for analyzing quality of raw fruit were collected immediately after cleaning and sizing, sealed in 211 × 400 J-enamel cans, and frozen for analysis.

The results of the analyses are shown in Table 3. Of the three cultivars evaluated, 'Cardinal' produced the highest yields, highest percent soluble solids, and highest titratable acidity. When the data were averaged across all other variables, the Arkansas breeding-line 'A-5344' was the firmest cultivar, as indicated by high shearpress values, followed by 'Cardinal.' The viscosity of puree was higher in 'Cardinal' and 'A-5344' than in 'Earlibelle.'

Most (79 to 84%) of the fruit of the three cultivars were large ripe fruit; therefore, the raw product quality of the composite fruit was similar to that of the large fruit. The slight raw product quality advantage that would be obtained by removing the small-sized fruit would probably not justify fruit sorting of these three cultivars for manufacture of puree.

Plant population (high or low density) had little or no effect on total fruit yield or raw puree quality.

In 1975, there was at least a 5 d period during which the strawberries could be once-over machine harvested without loss in quality or yield. During this 5 d period, the percentage of large ripe fruit increased from 68% to 92%. The slight reduction in percent soluble solids was attributed to 4.5 cm of rainfall that occurred after the first harvest dates. During the harvest sequence, there was an expected reduction in titratable acidity, green color (lower than b/a), fruit firmness, and puree viscosity.

Production of an acceptable raw product from a once-over harvest requires cultivars which have a concentrated ripening pattern. In addition, good fruit firmness is necessary for proper cleaning and handling. An acceptable raw product was obtained from all three of these highly-colored cultivars when they were harvested within the optimum 5 d period.

In the manufacture of preserves and jam from mechanically harvested fruit, combinations of green and ripe fruit were successfully used to produce high quality products.[70,105] Total anthocyanins (TACy) and 'a' values were higher on the riper fruit, but there was not much difference in color acceptance (Table 4). The color acceptance of the jam manufactured from 'Cardinal', 'Earlibelle' and 'A-5344' was rated much higher than jam from Tioga.[70] Color of jam from the darker colored cultivars declined much more during 9 months storage than Tioga. Ascorbic acid was lower in the 1/2 green and 1/2 ripe fruit blends than in blends with 7/8 ripe and 1/4 green fruit. Approximately 50% of the ascorbic acid was lost during 9 months storage of the jam at 21°C. The index of discoloration, which measures the extent of browning, was affected slightly by the percentage of green fruit.

Sistrunk et al.[109] showed that fruit harvested mechanically, in which both green and ripe fruit are picked, could be utilized in the manufacture of dessert pack (containing 4 parts fruit and 1 part sucrose). A fruit mixture of 50% green fruit was made into puree and mixed with sliced ripe fruit to make up the dessert pack. The frozen sliced fruit with green and ripe puree was lower in ascorbic acid, TACy, and color, but was higher in viscosity than fruit packs with 100% sliced ripe fruit. Mixtures of 40% green fruit puree and 60% sliced ripe fruit were rated acceptable in color and flavor by a sensory panel. Cultivars with sufficient red color were shown to be suitable for these blends of green and ripe fruit puree and ripe fruit. Higher cellulose and protopectin contents in the green fruit contributed to greater viscosity in the mixed packs.

B. TRANSPORT AND HOLDING

Rapid transport is also beneficial to shelf life with air shipment being the best. Cold fruit should not be placed in warm air for an extended period during loading due to the condensation that occurs on the fruit (dew point), thus necessitating additional cooling during transit. The major portion of fresh strawberries are transported by trucks with air shipment

TABLE 4
Effects of Cultivar and Fruit Maturity Mixtures on Sensory Evaluation, Total Anthocyanins, Reduced Ascorbic Acid, and Discoloration of Mechanically Harvested Strawberry Jam (Initally and After 9 mo at 21°C)

Cultivar/ maturity mixture[a]	Color intensity		Color acceptance		Flavor[b]		Total Acy (mg/100 g)		Ascorbic acid (mg/100 g)		Index of discoloration (A_{520}/A_{430})
	0	9 mo	0	9 mo	0	9 mo	0	9 mo	0	9 mo	9 mo
'Cardinal'											
1/2R, 1/2G	7.5	6.2	7.5	6.5	8.2	7.1	12.1	4.7	9.40	4.65	1.195
3/4R, 1/4G	8.6	7.3	8.2	6.4	8.1	7.2	13.2	5.6	9.80	5.55	1.147
7/8R, 1/8G	8.7	7.2	8.7	6.5	8.2	8.4	15.0	6.5	10.48	4.95	1.245
'Earlibelle'											
1/2R, 1/2G	7.2	7.3	7.4	6.1	6.5	6.9	11.0	5.3	9.05	5.25	1.375
3/4R, 1/4G	8.0	7.2	8.4	6.0	8.0	7.4	15.0	8.6	13.15	8.63	1.432
7/8R, 1/8G	8.8	8.4	8.9	6.9	8.1	7.7	17.2	9.0	13.33	9.00	1.361
'A-5344'											
1/2R, 1/2G	7.7	6.7	7.1	5.2	7.2	7.2	9.9	5.6	8.90	5.55	1.178
3/4R, 1/4G	7.9	7.3	8.3	5.6	7.9	7.4	11.2	5.6	8.90	5.55	1.205
7/8R, 1/8G	8.5	7.2	8.5	6.2	8.2	7.4	13.0	5.9	11.30	5.85	1.390
'Tioga'											
1/2R, 1/2G	3.3	5.0	4.5	4.0	5.3	6.0	3.7	2.6	8.95	2.60	0.787
3/4R, 1/4G	3.8	5.5	5.5	4.4	6.0	6.1	4.4	3.4	7.75	3.40	0.889
7/8R, 1/8G	4.8	5.3	6.3	4.2	7.4	7.6	5.3	4.7	7.90	4.65	0.931
1R, 0G	5.0	6.0	6.8	5.7	7.9	7.3	6.3	6.2	12.18	6.20	1.120

[a] R = large, ripe fruit; G = small, green fruit. Fraction indicates the proportion of each maturity used to prepare product.

[b] Attributes rated by a 10-member panel on a 10-member panel on a 1 to 10 scale: 10 = excellent, 5 = acceptable, and 1 = totally unacceptable.

From Morris, J. R., Sistrunk, W. A., Sims, C. A., Mair, G. L., and Wehurt, E. J., *J. Am. Soc. Hortic. Sci.*, in press, 1990. With permission.

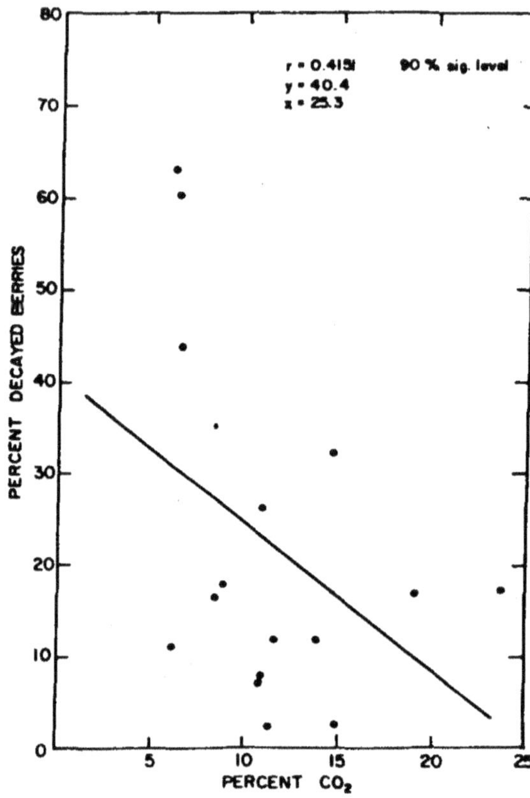

FIGURE 4. Effect of CO_2 concentration in transit on incidence of decay of California strawberries shipped by truck to eastern markets. (From Harvey, J. M., Harris, C. M., Tietjen, W. J., and Seriol, T., *USDA Sci. and Ed. Admin.*, AAT-W-12/July, 1980. With permission.)

providing a smaller percentage of transportation to market. Maintainance of low temperatures (0°C optimum) during transport is necessary for preventing deterioration. Decay losses have been found to increase directly as temperatures increase in transit.[43,62,128] In addition to mechanical refrigeration, dry ice or solid carbon dioxide has been used by shippers both for its cooling effect and its release of CO_2. The use of CO_2-enriched atmospheres in strawberry transport is helpful in reducing loss by decay.[21,41,128] A study in California has shown that strawberries shipped in pallets with average CO_2 levels above 10% had approximately one half as much decay as berries shipped in normal air.[128] In Figure 4, the effect of CO_2 concentration in transit on percent decayed berries shipped by truck from California to eastern markets is shown.

The use of CO_2, CO, and C_2H_4 at various concentrations and combinations in controlled atmospheres for holding strawberries has been studied by El-Kazzaz et al.[30] The most effective atmospheres in suppressing fruit rot were air + 15% CO_2 and a controlled atmosphere of 23.%O_2 + 10% CO. These researchers reported that 20 μl of C_2H_4 might enhance disease development. Borecka and Millikan[9] have found that storage at 4°C under CA's containing 20% CO_2 and 3% O_2 prevented the development of gray-mold rot up to 10 d without affecting the acidity and soluble solids of strawberries.

Another factor influencing quality of strawberries during transport or storage is relative humidity. Fruit held under low relative humidity will lose weight faster causing the fruit to look old and shriveled. Burton[14] recommends keeping the relative humidity between 90 to 95% at 0°C, especially if the fruit is to be held for an extended period of time.

The influence of different packaging treatments on strawberry quality has been studied by Miller et al.[64] No significant differences in pulp firmness were found among packaging treatments, but berries stored in rigid plastic baskets with solid plastic covers lost significantly less weight than those stored in mesh baskets with and without covers. Harvey et al.[44] found that berries shipped by truck in plastic-covered pallets versus non-covered pallets had a higher average load temperature but less decay when CO_2 atmospheres were maintained. A combination of low transit temperatures (3°C and below) and CO_2 atmospheres (over 10%) resulted in the least amount of decay.

When grown in regions to which they are well adapted, the following cultivars have been recommended as best for shipment: Albritton, Blakemore, Earlibelle, Headliner, Midway, Sunrise, Tennesse Beauty, Tioga, and Tufts.[88]

IV. FACTORS INFLUENCING THE QUALITY OF STRAWBERRIES FOR PROCESSING

Approximately 75% of all strawberries processed in the U.S. are produced in California. Strawberries which are frozen either for manufactured products or for dessert use total about 50% of the commercial strawberry crop in the U.S. The quality of frozen strawberries is based on color, character, and defects according to the U.S. Standards. Color is based on the outer surface of whole fruit and the entire surface area of sliced fruit, and not more than 15% of the surface may be materially darkened by being overripe or oxidized. Character is based on firmness, degree of wholeness, seediness, and disintegration of fruit.

Strawberries for commercial preserves, jam, and jelly manufacture are usually washed and capped, then frozen and stored with either 5:1 or 4:1 fruit: sugar ratio, and then thawed before processing.

A. FIRMNESS

The firmness of sliced frozen strawberries was improved by applying calcium, pectins and colloids to the prepared fruit before freezing.[6,7,33,45,63,127,129] Dipping in 2% cold low methoxyl pectin (LMP) increased the drained weight by 33%.[6] Neither $CaCl_2$ dips nor combinations of $CaCl_2$ and LMP were as effective in improving drip loss. An improvement in drained weight was obtained by using a vacuum of 26 in. to evacuate samples containing 0.1% high methoxyl pectin and sodium alginate.[7] Grab et al.[33] found that the use of both alginate and pectinate increased drained weight although a greater increase was obtained by adding 0.2 to 0.3% in the 60% syrup used to cover the fruit before freezing than by adding the dry powder to sugared fruit. The use of $CaCl_2$ was abandoned because of the off-flavor in the fruit. When comparing LMP, HMP and slow set pectin (SSP) for firming Missionary and Florida 90 cultivars of strawberries, fruit vacuumized in SSP had a better color than those treated in LMP and HMP. Fruit that was washed in water at 4°C was rated higher in color than fruit washed in water at 25°C. Also, fruit treated in 0.2 to 0.3% pectin with vacuum had a better color and higher drained weight than the control with no treatments. Recent studies by Morris et al.[71] demonstrated that dipping either sliced or whole fruit in 0.5% Ca lactate or a combination of 0.5% Ca lactate and 1% citric acid improved firmness and character. More firming occurred in sliced than in whole fruit.

Firmness of strawberries is dependent on the pectins and cellulose and the changes in these constituents during maturation, postharvest holding and processing. Water-soluble pectins increased with a delay in harvest and holding after harvest.[94] Also, the percent mushy slices increased and rate of syrup flow decreased under the same conditions, indicating that the syrup in thawed frozen fruit thinned out. There was a wide variation in pectins and cellulose among cultivars.[95,98] Cellulose content was higher in fruit from later harvests which could have been due to greater solubility rather than higher total cellulose although less mature fruit was higher in cellulose.[94,95] In the three cultivars analyzed, there appeared to be a close relationship between combined fractions of pectin and shearpress firmness. Sistrunk and Cash[98] found an increase in solubility of water-soluble pectin, hemicellulose and cellulose with holding time at 50°C in strawberry puree. Spayd and Morris[116] demonstrated that water soluble pectins increased while cellulose and protopectin decreased in strawberries during maturation from the small green berry stage to the full ripe stage. Viscosity decreased significantly during maturation. At the same time Calgon-soluble pectin and hemicellulose increased, then decreased sharply as the fruit ripened.

B. CULTIVAR

The firmness and maturity of genotypes have been shown to influence character of frozen strawberries.[86,88,120] Plunger-type testers and the shearpress have been used in determining firmness of fresh fruit.[71] Drained weight loss, percent mushy slices, and viscosity of syrup of thawed frozen fruit are other tests used to define character and firmness.[94,95,96,101] More mature fruit and a delay in harvest increased percent mushy slices and decreased viscosity and drained weight retention in thawed frozen fruit. No relationship was found between pH, free acid, ascorbic acid, and total solids and freezing adaptability of a large number of genotypes.[108] However, color was shown to be correlated with pH.[83,97,101,102]

C. ASCORBIC ACID

Field conditions, cultivar, maturity, post-harvest storage, and processing affect ascorbic acid in strawberries.[94,95,96,101,102] Ascorbic acid is one of the most thoroughly researched components of strawberries. The fresh fruit are a rich source of this vitamin, supplying 50 to 90 mg/100 g on a fresh weight basis.[83] Values varied widely among cultivars and growing conditions. Shading of the plants during maturation decreased the ascorbic acid as well as citric acid significantly, but total titratable acidity remained the same.[102] Shading the plants with plastic screening did not influence ascorbic acid as much as predicted since there was only a 3 mg/100 g difference between 30% and 73% shading. Highly significant correlations were found between color and ascorbic acid among large numbers of genotypes.[60] It appeared that there is a possibility of breeding and selecting for highly colored fruit with high ascorbic acid values.

D. PIGMENTS AND COLOR

Investigators have emphasized the importance of color to quality of strawberries.[1,25,75,97,98,115,116,117,133,134] Concentration and type of pigment are key factors that determine the intensity and stability of color.[16,17,94,134] While pelargonidin-3-glucoside (P-3-G) is the predominant pigment in strawberries, cyanidin-3-glucoside (C-3-G) is more stable, and consequently contributes to the stability of color after freezing and thawing. Some genotypes contain higher percentages of C-3-G than others, and thus may offer a means of breeding and selection of genotypes with more color stability.

The pH contributes more to stability of color in strawberries than most components as shown by research studies on many cultivars and selections.[83,97,134] Strawberries should have a pH below 3.51 and a pigment concentration of 450 to 700 mg/g of fresh fruit to be acceptable in color.[134] Any fruit with concentrations of pigment below 400 and above 850 mg/g were unacceptable.

Various chemical components of strawberries have been studied individually to determine their effect on purified P-3-G.[123] The rate of disappearance of P-3-G in a model system followed first order kinetics with some sugars, but in a sucrose system it was necessary for the reaction to occur at 90°C over an 120 min period to produce first-order kinetics. Other sugars, glucose, and fructose, were shown to degrade P-3-G more rapidly than sucrose. Sucrose had only a slight effect on color in strawberry juice, while fructose had a more pronounced effect on color loss during 5 months at 4°C.[29] Glucose appeared to improve color retention during the same storage period. There was only a 20 to 30% loss in ascorbic acid during storage of the juice. From the results obtained, it was suggested that anthocyanins may increase the stability of ascorbic acid.

The O_2 content and temperature of storage of strawberry juice and extracted pigments were demonstrated to be the most specific factors responsible for color loss.[74] Light caused very little change in color of juice, yet in extracted pigments the color was bleached by light. Small concentrations of sugar had little effect on either of the two forms studied, but increasing sugar and/or citric acid in fountain syrups increased the stability of color. Ascorbic acid at a concentration of 50 mg/100 ml of juice bleached the color, resulting in a loss of red color and development of a brown color when stored at 21 and 27°C.

Other researchers have shown varying changes in strawberry pigments in the presence of ascorbic acid, and concurrent development and precipitation of brown pigment.[10,81] In the presence of O_2, the loss of red pigment was dependent upon the pH.[36,59,97] During heat processing, strawberries with higher O_2 and lower H concentration lost more pigment.[36] The rate of oxidation is directly related to the amount of pigment that is in the form of the pseudobase when pigment solutions are heated.[59] The buffered pigments precipitated as either brown or red-brown pigments in the presence of O_2. Solutions that were buffered to a higher pH were more brown.

Ascorbic acid and fructose in strawberries are active components responsible for catalyzing the degradation of the pigments in the presence of O_2.[61] Regardless of the stage of processing in which O_2 is present during heating at high temperatures there was an increase in pigment loss and development of browning. Skrede[111] demonstrated that sucrose breaks down to glucose and fructose during thawing of frozen fruit, depending on thawing time. This could account for a rapid change in color of some genotypes during thawing. Glycine appeared to accelerate browning during heat processing of strawberries,[61] and in grape juice.[110] Glycine did not affect losses of ascorbic acid and anthocyanins when juice was heated at 43°C, yet rapid losses of both components occurred at 75°C.[110]

The two mechanisms of oxidation of strawberry anthocyanins by H_2O_2 proposed by Sondheimer and Kertesz[113] help to explain many of the changes in color during processing. Non-catalyzed reactions take place regardless of heating at high temperature (90°C) to inactivate enzymes prior to storage. At 20°C an increase in Fe^{++} ions and ascorbic acid accelerated the rate of degradation of anthocyanins. However, Fe^{++} catalyzed the destruction of pigments even when ascorbic acid was not present, yet ascorbic acid appeared to be needed to maintain the rate of reaction.

Other mechanisms of anthocyanin destruction in strawberry juice take place since losses occurred in the absence of O_2 when ascorbic acid was present.[114] Similar losses in anthocyanins did not occur in model systems. In other studies, there appeared to be a direct condensation mechanism that involved anthocyanins and ascorbic acid in the presence of both O_2 and N_2.[80]

Studies on Concord grape juice have shown that anthocyanins were degraded when ascorbic acid was stabilized with added cysteine and $SnCl_2$.[101,110] This indicates that anthocyanin pigments can be degraded independently of ascorbic acid changes. However, in juice without chemical additives increasing the ascorbic acid accelerated pigment losses.

Polyphenolases (PPO) have been found to degrade anthocyanins rapidly in fruit. The principal methods to prevent pigment changes by the enzyme are inactivation by heat, reduction in pH, lowering water activity, minimizing O_2 level and adding of antioxidants.[11] Heat is probably the simplest and most straight forward method of inactivating PPO since enzymes are proteins, and thus are easily denatured by heat. The temperature for optimum activity of PPO varies among different fruits and vegetables. In red tart cherries, the raction rate increased between 5° and 50°C, but it declined rapidly above 50°C.[79] Cash et al.[18]

demonstrated that Concord grape PPO was more active at 25 to 30°C and declined very rapidly with an increase in temperature above 30°C.

Activity of PPO is dependent on pH and the optimum activity of the enzyme varies in different fruit, depending on the number and kinds of isozymes. Strawberry PPO activity ceased at a pH of approximately 3.0, and the destruction of anthocyanins by PPO was accelerated when the pH was increased to 4.0.[17] While the PPO activity in strawberries was nil at pH 3.0, the PPO in Concord grapes exhibited approximately 50% activity at pH 3.0.[110]

There are a wide variations in PPO activity among cultivars of strawberries. The color of Tennessee Beauty was affected more by PPO than in other cultivars.[102] Also, the 'a' values were significantly lower in samples adjusted to pH 3.8 than at 3.0, indicating that a larger loss in color occurred due to the activity of the enzyme at higher pH. Total phenols were also higher in Tennessee Beauty than in other cultivars. These compounds may have some effect on color stability of anthocyanins.

Metal complexing of strawberry pigments has been studied by a number of investigators.[40,51,55,97,98,134] Since C-3-G has 2 vicinal phenolic groups it will complex with metal ions. In studies on strawberry puree, Sistrunk and Cash[98] demonstrated that red color was retained for 24 h at 50°C by the addition of low concentrations of $AlCl_3$, $SnCl_2$ and $SnCl_4$. Also, these metal ions, especially $SnCl_4$ prevented the oxidation of ascorbic acid. The blue colors in berry fruits are presumed to be due to metal complexes instead of co-pigmentation with other components.[40] The bathychromic shift from red to blue-purple takes place at pH 3.5 to 3.9, although maximum formation of complexed pigment occurred at pH 5.5.[51] Little color change in the strawberry puree occurred when $AlCl_3$ was added at pH 3.0 to 3.4,[98] but in the presence of Sn^{++} and Sn^{++++} ions a definite red-purple color developed that was stable. This conflicts with research by Asen et al.[5] in which the wavelength of a pigment solution of C-3-G shifted from 515 to 545 nm in a pH range of 3.0 to 3.5 when $AlCl_3$ was added. Research studies in our laboratories have not been able to demonstrate co-pigmentation of strawberry pigments with flavanoids and tannins to attain a more stable color.

The bathychromic shift that resulted from adding metal ions to strawberry puree was an acceptable color and not unlike the normal color of strawberries. The OD of extracted pigments was higher at pH 3.0 than at the other pH levels, a phenomenon reported by other investigators.[114,134] Most of the shift in OD at pH 3.8 occurred during the period of equilibrium before the samples were heated to 50°C.[98]

The changes in anthocyanins and the mechanism of color degradation in strawberry products such as preserves, jams, jellies, and concentrated juice have been studied extensively.[52,59,60,61,105,107,108] Rapid changes in color occurred during processing and storage, depending on the time and temperature.[1,24,105,107,108] The changes in color in concentrated strawberry products have been attributed to the degradation of monomeric anthocyanin pigment to polymeric pigments.[133] Polymeric pigments were soluble in 1*N* NaoH,[61,98] and chemical additives decreased the formation of these pigments in puree during storage.

The addition of $NaHSO_3$, citric acid and $SnCl_2$ to frozen mechanically harvested strawberries improved the color after freezing, storage and thawing.[55] Less browning occurred and sensory scores for color acceptability were higher. When these same treatments were applied to frozen whole and puree fruit for the manufacture of jam the effect on color stability was significant as compared to the control.[108] The use of ethylenediaminetetracetate-Na salt (EDTA) also provided protection from browning in the jam. Even after 9 months of storage at 24°C, jams made from fruit with the added chemicals were highly acceptable in color.

Similar studies on low-sugar strawberry spread (40% brix) indicated that these same chemicals did not protect the color beyond 3 to 4 months, especially when the product was stored at 24°C.[107] After 4 months, all spreads were low in quality except the product with added chemicals that was stored at 2°C.

Spayd and Morris[117] showed that color acceptability of jam was highly correlated with total anthocyanin content and more highly correlated with the log of the total anthocyanin content. A similar logarithmic relationship existed between sensory scores for lack of discoloration and the discoloration index. Prediction models were developed and total anthocyanin content of 2.9 OD units/g of jam and a discoloration index of 0.93 or higher were associated with acceptable sensory scores.

Using these mathematical relationships, models were developed for predicting color acceptability and lack of discoloration scores after 6 and 12 months storage of jam at 25°C, from initial TAcy and discoloration index values. For acceptable sensory color after 6 and 12 months at 25°, initial TAcy should be about 4.8 to 6.0 OD units/g of jam and the initial discoloration index should be about 1.3 to 1.6.

V. PROCESSING

Strawberry processors could use these objective color measurements in developing guidelines to suit their own standards for color acceptibility of acceptable strawberry products.

A. FREEZING

The rate of freezing of packaged berries has been shown to affect color, texture, and degree of oxidation.[35,132] During the freezing process, the cell-wall is damaged with the cell contents and cell wall becoming disorganized, especially during slow freezing.[4,135] The texture of frozen strawberries was retained better at -23.3°C than at higher or fluctuating temperatures.[136]

After freezing, the once turgid berry becomes soft and exudes a large amount of juice while thawing. The addition of colloidal materials such as pectin and sodium alginate or addition of sweetening agents prior to freezing can reduce drip losses and in some cases produce a firmer berry.[3,48,127]

During mixing of the fruit and sugar, an increase in mechanical breakage of sliced strawberries will occur.[95] Types of mixers for mixing the fruit with sugar during preparation for freezing can influence the texture of frozen sliced strawberries.[95,122]

During frozen storage, strawberries should be kept in air tight packages to prevent oxidative changes. Significant differences in quality retention due to package have been found in strawberries mixed with sugar (4 + 1).[35] Ascorbic acid content which was reduced by the presence of oxygen, probably influenced the subjective color and flavor retention.

VI. SUMMARY AND CONCLUSIONS

The numerous factors affecting quality of fresh and processed strawberries begin at the time of planting the plants and end with the consumer. Some of the more important factors influencing quality in the field are cultivar, maturity, harvest date, water relations, control of pests and diseases, and climate. Proper handling of fruit and maintenance of low temperatures during transport and during holding prior to either marketing or processing are extremely important.

The preparation and freezing process should be carried out rapidly by the use of equipment that avoid excessive physical damage by dropping, crushing, and mixing. Freezing immediately after mixing fruit with sugar and hydrocolloids reduces drip loss and retains more wholeness in whole and sliced fruit after thawing. Also, packaging in air-tight containers and holding at temperatures below -18°C reduces oxidation during storage of frozen fruit.

The most important quality attributes are processed or texture, assuming that defects are not objectionable. The bright orange-red color of strawberries is difficult to maintain because of the susceptibility of the main pigment in strawberries, P-3-G, to degradation. Color losses are accelerated by O_2, heat, metal ions, type of sugars present, ascorbic acid, total phenols, etc. In frozen strawberries, the presence of O_2 and storage temperature are the principal factors that influence color degradation. However, in strawberry products, the soluble solids, temperature and time of heating, and storage temperature have the greatest influence on quality. The changes in quality are caused by the degradation of the anthocyanins and development of brown polymeric compounds.

The character or texture of sliced and whole fruit has been one of the sources of complaints by consumers in the purchase of strawberry products such as sliced and whole frozen fruit, preserves, and toppings. The principal factors influencing character are cultivar, maturity, method of preparation for freezing, and method of cooking. The loss in character can be avoided by choosing a firm cultivar, harvest at optimum maturity, minimize holding time, and avoid physical damage during washing, slicing, and mixing with sugar.

The future of the strawberry industry in both fresh and processed products will depend heavily on the development of new cultivars that are high in yield, disease resistant, and high quality including good uniform color and firmness. The continuous improvement in harvesting and handling techniques that will be able to deliver higher quality fruit to the processing plant at a competitive price will enable processors to produce higher quality processed products. Also, the development of new products from strawberries with greater appeal to a larger number of consumers will expand the use of strawberries.

The producers of the raw product, processors, breeders, food scientists, engineers, plant pathologists, and marketing specialists must work together to attain the type of products and quality that meets the needs of the consumers in the new age.

REFERENCES

1. **Albers, J. E. and Wrolstad, R. E.,** Causitive factors of color deterioration in strawberry preserves during processing and storage, *J. Food Sci.*, 44, 75, 81, 1979.
2. **Adams, R. E., Keefer, R. F., and Hickman, C. E.,** Programs for improving yields of strawberries in West Virginia, *West Va. Agr. Exp. Sta. Bull.*, 618, 1973.
3. **Aref, M., Sidwilly, A. P., and Litwiller, E. M.,** The effects of various sweetening agents on frozen strawberries for preserve manufacture, *Food Technol.*, 10, 293, 1956.
4. **Armbruster, G.,** Cellular and textural changes in three varieties as a result of pre-freezing treatments, *Proc. Am. Soc. Hortic. Sci.*, 91, 876. 1967.
5. **Asen, S., Norris, K. H., and Stewart, A.N.,** Adsorption spectra and color of aluminum-cyanidin-3-glucoside complexes as influenced by pH, *Phytochem.*, 8, 653, 1969.
6. **Baker, G. L.,** Pectin as an aid in freezing fruit, I. Its application in the freezing preservation of strawberries, *Food Indust.*, 13, 1, 55, 1941.
7. **Barton, R. R.,** Improving the quality of frozen 'Premier' strawberries, *Proc. Am. Soc. Hortic. Sci.*, 58, 95, 1951.
8. **Bebeke, E. S., White, L. S., and Fabian, F. W.,** The incidence and pectolytic activity of fungi isolated from Michigan strawberry fruits, *Applied Microbiol.*, 2, 253, 1954.
9. **Borecka, H. and Millikan, D. F.,** Influence of storage conditions on the development of postharvest gray-mold rot of strawberries, *Phytoprotection*, 62, 1, 11, 1981.
10. **Beattie, H. G., Wheeler, K. A., and Pederson, C. S.,** Changes occurring in fruit juices during storage, *Food Res.*, 8, 395, 1943.
11. **Berk, Z.,** Phenolic compounds, in *Braverman's Introduction to the Biochemistry of Foods*, Calsevier Publ. Co, 1976.
12. **Bourne, M. C., Moyer, J. C., and Hand, D. B.,** Measurement of food texture by a universal testing machine, *Food Technol.*, 20, 522, 1966.
13. **Boyce, B. R. and Matlock, D. L.,** Strawberry nutrition, in *Fruit Nutrition*, Childers, N. F., Ed., Hort. Publ., 1966, 518.
14. **Burton, C. L.,** Factors effecting postharvest quality of fresh marketed strawberries in Michigan, *Hortic. Rep. No. 19* April 1972, Dept. of Hort., Michigan State University.

15. **Cannell, G. H., Voth, V., Bringhurst, R. S., and Proebsting, E. L.**, The influence of irrigation levels and application methods, polyethylene mulch, and nitrogen fertilizers on strawberry production in Southern California, 1961.
16. **Cash, J. and Sistrunk, W. A.**, Anthocyanin pigment concentration and type are important for color in strawberries, *Ark. Farm Res.* 19, 3, 8, 1970.
17. **Cash, J. and Sistrunk, W. A.**, Relationship of polyphenol oxidase to pigment degradation in strawberry juice, *Ark. Farm Res.* 20, 6, 2, 1971.
18. **Cash, J. N., Sistrunk, W. A., and Stutte, C. A.**, Characteristics of Concord grape polyphenoloxidase involved in juice color loss, *J. Food Sci.* 41, 1398.
19. **Childers, N. F.**, Modern Fruit Science, *Hortic. Publ.*, University of Fla., Gainesville, FL, 1983.
20. **Cochran, G. W. and Webster, J. E.**, The effect of fertilizers on the handling qualities and chemical analyses of strawberries and tomatoes, *Proc. Am. Soc. Hortic. Sci.*, 28, 236, 1931.
21. **Couey, H. M. and Wells, J. M.**, Low-oxygen or high-carbon dioxide atmospheres to control postharvest decay of strawberries, *Phytopathology*, 60, 47, 1970.
22. **Darrow, G. M., Ed.**, *The Strawberry: History, Breeding and Physiology*, Holt, Rinehart and Winston, New York, NY, 1966.
23. **Darrow, G. M., Scott, D. H., and Waldo, G. F.**, Strawberry varieties in the United States, *U.S.D.A. Farmer's Bull.* 1043, 1958.
24. **Darrow, G. M., Wilcox, M. S., Scott, D. H., and Hutchins, M. C.**, Breeding strawberries for vitamin C, *J. Hered.* 38, 363, 1947.
25. **Decareau, R. V., Livingston, G. E., and Fellers, C. R.**, Color changes in strawberry jellies, *Food Technol.* 10, 125, 1956.
26. **Degman, E. S. and Weinberger, J. H.**, Studies on firmness and keeping quality of certain fruits, I. Effects of nitrogen fertilization, II. Effects of Potassium fertilization. *Md. Agric. Expt. Sta. Bull.*, 366, 1934.
27. **Dennison, R. A. and Hall, C. B.**, Influence of nitrogen, phosphorus, potash and lime on the growth and yield of strawberries, *Proc. Fla. St. Hortic. Soc.*, 69, 224, 1957.
28. **Duewer, R. G. and Zych, C. C.**, Heritability of soluble solids and acids in progenies of the cultivated strawberry *(Fragaria X Ananassa Duch). Proc. Am. Soc. Hortic. Sci.* 90, 153, 1967.
29. **El-kady, S. and Ammar, K.**, Stability of color in plum and strawberry juices during storage, *J. Agri. Res.*, 3, 99, 1977.
30. **El-Kazzaz, M. K., Sommer, N. F., and Fortlage, R. J.**, Effect of different atmospheres on postharvest decay and quality of fresh strawberries. *Phytopathology*, 73, (2) 282, 1983.
31. **Ezell, B. D., Darrow, G. M., Wilcox, M. S., and Scott, D. H.**, Ascorbic acid content of strawberries, *Food Res.* 12, 510, 1947.
32. Federal Register, 1983, United States Standards for Grades of Strawberries, Office of the Federal Register, Publ. 7CFR51. 416.
33. **Grab, E. G., Jr., Wegener, J. B., and Baer, B. H.**, Firming strawberries before freezing, *Food Packer*, 29, 11, 39, 42, 1948.
34. **Guadagni, D. G.**, Time-temperature tolerance of frozen foods, VIII. Organeoleptic evaluation of frozen strawberries, raspberries and peaches, *Food Technol.*, 9, 471, 1954.
35. **Guadagni, G. C., Nimmo, C. C., and Jansen, E. F.**, Time-temperature tolerance of frozen foods, VI. Retail packages of frozen strawberries, *Food Technol.*, 9, 389, 1957.
36. **Habib, A. T. and Brown, H. D.**, The effect of oxygen and hydrogen-ion concentration on color changes in processed beets, strawberries and raspberries, *Proc. Am. Soc. Hortic. Sci.*, 68, 482, 1956.
37. **Haller, M. H., Harding, P. L., and Rose, D. H.**, The interrelation of firmness, dry weight and respiration in strawberries, *Proc. Am. Soc. Hortic. Sci.*, 29, 330, 1932.
38. **Hansche, P. E., Bringhurst, R. S., and Voth, V.**, Estimates of genetic and environmental parameters in the strawberry. *Proc. Am. Soc. Hortic. Sci.*, 92, 338, 1968.
39. **Hansen, E. and Waldo, G. F.**, Ascorbic acid content of small fruits in relation to genetic and environmental factors, *Food Res.*, 9, 453, 1944.
40. **Harborne, J. B.**, Flavonoids: Distribution and contribution to plant color, in *Chemistry and Biochemistry of Plant Pigments*, Goodwin, T. W., Ed., Academic Press, New York, 1965, 247.
41. **Harris, C. M. and Harvey, J. M.**, Quality and decay of California strawberries stored in CO_2-enriched atmospheres, *Plant Dis. Rep.*, 57, 1, 44, 1973.
42. **Hartman, J. D., White-Steveni, R. H., and Hoffman, M. B.**, Phosphorus and nitrogen fertilization for strawberries on Long Island, New York, *Proc. Am. Soc. Hortic. Sci.*, 36, 473, 1938.
43. **Harvey, J. M., Couey, H. M., Harris, C. M., and Porter, F. M.**, Air Transport of California strawberries: factors affecting market quality in summer shipments, *Mkt. Res. Rpt.*, 751, 11, 1965.
44. **Harvey, J. M., Harris, C. M., Tietjen, W. J., and Seriol, T.**, Quality maintenance in truck shipments of California strawberries, *USDA Sci. and Ed. Admin.*, AAT-W-12/July, 1980.
45. **Haut, I. C., Webster, J. E., and Cochran, G. W.**, The influence of commercial fertilizers upon the firmness and chemical composition of strawberries and tomatoes, *Proc. Am. Soc. Hortic. Sci.*, 33, 405, 1935.

46. **Hawkins, L. A. and Sando, C. E.,** Effect of temperature on the resistance to wounding of certain small fruits and cherries, *U.S.D.A. Bull.*, 830, 1920.
47. **Haynes, R. D., Harlin, H., Mundt, J. O., and Stokes, R.,** The use of detergents in removing mold from strawberries, *Quick Frozen Foods,* 15, (2) 53, 1953.
48. **Hoover, M. M. and Dennison, R. A.,** Some treatments influencing the quality of Florida frozen strawberries, *Proc. Am. Soc. Hortic. Sci.*, 65, 188, 1955.
49. **Hull, J., Moulton, J., and Flore, J.,** Commercial strawberry culture in Michigan, Farm Science Series June 1977 (Revision) *Extention Bull.*, E-682.
50. **Jahn, O. L. and Crosby, E. A.,** Growth of the strawberry plant as affected by several nitrogen carriers, *Proc. Am. Soc. Hortic. Sci.*, 71, 207, 1957.
51. **Jurd, L. and Asen, S.,** The formation of metal and co-pigment complexes of cyanidin-3-glucoside, *Phytochem.*, 5, 1966.
52. **Hertesz, Z. I. and Sondheimer, E.,** To reduce loss in strawberry preserves, *Food Indust.*, 20, 1300, 1408, 1948.
53. **Kimbrough, W. D.,** The quality of strawberries as influenced by rainfall, soil moisture and fertilizer treatments, *Proc. Am. Soc. Hortic. Sci.*, 27, 184, 1930.
54. **Kirsh, R. K.,** The importance of interaction effects in fertilizer and lime studies with strawberries, *Proc. Am. Soc. Hortic. Sci.*, 73, 181, 1959.
55. **Kozup, J., Sistrunk, W. A., and Morris, J. R.,** Chemical stabilization of color in machine-harvested strawberries, *Ark. Farm Res.*, 29, (4) 5, 1980.
56. **Long, J. H.,** The use of certain nutrient elements at the time of flower formation in the strawberry, *Proc. Am. Soc. Hortic. Sci.*, 37, 553, 1939.
57. **Long, J. H. and Murneek, A. E.,** Nitrogen and carbohydrate content of the strawberry plant, seasonal changes, and effects of fertilizers, *Mo. Agr. Exp. Sta. Res. Bul.*, 252, 1937.
58. **Loree, R. E.,** The nutrient requirements of the strawberry, *Mich. Agr. Exp. Sta. Tech. Bul.*, 70, 1925.
59. **Lukton, A., Chichester, C. O., and Mackinney, G.,** The breakdown of strawberry anthocyanin pigment, *Food Technol.*, 10, 427, 1956.
60. **Lundergan, V. A. and Moore, J. N.,** Inheritance of ascorbic acid content and color intensity in fruits of strawberry *(Fragaria × Ananassa Duch.) J. Am. Soc. Hortic. Sci.*, 100, 633, 1975.
61. **Markakis, P., Livingston, G. E., and Fellers, C. R.,** Quantitative aspects of strawberry pigment degradation, *Food Res.*, 22, 117, 1957.
62. **Maxie, E. C., Mitchell, F. G., and Greathead, A.,** Studies on strawberry quality, *Calif. Agric.*, 13, 2, 1, 16, 1959.
63. **Meschter, E. E.,** Effects of carbohydrates and other factors on strawberry products, *J. Agri. Food Chem.*, 1, 574, 1953.
64. **Miller, W. R., Davis, P. L., Dow, A., and Bongers, A. J.,** Quality of strawberries packed in different consumer units and stored under simulated air-freight shipping conditions, *Hort. Science,* 18, 3, 310, 1983.
65. **Mitchell, F. G., Maxie, E. C., and Greathead, A. S.,** Handling strawberries for fresh market., *Calif. Agric. Expt. Sta. Ext. Ser. Circ.*, 527, 1964.
66. **Moore, J. N. and Sistrunk, W. A.,** Breeding strawberries for superior fruit quality, in *The Strawberry,* Childers, N. F., Ed., Rutgers University, New Brunswick, NJ, 1981, 149.
67. **Morrison, W. W.,** Preparing Strawberries for Market, *Farmers' Bull. No.* 1560 U.S. Dept. of Agric., 1959.
68. **Morrow, E. B. and Darrow, G. M.,** Inheritance of some characteristics in strawberry varieties, *Proc. Am. Soc. Hortic. Sci.*, 39, 362, 1941.
69. **Morris, J. R., Kattan, A. A., Nelson, G. S., and Cawthon, D. L.,** Developing a mechanized system for production, harvesting and handling of strawberries, *HortScience,* 13, 413, 1978.
70. **Morris, J. R., Sistrunk, W. A., Kattan, A. A., Nelson, G. S., and Cawthon, D. L.** Quality of mechanically harvested strawberries for processing, *Food Technol.*, 32, 92, 1979.
71. **Morris, J. R., Sistrunk, W. A., Sims, C. A., Main, G. L., and Wehunt, E. J.,** Effect of cultivar, post-harvest holding, pre-processing treatments and style of pack on the processing quality of strawberries, *J. Am. Soc. Hortic. Sci.*, in press.
72. **Morris, J. R., Spayd, S. E., Cawthon, D. L., Kattan, A. A., and Nelson, G. S.,** Strawberry clonal fruit yield and quality responses to hand picking prior to once-over machine harvest, *J. Am. Soc. Hortic. Sci.*, 10, 864, 1979.
73. **Mundt, J. O.,** Microbiology of strawberries during harvest, *Quick Frozen Foods,* 12, 9, 53, 1950.
74. **Nebesky, E. A., Esselen, Jr., W. B., McConnell, J. E. W., and Fellers, C. R.,** Stability of color in fruit juices, *Food Res.*, 14, 261, 1949.
75. **Nelson, J. W., Barritt, B. H., and Wolford, E. R.,** Influence of location and cultivar on color and chemical composition of strawberry fruit, *Washington Agr. Exp. Sta. Tech. Bull.*, 74, 1972.
76. **Ourecky, D. K. and Bourne, M. C.,** Measurement of strawberry texture with the Instron machine, *Proc. Am. Soc. Hortic. Sci.*, 93, 317, 1968.

77. **Overholser, E. L. and Claypool, L. L.,** The relation of fertilizer to the respiration and certain physical properties of strawberries, *Proc. Am. Soc. Hortic. Sci.*, 28, 220, 1931.
78. **Overholser, E. L., Hardy, M. B., Locklin, H. D.,** Respiration studies of strawberries, *Plant Physiol.*, 6, 549, 1931.
79. **Peng, C. Y. and Markakis, P.,** Effect of phenolase on anthocyanins, *Nature*, 199, 597, 1963.
80. **Poei-Langston, M. S. and Wrolstad, R. E.,** Color degradation in an ascorbic acid-anthocyanin-flavanol model system, *J. Food Sci.*, 46, 1218, 1236, 1981.
81. **Pratt, D. E., Balkom, C. M., Powers, J. J., and Mills, L. W.,** Fruit color stability: interactions of ascorbic acid, riboflavin and anthoyanin pigments, *J. Agri. Food Chem.*, 2, 367, 1954.
82. **Ricketson, C. L.,** Plant spacing in solid-bed strawberry plantings, *Rep. Hortic. Res. Inst.*, Ontario, Canada, 1969, 67.
83. **Robinson, W. B., Lee, F. A., Slate, G. L., and Pederson, C. S.,** Chemical composition and freezing adaptability of strawberries, *NY State Agr. Exp. Sta. Bull.*, 726 Cornell, 1947.
84. **Saxena, G. K. and Locascio, S. J.,** Fruit quality of fresh strawberries as influenced by nitrogen and potassium nutrition, *Proc. Am. Soc. Hortic. Sci.*, 92, 354, 1968.
85. **Schrader, A. L. and Scott, D. H.,** Quality comparisons of strawberry varieties as affected by processing with the freezing method, *Proc. Am. Soc. Hortic. Sci.*, 50, 246, 1947.
86. **Schrader, A. L. and Scott, D. H.,** Quality comparisons of strawberry varieties as affected by processing with the freezing method, *Proc. Am. Soc. Hortic. Sci.*, 50, 246, 1947.
87. **Scott, D. H. and Lawrence, F. J.,** Strawberries. in *Advances in Fruit Breeding*, Janick, J., and Moore, J. N., Eds., Purdue Press, Lafayette, IN, 1975, 71.
88. **Scott, D. H., Lawrence, F. J., and Draper, A. D.,** Strawberry variety suggestions, USA in *The Strawberry-Cultivar to Marketing* Childers, N. F., Ed., 1981, 180.
89. **Shah, J. N. and Worthington, O. J.,** Comparison of several methods and instruments for specifying the color of frozen strawberries, *Food Technol.*, 8, 121, 1953.
90. **Shoemaker, J. S.,** The strawberry of Ohio, *Ohio Agr. Exp. Sta. Bull.* 444, 1929.
91. **Shoemaker, J. S.,** Strawberries, *in Small Fruit Culture*, 5th ed., AVI Publishing, Westport, CT, 1978, 103.
92. **Shoemaker, J. S. and Greve, E. W.,** Relation of nitrogen fertilizer to the firmness and composition of strawberries, *Ohio Agric. Expt. Sta. Bull.*, 466, 1930.
93. **Shrader, A. L. and Scott, D. H.,** Quality comparisons of strawberry varieties as affected by processing with the freezing method, *Proc. Am. Soc. Hortic. Sci.*, 50, 246, 1947.
94. **Sistrunk, W. A.,** Field conditions and processing practices related to frozen strawberry quality, *Proc. Am. Soc. Hortic. Sci.*, 83, 440, 1963.
95. **Sistrunk, W A., Cain, R. F., Vaughn, E. K., and Lagerstedt, H. B.,** Factors contributing to the breakdown of frozen sliced strawberries, *Food Technol.*, 14, 640, 1960.
96. **Sistrunk, W. A., Cain, R. F., Kirk, D. E., Vaughn, E. K., and Lagerstedt, H. B.,** Relationship of field and processing factors to the quality of frozen strawberries, *Ore. Agr. Exp. Sta., Spec. Rept.*, 138, 23, 1962.
97. **Sistrunk, W. A. and Cash, J. N.,** Stabilizing the color of frozen strawberries, *Ark. Farm Res.*, 17, 3, 2, 1968.
98. **Sistrunk, W. A. and Cash, J. N.,** The effect of certain chemicals on color and polysaccharides of strawberry puree, *Food Technol.*, 23, 169, 1971.
99. **Sistrunk, W. A. and Gascoigne, H. L.,** Stability of color in 'Concord' grape juice and expression of color, *J. Food Sci.*, 48, 430, 1983.
100. **Sistrunk, W. A., Junek, J., and Morris, J. R.,** Effects of mechanization on product quality and utilization of strawberries, in *Strawberry Mechanization*, Martin, L. W. and Morris, J. R., Eds., *Ore. Agri. Exp. Sta. Bull.*, 645, 160, 1980.
101. **Sistrunk, W. A. and Moore, J. N.,** Assessment of strawberry quality-fresh and frozen, *Food Technol.*, 21, 449, 1967.
102. **Sistrunk, W. A. and Moore, J. N.,** Strawberry quality studies in relation to new variety development, *Ark. Agr. Exp. Sta. Bull.*, 761, 1971.
103. **Sistrunk, W. A. and Moore, J. N.,** Evaluating selections and cultivars for processing, *Ark. Farm Res.*, 26, 1, 7, 1977
104. **Sistrunk, W. A. and Moore, J. N.,** Evaluation of new strawberry selections and cultivars for processing, *Ark. Farm Res.*, 28, 1, 7, 1979.
105. **Sistrunk, W. A. and Morris, J. R.,** Storage stability of strawberry products manufactured from mechanically harvested strawberries, *J. Am. Soc. Hortic. Sci.*, 103, 616, 1978.
106. **Sistrunk, W. A. and Moore, J. N.,** Evaluating strawberry selections for mechanization and high quality, in *Strawberry Mechanization*, Martin, L. W., and Morris, J. R., Eds., *Ore. Agr. Exp. Sta. Bull.*, 645, 133, 1980.
107. **Sistrunk, W. A., Morris, J. R., and Gascoigne, H. L.,** Effect of storage temperature and treatment on quality changes in strawberry spread, *Ark. Farm Res.*, 32, 2, 7, 1982.

108. **Sistrunk, W. A., Morris, J. R., and Kozup, J.,** The effect of chemical treatments and heat on color stability of frozen machine-harvested strawberries for jam, *J. Am. Soc. Hortic. Sci.*, 107, 693, 1982.
109. **Sistrunk, W. A., Wang, R. C., and Morris, J. R.,** Effect of combining mechanically harvested green and ripe puree and sliced fruit, processing methodology and frozen storage on quality of strawberries, *J. Food Sci.*, 48, 1609, 1983.
110. **Skalski, C. and Sistrunk, W. A.,** Factors influencing color degradation in 'Concord' grape juice, *J. Food Sci.*, 38, 1060, 1973.
111. **Skrede, G.,** Changes in sucrose, fructose and glucose content of frozen strawberries with thawing, *J. Food Sci.*, 48, 1094, 1983.
112. **Skrede, G.,** Strawberry varieties for industrial jam production, *J. Sci. Food Agric.*, 31, 670, 1980.
113. **Sondheimer, E. and Kertesz, Z. I.,** The kinetics of the oxidation of strawberry anthocyanin by hydrogen peroxide, *Food Res.*, 17, 288, 1952.
114. **Sondheimer, E. and Kertesz, Z. I.,** Participation of ascorbic acid in the destruction of anthocyaninin strawberry juice and model systems, *Food Res.*, 18, 475, 1953.
115. **Spayd, S. E. and Morris, J. R.,** Prediction of sensory color of strawberry jam, *Ark. Farm Res.*, 29, 5, 12, 1980.
116. **Spayd, S. E. and Morris, J. R.,** Influence of immature fruits on strawberry jam quality and storage stability, *J. Food Sci.*, 46, 414, 1981.
117. **Spayd, S. E. and Morris, J. R.,** Changes in strawberry quality during Maturation, *Ark. Farm Res.*, 30, 1, 6, 1981.
118. **Stadelbacher, I. J.,** Flower initiation and fruiting response of the strawberry as related to nitrogen nutrient levels, *Diss. Abstr.*, 24, 1780, 1963.
119. **Staudt, G.,** (The geographical distribution of the genus Fragria and the importance of the genus) *Ber. Deutsch. Bot. Ges.*, 66, 236, Biol. Abstr., 28, 10128, 1953.
120. **Szczesniak, A. S. and Smith, B. J.,** Observations on strawberry texture, a three pronged approach, *J. Texture studies*, 1, 65, 1969.
121. **Szymoniak, B.,** Fertilizer test for strawberry, *La. Agric. Expt. Sta. Rpt.*, 1925.
122. **Talburt, W. F., Leinbach, L. R., Brekke, J. E., and McHenry, R. O.,** Factors affecting character grade of frozen strawberries, *Food Technol.*, 9, 111, 1955.
123. **Tinsley, I. J. and Bockian, A. H.,** Some affects of sugars on the breakdown of pelargonidin-3-glucoside in model systems at 90°C., *Food Res.*, 25, 161, 1960.
124. **Voth, V., Bringhurst, R. S., Bowen, Jr., H. J.,** Effect of bed system, bed height and clear polyethylene mulch on yield, salt accumulation and soil temperature in California strawberries. *Proc. Am. Soc. Hortic. Sci.*, 91, 242, 1967.
125. **Voth, V., Urin, K., and Bringhurst, R. S.,** Effect of high nitrogen applications on yield, earliness, fruit quality, and leaf composition of California strawberries, *Proc. Am. Soc. Hortic. Sci.*, 91, 249, 1967.
126. **Waltman, C. S.,** Nitrogen and phosphorus relationships in strawberries, *Ky. Agr. Exp. Sta. Bull.*, 562, 1951.
127. **Wegener, J. B., Baer, B. A., and Rodgers, P. D.,** Improving quality of frozen strawberries with added colloids, *Food Technol.*, 5, 76, 1951.
128. **Wells, J. M.,** Modified atmospheres, chemical, and heat treatments to control postharvest decay of California strawberries, *Plant Dis. Rep.*, 54, 431, 1970.
129. **Whitney, E. N. and Hamilton, E. M., Eds.,** *Understanding Nutrition*, West Publ. Co., 1977, 607.
130. **Wilhelm, S.,** Diseases of strawberry, *Calif. Agr. Ext. Circ.*, 494, 1961.
131. **Wilson, W. F., Jr. and Giamalva, M. M.,** Days from blossom to harvest of Louisiana strawberries, *Proc. Am. Soc. Hortic. Sci.*, 63, 201, 1954.
132. **Winter, J. D. and Trantanella, S. R.,** Effect of packaging on palatability and weight loss of frozen strawberries, *Proc. Am. Soc. Hortic. Sci.*, 73, 189, 1959.
133. **Wrolstad, R. E., Lee, D. D., and Poei, M. S.,** The effect of microwave blanching on the color and composition of strawberry concentrate, *J. Food Sci.*, 45, 1573, 1980.
134. **Wrolstad, R. E., Putnam, T. E., and Varseveld, G. W.,** Color quality of frozen strawberries: effect of anthocyanins, pH, total acidity, and ascorbic acid variability, *J. Food Sci.*, 35, 448, 1970.
135. **Woodroof, J. G.,** Microscopic studies of frozen fruits and vegetables, *GA Agr. Exp. Sta. Bull.*, No. 201, 1938.
136. **Woodroof, J. G. and Shelor, E.,** Effect of Freezing storage on strawberries, blackberries, raspberries and peaches, *Food Freezing*, 2, 206, 1947.
137. **Zych, C. C.,** Fruit maturation times of strawberry varieties, *Fruit. Var. Hortic. Dig.*, 20, 51, 1966.

INDEX

N

O

P

Q

R

S

T

U

V

W

X

Y